STRUCTURAL LOAD DETERMINATION
UNDER 2009 IBC® AND ASCE/SEI 7-05

DAVID A. FANELLA, PH.D., S.E., P.E., F.ASCE

**Structural Load Determination
under 2009 IBC and ASCE/SEI 7-05
Second Edition**

ISBN: 978-1-58001-924-8

Cover Art Director:	Dianna Hallmark
Cover Design:	Dianna Hallmark
Project Editor/Typesetting:	Jodi Tahsler
Project Head:	John Henry
Publications Manager:	Mary Lou Luif

COPYRIGHT © 2009

ALL RIGHTS RESERVED. This publication is a copyrighted work owned by the International Code Council, Inc. Without advance written permission from the copyright owner, no part of this book may be reproduced, distributed or transmitted in any form or by any means, including, without limitation, electronic, optical or mechanical means (by way of example, and not limitation, photocopying or recording by or in an information storage retrieval system). For information on permission to copy material exceeding fair use, please contact: Publications, 4051 West Flossmoor Road, Country Club Hills, IL 60478. Phone 1-888-ICC-SAFE (422-7233).

The information contained in this document is believed to be accurate; however, it is being provided for informational purposes only and is intended for use only as a guide. Publication of this document by the ICC should not be construed as the ICC engaging in or rendering engineering, legal or other professional services. Use of the information contained in this book should not be considered by the user to be a substitute for the advice of a registered professional engineer, attorney or other professional. If such advice is required, it should be sought through the services of a registered professional engineer, licensed attorney or other professional.

Trademarks: "International Code Council" and the "International Code Council" logo and the "International Building Code" are trademarks of International Code Council, Inc.

Errata on various ICC publications may be available at www.iccsafe.org/errata.

First Printing: November 2009

PREFACE

The purpose of *Structural Load Determination under 2009 IBC and ASCE/SEI 7-05* is to provide a detailed guide to the proper determination of structural loads in accordance with the 2009 *International Building Code®* (IBC®) and *Minimum Design Loads for Buildings and Other Structures* (ASCE/SEI 7-05) with Supplement No. 2. The 2009 IBC references the 2005 edition of the ASCE/SEI 7 standard for many code-prescribed loads, most notably environmental loads such as flood, snow, wind and seismic load effects. In general, the IBC contains only the structural design criteria for environmental loads, while the technical design provisions for these loads are contained in the ASCE/SEI 7 standard.

This book is an essential resource for civil and structural engineers, architects, plan check engineers and students who need an efficient and practical approach to load determination under the 2009 IBC and ASCE/SEI 7-05 standard. The book is especially valuable to code users who are familiar with the structural load provisions of the previous legacy codes such as the *Uniform Building Code* (UBC). It has been reported that one of the most significant changes for code users transitioning from the UBC to the IBC is the way snow loads, wind pressures and earthquake ground motion load effects are determined under the IBC and ASCE/SEI 7-05 compared to previous legacy codes. *Structural Load Determination under 2009 IBC and ASCE/SEI 7-05* is a practical resource that will help code users make the transition quickly.

The book illustrates the application of code provisions and methodology for determining structural loads through the use of numerous flowcharts and practical design examples. Included are load combinations for allowable stress design, load and resistance factor (strength) design, seismic load combinations with vertical load effect and special seismic load combinations; dead loads, live loads and rain loads; snow loads, flood loads, wind loads and earthquake load effects. For wind load determination, flowcharts and design examples are presented for the simplified procedure (Method 1), the analytical procedure (Method 2) and the new alternate all-heights method in the 2009 IBC. Seismic design criteria, determination of seismic design category, the simplified method, equivalent lateral force procedure and nonbuilding structures are some of the topics illustrated through flowcharts and design examples.

This publication is an update to the previous publication, *Structural Load Determination under 2006 IBC and ASCE/SEI 7-05*. A new section has been added to Chapter 5 that covers the alternate all-heights wind design method in 2009 IBC Section 1609.6. This method is a simplified procedure based on Method 2 of ASCE/SEI 7-05 that applies to regularly-shaped buildings and structures that meet the five conditions given in IBC 1609.6.1. Net wind pressures p_{net} are calculated using design pressure coefficients C_{net}, which are given in IBC Table 1609.6.2(2) for main wind-force-resisting systems and components and cladding. A flowchart on how to determine p_{net} is provided along with examples on how to apply the alternate all-heights method.

A new Chapter 7 was added covering the determination of flood loads in accordance with IBC Section 1612, ASCE/SEI 7-05 and ASCE/SEI 24-05. Section 1612 of the IBC requires all structures sited in designated flood hazard areas to be designed and constructed to resist the effects of flood hazards and flood loads. Flood hazards may include erosion and scour whereas flood loads include flotation, lateral hydrostatic pressures, hydrodynamic pressures (due to moving water), wave impact and debris impact. Chapter 7 covers (1) identification of the various types of flood hazard areas and zones, (2) design and construction requirements and (3) determination of flood loads in accordance with IBC 1612, Chapter 5 of ASCE/SEI 7-05 and ASCE/SEI 24-05. Examples that clearly illustrate the provisions are included for residential structures in a Non-Coastal A Zone, a Coastal A Zone and a V Zone.

Load Determination under 2009 IBC and ASCE/SEI 7-05 is a multipurpose resource for civil and structural engineers, architects and plan check engineers because it can be used as a self-learning guide as well as a reference manual.

About the International Code Council

The International Code Council® (ICC®) is a nonprofit membership association dedicated to protecting the health, safety, and welfare of people by creating better buildings and safer communities. The mission of ICC is to provide the highest quality codes, standards, products and services for all concerned with the safety and performance of the built environment. ICC is the publisher of the family of the International Codes® (I-Codes®), a single set of comprehensive and coordinated model codes. This unified approach to building codes enhances safety, efficiency and affordability in the construction of buildings. The Code Council is also dedicated to innovation, sustainability and energy efficiency. Code Council subsidiary, ICC Evaluation Service, issues Evaluation Reports for innovative products and reports of Sustainable Attributes Verification and Evaluation (SAVE).

Headquarters: 500 New Jersey Avenue, NW, 6th Floor, Washington, DC 20001-2070
District Offices: Birmingham, AL; Chicago. IL; Los Angeles, CA
1-888-422-7233

www.iccsafe.org

About the Author

David A. Fanella, Ph.D., S.E., P.E., F.ASCE, is Associate Principal and Director of New Structures at Klein and Hoffman Inc., Chicago, Illinois. Dr. Fanella holds a Ph.D. in structural engineering from the University of Illinois at Chicago and is a licensed Structural Engineer in the State of Illinois and a licensed Professional Engineer in numerous states. He was formerly with the Portland Cement Association in Skokie, Illinois, where he was responsible for the buildings and special structures market. Dr. Fanella is an active member of a number of American Concrete Institute (ACI) Committees and is an Associate Member of the ASCE 7 Committee. He is currently President-Elect of the Structural Engineers Association of Illinois. Dr. Fanella has authored or coauthored many structural publications, including a series of articles on time-saving methods for reinforced concrete design.

TABLE OF CONTENTS

CHAPTER 1 – INTRODUCTION 1-1
  1.1 OVERVIEW 1-1
  1.2 SCOPE 1-2
  1.3 REFERENCES 1-4

CHAPTER 2 – LOAD COMBINATIONS 2-1
  2.1 INTRODUCTION 2-1
  2.2 LOAD EFFECTS 2-1
  2.3 LOAD COMBINATIONS USING STRENGTH DESIGN OR LOAD AND RESISTANCE FACTOR DESIGN 2-2
  2.4 LOAD COMBINATIONS USING ALLOWABLE STRESS DESIGN 2-4
  2.5 LOAD COMBINATIONS WITH OVERSTRENGTH FACTOR 2-8
  2.6 LOAD COMBINATIONS FOR EXTRAORDINARY EVENTS 2-11
  2.7 EXAMPLES 2-11

      2.7.1 Example 2.1 – Column in Office Building, Strength Design Load Combinations for Axial Loads 2-12

      2.7.2 Example 2.2 – Column in Office Building, Strength Design Load Combinations for Axial Loads and Bending Moments 2-13

      2.7.3 Example 2.3 – Beam in University Building, Strength Design Load Combinations for Shear Forces and Bending Moments 2-14

      2.7.4 Example 2.4 – Beam in University Building, Basic Allowable Stress Design Load Combinations for Shear Forces and Bending Moments 2-16

      2.7.5 Example 2.5 – Beam in University Building, Alternative Basic Allowable Stress Design Load Combinations for Shear Forces and Bending Moments 2-18

      2.7.6 Example 2.6 – Collector Beam in Residential Building, Load Combinations using Strength Design and Basic Load Combinations for Strength Design with Overstrength Factor for Axial Forces, Shear Forces, and Bending Moments 2-19

      2.7.7 Example 2.7 – Collector Beam in Residential Building, Load Combinations using Allowable Stress Design (Basic Load Combinations) and Basic Combinations for Allowable Stress Design with Overstrength Factor for Axial Forces, Shear Forces, and Bending Moments 2-21

      2.7.8 Example 2.8 – Collector Beam in Residential Building, Load Combinations using Allowable Stress Design (Alternative Basic Load Combinations) and Basic Combinations for Allowable Stress Design with Overstrength Factor for Axial Forces, Shear Forces, and Bending Moments 2-22

      2.7.9 Example 2.9 – Timber Pile in Residential Building, Basic Allowable Stress Design Load Combinations for Axial Forces 2-23

CHAPTER 3 – DEAD, LIVE, AND RAIN LOADS 3-1

3.1 DEAD LOADS 3-1
3.2 LIVE LOADS 3-2
3.2.1 General 3-2
3.2.2 Reduction in Live Loads 3-3
3.2.3 Distribution of Floor Loads 3-9
3.2.4 Roof Loads 3-10
3.2.5 Crane Loads 3-11
3.2.6 Interior Walls and Partitions 3-12
3.3 RAIN LOADS 3-12
3.4 EXAMPLES 3-13
3.4.1 Example 3.1 – Live Load Reduction, General Method of IBC 1607.9.1 3-14
3.4.2 Example 3.2 – Live Load Reduction, Alternate Method of IBC 1607.9.2 3-24
3.4.3 Example 3.3 – Live Load Reduction on a Girder 3-32
3.4.4 Example 3.4 – Rain Load, IBC 1611 3-33

CHAPTER 4 – SNOW LOADS 4-1

4.1 INTRODUCTION 4-1
4.2 FLOWCHARTS 4-4
4.3 EXAMPLES 4-14
4.3.1 Example 4.1 – Warehouse Building, Roof Slope of 1/2 on 12 4-14
4.3.2 Example 4.2 – Warehouse Building, Roof Slope of 1/4 on 12 4-19
4.3.3 Example 4.3 – Warehouse Building (Roof Slope of 1/2 on 12) and Adjoining Office Building (Roof Slope of 1/2 on 12) 4-20
4.3.4 Example 4.4 – Six-Story Hotel with Parapet Walls 4-28
4.3.5 Example 4.5 – Six-Story Hotel with Rooftop Unit 4-33
4.3.6 Example 4.6 – Agricultural Building 4-35
4.3.7 Example 4.7 – University Facility with Sawtooth Roof 4-39
4.3.8 Example 4.8 – Public Utility Facility with Curved Roof 4-42

CHAPTER 5 – WIND LOADS 5-1

5.1 INTRODUCTION 5-1
5.2 FLOWCHARTS 5-3
5.2.1 Allowed Procedures 5-5
5.2.2 Method 1 – Simplified Procedure 5-7
5.2.3 Method 2 – Analytical Procedure 5-11
5.2.4 Alternate All-heights Method 5-30
5.3 EXAMPLES 5-32
5.3.1 Example 5.1 – Warehouse Building using Method 1, Simplified Procedure 5-32

	5.3.2	Example 5.2 – Warehouse Building using Low-rise Building Provisions of Method 2, Analytical Method...	5-43
	5.3.3	Example 5.3 – Warehouse Building using Provisions of Method 2, Analytical Procedure	5-52
	5.3.4	Example 5.4 – Warehouse Building using Alternate All-heights Method................	5-60
	5.3.5	Example 5.5 – Residential Building using Method 2, Analytical Procedure............	5-69
	5.3.6	Example 5.6 – Six-Story Hotel using Method 2, Analytical Procedure	5-87
	5.3.7	Example 5.7 – Six-Story Hotel Located on an Escarpment using Method 2, Analytical Procedure	5-98
	5.3.8	Example 5.8 – Six-Story Hotel using Alternate All-heights Method	5-105
	5.3.9	Example 5.9 – Fifteen-Story Office Building using Method 2, Analytical Procedure	5-110
	5.3.10	Example 5.10 – Agricultural Building using Method 2, Analytical Procedure	5-125
	5.3.11	Example 5.11 – Freestanding Masonry Wall using Method 2, Analytical Procedure	5-131

CHAPTER 6 – EARTHQUAKE LOADS 6-1

6.1	INTRODUCTION	6-1
6.2	SEISMIC DESIGN CRITERIA..	6-2
	6.2.1 Seismic Ground Motion Values	6-2
	6.2.2 Occupancy Category and Importance Factor	6-4
	6.2.3 Seismic Design Category	6-5
	6.2.4 Design Requirements for SDC A	6-5
6.3	SEISMIC DESIGN REQUIREMENTS FOR BUILDING STRUCTURES	6-6
	6.3.1 Basic Requirements	6-6
	6.3.2 Seismic Force-Resisting Systems	6-6
	6.3.3 Diaphragm Flexibility, Configuration Irregularities, and Redundancy	6-7
	6.3.4 Seismic Load Effects and Combinations..	6-12
	6.3.5 Direction of Loading	6-13
	6.3.6 Analysis Procedure Selection	6-15
	6.3.7 Modeling Criteria	6-15
	6.3.8 Equivalent Lateral Force Procedure	6-15
	6.3.9 Modal Response Spectral Analysis	6-18
	6.3.10 Diaphragms, Chords, and Collectors...	6-18
	6.3.11 Structural Walls and Their Anchorage	6-20
	6.3.12 Drift and Deformation	6-20
	6.3.13 Foundation Design	6-20
	6.3.14 Simplified Alternative Structural Design Criteria for Simple Bearing Wall or Building Frame Systems	6-21
6.4	SEISMIC DESIGN REQUIREMENTS FOR NONSTRUCTURAL COMPONENTS	6-22

	6.4.1	General	6-22
	6.4.2	Seismic Demands on Nonstructural Components	6-23
	6.4.3	Nonstructural Component Anchorage	6-23
	6.4.4	Architectural Components	6-23
	6.4.5	Mechanical and Electrical Components	6-24
6.5	SEISMIC DESIGN REQUIREMENTS FOR NONBUILDING STRUCTURES		6-24
	6.5.1	General	6-24
	6.5.2	Reference Documents	6-25
	6.5.3	Nonbuilding Structures Supported by Other Structures	6-25
	6.5.4	Structural Design Requirements	6-25
	6.5.5	Nonbuilding Structures Similar to Buildings	6-25
	6.5.6	Nonbuilding Structures Not Similar to Buildings	6-26
	6.5.7	Tanks and Vessels	6-26
6.6	FLOWCHARTS		6-26
	6.6.1	Seismic Design Criteria	6-28
	6.6.2	Seismic Design Requirements for Building Structures	6-38
	6.6.3	Seismic Design Requirements for Nonstructural Components	6-55
	6.6.4	Seismic Design Requirements for Nonbuilding Structures	6-57
6.7	EXAMPLES		6-61
	6.7.1	Example 6.1 – Residential Building, Seismic Design Category	6-61
	6.7.2	Example 6.2 – Residential Building, Permitted Analytical Procedure	6-64
	6.7.3	Example 6.3 – Office Building, Seismic Design Category	6-70
	6.7.4	Example 6.4 – Office Building, Permitted Analytical Procedure	6-74
	6.7.5	Example 6.5 – Office Building, Allowable Story Drift	6-84
	6.7.6	Example 6.6 – Office Building, P-delta Effects	6-86
	6.7.7	Example 6.7 – Health Care Facility, Diaphragm Design Forces	6-88
	6.7.8	Example 6.8 – Health Care Facility, Nonstructural Component	6-93
	6.7.9	Example 6.9 – Residential Building, Vertical Combination of Structural Systems	6-95
	6.7.10	Example 6.10 – Warehouse Building, Design of Roof Diaphragm, Collectors, and Wall Panels	6-102
	6.7.11	Example 6.11 – Retail Building, Simplified Design Method	6-112
	6.7.12	Example 6.12 – Nonbuilding Structure	6-121

CHAPTER 7 – FLOOD LOADS 7-1

7.1	INTRODUCTION	7-1
7.2	FLOOD HAZARD AREAS	7-1
7.3	DESIGN AND CONSTRUCTION	7-3

	7.3.1	General	7-3
	7.3.2	Flood Loads	7-4
7.4	EXAMPLES		7-10
	7.4.1	Example 7.1 – Residential Building Located in a Non-Coastal A Zone	7-10
	7.4.2	Example 7.2 – Residential Building Located in a Coastal A Zone	7-14
	7.4.3	Example 7.3 – Residential Building Located in a V Zone	7-18

ACKNOWLEDGMENTS

The writer is deeply grateful to John R. Henry, P.E., Principal Staff Engineer, International Code Council, Inc., for his thorough review of the second edition of this publication. His insightful comments and suggestions for improvement have added significant value to this edition.

Thanks are also due to Adugna Fanuel, S.E., LEED AP, Christina Harber, S.E. and Majlinda Agojci, all of Klein and Hoffman, Inc., for their contributions. Their help in modeling and analyzing some of the example buildings and their review of the text and example problems were invaluable.

CHAPTER 1　　　　　　　　　　　　　　　　　　　　*INTRODUCTION*

1.1　OVERVIEW

The purpose of this publication is to assist in the proper determination of structural loads in accordance with the 2009 edition of the *International Building Code*® (IBC®) [1.1] and the 2005 edition of ASCE/SEI 7 *Minimum Design Loads for Buildings and Other Structures* [1.2].[1] Chapter 16 of the 2009 IBC, Structural Design, prescribes minimum structural loading requirements that are to be used in the design of all buildings and structures. The intent is to subject buildings and structures to loads that are likely to be encountered during their life span, thereby minimizing hazard to life and improving performance during and after a design event.

The snow load provisions in Section 1608, the wind load provisions in Section 1609, the flood load provisions in Section 1612 and the earthquake load provisions in Section 1613 are based on the provisions of Chapter 7, Chapter 6, Chapter 5, and Chapters 11 through 23 (with some exceptions) of ASCE/SEI 7, respectively. These ASCE/SEI 7 chapters are referenced in the aforementioned sections of the IBC.[2]

The seismic requirements of the 2009 IBC and ASCE/SEI 7 are based primarily on those in the 2003 edition of *NEHRP Recommended Provisions for Seismic Regulations for New Buildings and Other Structures, Part 1: Provisions* [1.3]. The NEHRP document, which has been updated every three years since the first edition in 1985, contains state-of-the-art criteria for the design and construction of buildings anywhere in the U.S. and its territories that are subject to the effects of earthquake ground motion. Life safety is the primary goal of the provisions. The requirements are also intended to enhance the performance of high-occupancy buildings and to improve the capability of essential facilities to function during and after a design-basis earthquake.

In addition to minimum design load requirements, Chapter 16 contains other important criteria that have a direct impact on the design of buildings and structures, including permitted design methodologies and design load combinations. For example, new Section 1614 contains provisions for structural integrity, which are applicable to high-rise buildings that are assigned to Occupancy Category III or IV and that are bearing wall structures or frame structures.[3]

---

[1] Numbers in brackets refer to references listed in Section 1.3 of this publication.
[2] ASCE/SEI 7-05 is one of a number of codes and standards that are referenced by the IBC. These documents, which can be found in Chapter 35 of the 2009 IBC, are considered part of the requirements of the IBC to the prescribed extent of each reference (see Section 101.4 of the 2009 IBC).
[3] High-rise buildings are defined in Section 202 as a building with an occupied floor located more than 75 ft above the lowest level to fire department vehicle access. Occupancy categories are defined in IBC Table 1604.5. Definitions of bearing wall structures and frame structures are given in Section 1614.2.

1.2 SCOPE

The content of this publication is geared primarily to practicing structural engineers. The load requirements of the 2009 IBC and ASCE/SEI 7-05 are presented in a straightforward manner with emphasis placed on the proper application of the provisions in everyday practice.

Code provisions have been organized in comprehensive flowcharts, which provide a road map that guides the reader through the requirements. Included in the flowcharts are the applicable section numbers and equation numbers from the 2009 IBC and ASCE/SEI 7-05 that pertain to the specific requirements. A basic description of flowchart symbols used in this publication is provided in Table 1.1.

Table 1.1 Summary of Flowchart Symbols

Symbol		Description
(rounded rectangle)	Terminator	The terminator symbol represents the starting or ending point of a flowchart.
(rectangle)	Process	The process symbol indicates a particular step or action that is taken within a flowchart.
(diamond)	Decision	The decision symbol represents a decision point, which requires a "yes" or "no" response.
(pentagon)	Off-page Connector	The off-page connector symbol is used to indicate continuation of the flowchart on another page.
(circle with cross)	Or	The logical "Or" symbol is used when a process diverges in two or more branches. Any one of the branches attached to this symbol can be followed.
(arrow)	Connector	The connector symbol indicates the sequence and direction of a process.

Numerous completely worked-out design examples are included in the chapters that illustrate the proper application of the code requirements. These examples follow the steps provided in the referenced flowcharts.

Readers who are interested in the history and design philosophy of the requirements can find detailed discussions in the commentary of ASCE/SEI 7 [1.2] and in *NEHRP*

Recommended Provisions for Seismic Regulations for New Buildings and Other Structures, Part 2: Commentary [1.4].

In addition to practicing structural engineers, engineers studying for licensing exams, structural plan checkers and others involved in structural engineering, such as advanced undergraduate students and graduate students, will find the flowcharts and the worked-out design examples to be very useful.

Throughout this publication, section numbers from the 2009 IBC are referenced as illustrated by the following: Section 1613 of the 2009 IBC is denoted as IBC 1613. Similarly, Section 11.4 from the 2005 ASCE/SEI 7 is referenced as ASCE/SEI 11.4 or as 11.4.

Chapter 2 outlines the required load combinations that must be considered when designing a building or its members for a variety of load effects. Load combinations using strength design or load and resistance factor design and load combinations using allowable stress design are both covered. Examples are provided that illustrate the strength design and allowable stress design load combinations for different types of members subject to different types of load effects.

Dead, live and rain loads are discussed in Chapter 3. The general method and an alternate method of live load reduction are covered, and flowcharts and examples illustrate both methods. The rain load provisions of IBC 1611 are also described, and an example demonstrates the calculation of a design rain load for a roof with scuppers.

Design provisions for snow loads are given in Chapter 4. A series of flowcharts highlight the requirements, and examples show the determination of flat roof snow loads, sloped roof snow loads, unbalanced roof snow loads and snow drift loads on a variety of flat and sloped roofs, including gable roofs, monoslope roofs, sawtooth roofs and curved roofs. Examples are also given that illustrate design snow loads for parapets and rooftop units.

Chapter 5 presents the design requirements for wind loads. Flowcharts are provided for the procedures that are allowed to be used when analyzing main wind-force-resisting systems and components and cladding. Other flowcharts give step-by-step procedures on how to determine design wind pressures on main wind-force-resisting systems and components and cladding of enclosed, partially enclosed and open buildings using the simplified procedure and the analytical procedure. A number of worked-out examples illustrate the design requirements for a variety of buildings and structures.

Earthquake loads are presented in Chapter 6. Information on how to determine design ground accelerations, site class and the seismic design category (SDC) of a building or structure is included, as are the various methods of analysis and their applicability for regular and irregular building and structures. Flowcharts and examples are provided that cover seismic design criteria, seismic design requirements for building structures, seismic design requirements for nonstructural components and seismic design requirements for nonbuilding structures.

Chapter 7 contains the requirements for flood loads. Included is information on flood hazard areas and flood hazard zones. Equations are provided for the following types of flood loads: hydrostatic loads, hydrodynamic loads, wave loads (breaking wave loads on vertical pilings and columns, breaking wave loads on vertical and nonvertical walls, and breaking wave loads from obliquely incident waves), and impact loads. Examples illustrate load calculations for a residential building in a Non-Coastal A Zone, a Coastal A Zone, and a V Zone.

1.3 REFERENCES

1.1 International Code Council, *International Building Code*, Washington, DC, 2009.

1.2 Structural Engineering Institute of the American Society of Civil Engineers, *Minimum Design Loads for Buildings and Other Structures*, ASCE/SEI 7-05, Reston, VA, 2006.

1.3 Building Seismic Safety Council, *NEHRP Recommended Provisions for Seismic Regulations for New Buildings and Other Structures, Part 1: Provisions*, FEMA 450-1/2003 edition, Washington, DC, 2004.

1.4 Building Seismic Safety Council, *NEHRP Recommended Provisions for Seismic Regulations for New Buildings and Other Structures, Part 2: Commentary*, FEMA 450-2/2003 edition, Washington, DC, 2004.

CHAPTER 2 — LOAD COMBINATIONS

2.1 INTRODUCTION

In accordance with IBC 1605.1, structural members of buildings and other structures must be designed to resist the load combinations of IBC 1605.2, 1605.3.1 or 1605.3.2. Load combinations that are specified in Chapters 18 through 23 of the IBC, which contain provisions for soils and foundations, concrete, aluminum, masonry, steel and wood, must also be considered. The structural elements identified in ASCE/SEI 12.2.5.2, 12.3.3.3 and 12.10.2.1 must be designed for the load combinations with overstrength factor of ASCE/SEI 12.4.3.2. These load combinations and their applicability are examined in Section 2.5 of this publication.

IBC 1605.2 contains the load combinations that are to be used when strength design or load and resistance factor design is utilized. Load combinations using allowable stress design are given in IBC 1605.3. Both sets of combinations are examined in Sections 2.3 and 2.4 of this publication, respectively. In addition to design for strength, the combinations of IBC 1605.2 or 1605.3 can be used to check overall structural stability, including stability against overturning, sliding and buoyancy (IBC 1605.1.1).

According to IBC 1605.1, load combinations must be investigated with one or more of the variable loads set equal to zero.[1] It is possible that the most critical load effects on a member occur when variable loads are not present.

ASCE/SEI 2.3 and 2.4 contain load combinations using strength design and allowable stress design, respectively. The load combinations are essentially the same as those in IBC 1605.2 and 1605.3 with some exceptions. Differences in the IBC and ASCE/SEI 7 load combinations are covered in the following sections.

Prior to examining the various load combinations, a brief introduction on load effects is given in Section 2.2.

2.2 LOAD EFFECTS

The load effects that are included in the IBC and ASCE/SEI 7 load combinations are summarized in Table 2.1. More details on these load effects can be found in the IBC and ASCE/SEI 7, as well as in subsequent chapters of this publication, as noted in the table.

---

[1] By definition, a "variable load" is a load that is not considered to be a permanent load (see IBC 1602). Permanent loads are those loads that do not change or that change very slightly over time, such as dead loads. Live loads, roof live loads, snow loads, rain loads, wind loads and earthquake loads are all examples of variable loads.

Table 2.1 Summary of Load Effects

Notation	Load Effect	Notes
D	Dead load	See IBC 1606 and Chapter 3 of this publication
D_i	Weight of ice	See Chapter 10 of ASCE/SEI 7
E	Combined effect of horizontal and vertical earthquake-induced forces as defined in ASCE/SEI 12.4.2	See IBC 1613, ASCE/SEI 12.4.2 and Chapter 6 of this publication
E_m	Maximum seismic load effect of horizontal and vertical forces as set forth in ASCE/SEI 12.4.3	See IBC 1613, ASCE/SEI 12.4.3 and Chapter 6 of this publication
F	Load due to fluids with well-defined pressures and maximum heights	---
F_a	Flood load	See IBC 1612 and Chapter 7 of this publication
H	Load due to lateral earth pressures, ground water pressure or pressure of bulk materials	See IBC 1610 for soil lateral loads
L	Live load, except roof live load, including any permitted live load reduction	See IBC 1607 and Chapter 3 of this publication
L_r	Roof live load including any permitted live load reduction	See IBC 1607 and Chapter 3 of this publication
R	Rain load	See IBC 1611 and Chapter 3 of this publication
S	Snow load	See IBC 1608 and Chapter 4 of this publication
T	Self-straining force arising from contraction or expansion resulting from temperature change, shrinkage, moisture change, creep in component materials, movement due to differential settlement or combinations thereof	---
W	Load due to wind pressure	See IBC 1609 and Chapter 5 of this publication
W_i	Wind-on-ice load	See Chapter 10 of ASCE/SEI 7

2.3 LOAD COMBINATIONS USING STRENGTH DESIGN OR LOAD AND RESISTANCE FACTOR DESIGN

The basic load combinations where strength design or load and resistance factor design is used are given in IBC 1605.2 and are summarized in Table 2.2. These combinations of factored loads establish the required strength that needs to be provided in the structural members of a building or structure.

Factored loads are determined by multiplying nominal loads (i.e., loads specified in Chapter 16 of the IBC) by a load factor, which is typically greater than or less than 1.0. Earthquake load effects are an exception to this: a load factor of 1.0 is used to determine the maximum effect, since an earthquake load is considered a strength-level load.

Table 2.2 Summary of Load Combinations Using Strength Design or Load and Resistance Factor Design (IBC 1605.2.1)

Equation No.	Load Combination
16-1	$1.4(D + F)$
16-2	$1.2(D + F + T) + 1.6(L + H) + 0.5(L_r \text{ or } S \text{ or } R)$
16-3	$1.2D + 1.6(L_r \text{ or } S \text{ or } R) + (f_1 L \text{ or } 0.8W)$
16-4	$1.2D + 1.6W + f_1 L + 0.5(L_r \text{ or } S \text{ or } R)$
16-5	$1.2D + 1.0E + f_1 L + f_2 S$
16-6	$0.9D + 1.6W + 1.6H$
16-7	$0.9D + 1.0E + 1.6H$

f_1 = 1 for floors in places of public assembly, for live loads in excess of 100 psf, and for parking garage live load
= 0.5 for other live loads

f_2 = 0.7 for roof configurations (such as sawtooth) that do not shed snow off the structure
= 0.2 for other roof configurations

Load combinations are constructed by adding to the dead load one or more of the variable loads at its maximum value, which is typically indicated by a load factor of 1.6. Also included are other variable loads with load factors less than 1.0; these are companion loads that represent arbitrary point-in-time values for those loads. Certain types of variable loads, such as wind and earthquake loads, act in more than one direction on a building or structure, and the appropriate sign of the variable load must be considered in the load combinations.

According to the exception in this section, factored load combinations that are specified in other provisions of the IBC take precedence to those listed in IBC 1605.2.

The load combinations given in IBC 1605.2.1 are the same as those in ASCE/SEI 2.3.2 with the following exceptions:

- The variable f_1 that is present in IBC Eqs. 16-3, 16-4 and 16-5 is not found in ASCE/SEI combinations 3, 4 and 5. Instead, the load factor on the live load L in the ASCE/SEI combinations is equal to 1.0 with the exception that the load factor on L is permitted to equal 0.5 for all occupancies where the live load is less than or equal to 100 psf, except for parking garages or areas occupied as places of public assembly (see exception 1 in ASCE/SEI 2.3.2). This exception makes these load combinations the same in ASCE/SEI 7 and the IBC.

- The variable f_2 that is present in IBC Eq. 16-5 is not found in ASCE/SEI combination 5. Instead, a load factor of 0.2 is applied to S in the ASCE/SEI combination. The third exception in ASCE/SEI 2.3.2 states that in combinations

2, 4 and 5, S shall be taken as either the flat roof snow load p_f or the sloped roof snow load p_s. This essentially means that the balanced snow load defined in ASCE/SEI 7.3 for flat roofs and 7.4 for sloped roofs can be used in combinations 2, 4 and 5. Drift loads and unbalanced snow loads are covered by combination 3. More information on snow loads can be found in Chapter 4 of this publication.

According to IBC 1605.2.2, the load combinations of ASCE/SEI 2.3.3 are to be used where flood loads F_a must be considered in the design.[2] In particular, $1.6W$ in IBC Eqs. 16-4 and 16-6 shall be replaced by $1.6W + 2.0F_a$ in V Zones or Coastal A Zones.[3] In Non-Coastal A Zones, $1.6W$ in IBC Eqs. 16-4 and 16-6 shall be replaced by $0.8W + 1.0F_a$.

ASCE/SEI 2.3.4 provides load combinations that include atmospheric ice loads, which are not found in the IBC. The following load combinations must be considered when a structure is subjected to atmospheric ice and wind-on-ice loads:[4]

- $0.5(L_r$ or S or $R)$ in ASCE/SEI combination 2 (IBC Eq. 16-2) shall be replaced by $0.2D_i + 0.5S$

- $1.6W + 0.5(L_r$ or S or $R)$ in ASCE/SEI combination 4 (IBC Eq. 16-4) shall be replaced by $D_i + W_i + 0.5S$

- $1.6W$ in ASCE/SEI combination 6 (IBC Eq. 16-6) shall be replaced by $D_i + W_i$

2.4 LOAD COMBINATIONS USING ALLOWABLE STRESS DESIGN

The basic load combinations where allowable stress design (working stress design) is used are given in IBC 1605.3. A set of basic load combinations is given in IBC 1605.3.1, and a set of alternative basic load combinations is given in IBC 1605.3.2. Both sets are examined below.

The basic load combinations of IBC 1605.3.1 are summarized in Table 2.3. A factor of 0.75 is applied where these combinations include more than one variable load, since the probability is low that two or more of the variable loads will reach their maximum values at the same time.

A factor of 0.6 is applied to the dead load D in IBC Eqs. 16-14 and 16-15. This factor limits the dead load that resists horizontal loads to approximately two-thirds of its actual value.[5] These load combinations apply to the design of all members in a structure and also provide for overall stability of a structure.

---

[2] Flood loads are determined by Chapter 5 of ASCE/SEI 7-05 and are covered in Chapter 7 of this publication.
[3] Definitions of Coastal High Hazard Areas (V Zones) and Coastal A Zones are given in ASCE/SEI 5.2.
[4] Atmospheric and wind-on-ice loads are determined by Chapter 10 of ASCE/SEI 7.
[5] Previous editions of the legacy building codes specified that the overturning moment and sliding due to wind load could not exceed two-thirds of the dead load stabilizing moment. This provision was not typically applied to all members in the building.

Table 2.3 Summary of Basic Load Combinations using Allowable Stress Design (IBC 1605.3.1)

Equation No.	Load Combination
16-8	$D + F$
16-9	$D + H + F + L + T$
16-10	$D + H + F + (L_r \text{ or } S \text{ or } R)$
16-11	$D + H + F + 0.75(L + T) + 0.75(L_r \text{ or } S \text{ or } R)$
16-12	$D + H + F + (W \text{ or } 0.7E)$
16-13	$D + H + F + 0.75(W \text{ or } 0.7E) + 0.75L + 0.75(L_r \text{ or } S \text{ or } R)$
16-14	$0.6D + W + H$
16-15	$0.6D + 0.7E + H$

As noted in Section 2.3 of this document, the combined effect of horizontal and vertical earthquake-induced forces E is a strength-level load. A factor of 0.7 (which is approximately equal to 1/1.4) is applied to E in IBC Eqs. 16-12, 16-13 and 16-15 to convert the strength-level load to a service-level load.

Two exceptions are given in IBC 1605.3.1. The first exception states that crane hook loads need not be combined with roof live load or with more than three-fourths of the snow load or one-half of the wind load. It is important to note this exception does not eliminate the need to combine live loads other than crane live loads with wind and snow loads in the prescribed manner. In other words, the load combinations in IBC Eqs. 16-11 and 16-13 must be investigated without the crane live load and with the crane live load using the criteria in the exception. In particular, the following load combinations must be investigated where crane live loads L_c are present:[6]

- IBC Eq. 16-11:

    $D + H + F + 0.75L + 0.75(L_r \text{ or } S \text{ or } R)$

    and

    $D + H + F + 0.75(L + L_c) + 0.75(0.75S \text{ or } R)$

- IBC Eq. 16-13 with E:

    $D + H + F + 0.75(0.7E) + 0.75L + 0.75(L_r \text{ or } S \text{ or } R)$

---

[6] Load effects T are not considered here for simplicity.

and

$$D + H + F + 0.75(0.7E) + 0.75(L + L_c) + 0.75(0.75S \text{ or } R)$$

- IBC Eq. 16-13 with W:

$$D + H + F + 0.75W + 0.75L + 0.75(L_r \text{ or } S \text{ or } R)$$

and

$$D + H + F + 0.75(0.5W) + 0.75(L + L_c) + 0.75(0.75S \text{ or } R)$$

The second exception in IBC 1605.3.1 states that flat roof snow loads p_f that are less than or equal to 30 psf and roof live loads that are less than or equal to 30 psf need not be combined with seismic loads. Also, where p_f is greater than 30 psf, 20 percent of the snow load must be combined with seismic loads.

Increases in allowable stresses that are given in the materials chapters of the IBC or in referenced standards are not permitted when the load combinations of IBC 1605.3.1 are used (IBC 1605.3.1.1). However, it is permitted to use the duration of load factor when designing wood structures in accordance with Chapter 23 of the IBC, which references the 2005 edition of the *National Design Specification for Wood Construction* (NDS-05).

According to IBC 1605.3.1.2, the load combinations of ASCE/SEI 2.4.2 are to be used where flood loads F_a must be considered in design.[7] In particular, $1.5F_a$ must be added to the other loads in IBC Eqs. 16-12, 16-13 and 16-14, and E is set equal to zero in IBC Eqs. 16-12 and 16-13 in V Zones or Coastal A Zones.[8] In noncoastal A Zones, $0.75F_a$ must be added to the other loads in IBC Eqs. 16-12, 16-13 and 16-14, and E is set equal to zero in IBC Eqs. 16-12 and 16-13.

The load combinations of IBC 1605.3.1 and ASCE/SEI 2.4.1 are the same except for the following:

- There is no specific exception for crane loads in ASCE/SEI 2.4.1.

- The exception in ASCE/SEI 2.4.1 states that in combinations 4 and 6, S shall be taken as either the flat roof snow load p_f or the sloped roof snow load p_s. The balanced snow load defined in ASCE/SEI 7.3 for flat roofs and 7.4 for sloped roofs can be used in combinations 4 and 6, and drift loads and unbalanced snow loads are covered by combination 3. More information on snow loads can be found in Chapter 4 of this publication.

---

[7] Flood loads are determined by Chapter 5 of ASCE/SEI 7-05 and are covered in Chapter 7 of this publication.
[8] Definitions of Coastal High Hazard Areas (V Zones) and Coastal A Zones are given in ASCE/SEI 5.2.

ASCE/SEI 2.4.3 provides load combinations including atmospheric ice loads, which are not found in the IBC. The following load combinations must be considered when a structure is subjected to atmospheric ice and wind-on-ice loads:[9]

- $0.7D_i$ shall be added to combination 2 (IBC Eq. 16-9)

- $(L_r$ or S or $R)$ in combination 3 (IBC Eq. 16-10) shall be replaced by $0.7D_i + 0.7W_i + S$

- W in combination 7 (IBC Eq. 16-14) shall be replaced by $0.7D_i + 0.7W_i$

As noted previously, a second set of load combinations are provided in the IBC for allowable stress design. The alternative basic load combinations can be found in IBC 1605.3.2 and are summarized in Table 2.4.

Table 2.4 Summary of Alternative Basic Load Combinations using Allowable Stress Design (IBC 1605.3.2)

Equation No.	Load Combination
16-16	$D + L + (L_r$ or S or $R)$
16-17	$D + L + \omega W$
16-18	$D + L + \omega W + S/2$
16-19	$D + L + S + \omega W/2$
16-20	$D + L + S + E/1.4$
16-21	$0.9D + E/1.4$

These load combinations are based on the allowable stress load combinations that appeared in the *Uniform Building Code* for many years.

It should be noted that the alternative allowable stress design load combinations do not include a load combination comparable to IBC Eq. 16-14 for dead load counteracting wind load effects. Instead of a specific load combination, IBC 1605.3.2 states that for load combinations that include counteracting effects of dead and wind loads, only two-thirds of the minimum dead load that is likely to be in place during a design wind event is to be used in the load combination. This is equivalent to a load combination of $0.67D + W$.

As was discussed previously in this section, the combined effect of horizontal and vertical earthquake-induced forces E is a strength-level load. This strength-level load is divided by 1.4 in IBC Eqs. 16-20 and 16-21 to convert it to a service-level load.

ASCE/SEI 12.13.4 permits a reduction of foundation overturning due to earthquake forces, provided that the criteria of that section are satisfied. Such a reduction is not

---

[9] Atmospheric and wind-on-ice loads are determined by Chapter 10 of ASCE/SEI 7.

permitted when the alternative basic load combinations are used to evaluate sliding, overturning and soil bearing at the soil-structure interface. Also, the vertical seismic load effect E_v in ASCE/SEI Eq. 12.4-4 may be taken as zero when proportioning foundations using these load combinations.

The coefficient ω in IBC Eqs. 16-17, 16-18 and 16-19 is equal to 1.3 where wind loads are calculated in accordance with ASCE/SEI Chapter 6.[10]

Unlike the basic load combinations of IBC 1605.3.1, allowable stresses are permitted to be increased or load combinations are permitted to be reduced where permitted by the material chapters of the IBC (Chapters 18 through 23) or by referenced standards when the alternative basic load combinations of IBC 1605.3.2 are used. This applies to those load combinations that include wind or earthquake loads.

The two exceptions in IBC 1605.3.2 for crane hook loads and for combinations of snow loads, roof live loads, and earthquake loads are the same as those in IBC 1605.3.1, which were discussed previously.

IBC 1605.3.2.1 requires that where F, H or T must be considered in design, each applicable load is to be added to the load combinations in IBC Eqs. 16-16 through 16-21.

ASCE/SEI 7-05 does not contain provisions for the alternative basic load combinations of IBC 1605.3.2.

2.5 LOAD COMBINATIONS WITH OVERSTRENGTH FACTOR

The following load combinations, which are given in ASCE/SEI 12.4.3.2, must be used where required by ASCE 12.2.5.2, 12.3.3.3 or 12.10.2.1 instead of the corresponding load combinations in IBC 1605.2 and 1605.3 (IBC 1605.1, item 3):

- Basic Combinations for Strength Design with Overstrength Factor[11]

    IBC Eq. 16-5: $(1.2 + 0.2S_{DS})D + \Omega_o Q_E + L + 0.2S$

    IBC Eq. 16-7: $(0.9 - 0.2S_{DS})D + \Omega_o Q_E + 1.6H$

- Basic Combinations for Allowable Stress Design with Overstrength Factor

    IBC Eq. 16-12: $(1.0 + 0.14S_{DS})D + H + F + 0.7\Omega_o Q_E$

---

[10] It is shown in Chapter 5 of this publication that the wind directionality factor, which is equal to 0.85 for building structures, is explicitly included in the velocity pressure equation for wind. In earlier editions of ASCE/SEI 7 and in the legacy codes, the directionality factor was part of the load factor, which was equal to 1.3 for wind. Thus, for allowable stress design, $\omega = 1.3 \times 0.85 \approx 1.0$, and for strength design, $\omega = 1.6 \times 0.85 \approx 1.3$.

[11] See Notes 1 and 2 in ASCE/SEI 12.4.3.2 that pertain to these load combinations.

IBC Eq. 16-13:
$(1.0 + 0.105 S_{DS})D + H + F + 0.525\Omega_o Q_E + 0.75L + 0.75(L_r \text{ or } S \text{ or } R)$

IBC Eq. 16-15: $(0.6 - 0.14 S_{DS})D + 0.7\Omega_o Q_E + H$

- Alternative Basic Combinations for Allowable Stress Design with Overstrength Factor

IBC Eq. 16-20: $\left(1.0 + \dfrac{0.2 S_{DS}}{1.4}\right)D + \dfrac{\Omega_o Q_E}{1.4} + L + S$

IBC Eq. 16-21: $\left(0.9 - \dfrac{0.2 S_{DS}}{1.4}\right)D + \dfrac{\Omega_o Q_E}{1.4}$

where $E_m = E_{mh} + E_v = \Omega_o Q_E + 0.2 S_{DS} D$ for use in IBC Eqs. 16-5, 16-12, 16-13 and 16-20

$ = E_{mh} - E_v = \Omega_o Q_E - 0.2 S_{DS} D$ for use in IBC Eqs. 16-7, 16-15 and 16-21

Ω_o = system overstrength factor obtained from ASCE/SEI Table 12.2-1 for a particular seismic-force-resisting system

Q_E = effects of horizontal seismic forces on a building or structure

S_{DS} = design spectral response acceleration parameter at short periods determined by IBC 1613.5 or ASCE/SEI 11.4

ASCE/SEI 12.4.3.3 permits allowable stresses to be increased by a factor of 1.2 where allowable stress design is used with seismic load effect including overstrength factor. This increase is not to be combined with increases in allowable stresses or reductions in load combinations that are otherwise permitted in ASCE/SEI 7 or in other referenced materials standards. However, the duration of load factor is permitted to be used when designing wood members in accordance with the referenced standard.

Provisions for cantilever column systems are given in ASCE/SEI 12.2.5.2. In addition to the design requirements of that section, the members in such systems must be designed to resist the strength or allowable stress load combinations of IBC 1605.2 or 1605.3 and the applicable load combinations with overstrength factor specified in ASCE/SEI 12.4.3.2.

The provisions of ASCE/SEI 12.3.3.3 apply to structural members that support discontinuous frames or shear wall systems where the discontinuity is severe enough to be deemed a structural irregularity. In particular, columns, beams, trusses or slabs that support discontinuous walls or frames having horizontal irregularity Type 4 of ASCE/SEI Table 12.3-1 or vertical irregularity Type 4 of ASCE/SEI Table 12.3-2 must be designed to resist the load combinations with overstrength factor specified in ASCE/SEI 12.4.3.2

in addition to the strength design or allowable stress design load combinations described in previous sections of this publication.[12]

An example of columns supporting a shear wall that has been discontinued at the first floor of a multistory building is illustrated in Figure 2.1. The columns in this case must be designed to resist the load combinations with overstrength factor.

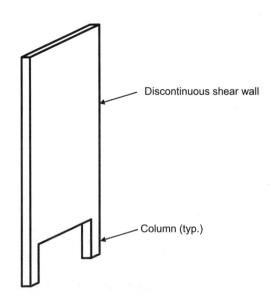

Figure 2.1 Example of Columns Supporting Discontinuous Shear Wall

ASCE/SEI 12.10.2.1 applies to collector elements in structures assigned to Seismic Design Category (SDC) C and higher.[13] Collectors, which are also commonly referred to as drag struts, are elements in a structure that are used to transfer the loads from a diaphragm to the elements of the lateral-force-resisting system (LFRS) where the lengths of the vertical elements in the LFRS are less than the length of the diaphragm at that location. An example of collector beams and a shear wall is shown in Figure 2.2. The collector beams collect the force from the diaphragm and distribute it to the shear wall.

In general, collector elements, splices and connections to resisting elements must all be designed to resist the load combinations with overstrength factor in addition to the strength design or allowable stress design load combinations presented previously. The exception in ASCE/SEI 12.10.2.1 permits collectors, splices and connections in structures braced entirely by light frame shear walls to be only designed to resist the forces prescribed in ASCE/SEI 12.10.1.1. Light frame construction is defined in ASCE/SEI 11.2 and includes systems composed of repetitive wood and cold-formed steel framing.

---

[12] Additional information on structural irregularities can be found in ASCE/SEI 12.3 and Chapter 6 of this publication.
[13] More information on how to determine the Seismic Design Category of a building or structure is given in IBC 1613.5.6, ASCE/SEI 11.6 and Chapter 6 of this publication.

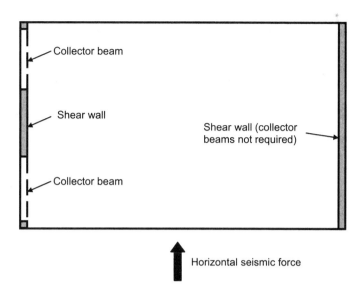

Figure 2.2 Example of Collector Beams and Shear Walls

2.6 LOAD COMBINATIONS FOR EXTRAORDINARY EVENTS

ASCE/SEI 2.5 requires that the strength and stability of a structure be checked to ensure that it can withstand the effects from extraordinary events (i.e., low-probability events) such as fires, explosions and vehicular impact. More information on these types of events and recommended load combinations that include the effects of such events can be found in ASCE/SEI C2.5. In that section, reference is made to ASCE/SEI 1.4 and C1.4, which address general structural integrity requirements. Included is a discussion on resistance to progressive collapse.

New provisions for structural integrity are contained in IBC 1614. These are applicable to buildings classified as high-rise buildings in accordance with IBC 403 and assigned to Occupancy Category III or IV with bearing wall structures or frame structures.[14] Specific load combinations are not included in these prescriptive requirements; they are meant to improve the redundancy and ductility of these types of framing systems in the event of damage due to an abnormal loading event.

2.7 EXAMPLES

The following examples illustrate the load combinations that were discussed in the previous sections of this publication.

---

[14] A high-rise building is defined in IBC 202 as a building with an occupied floor located more than 75 ft above the lowest level of fire department vehicle access. Occupancy Categories III and IV are defined in IBC Table 1604.5.

2.7.1 Example 2.1 – Column in Office Building, Strength Design Load Combinations for Axial Loads

Determine the strength design load combinations for a column in a multistory office building using the nominal axial loads in the design data. The live load on the floors is less than 100 psf and the roof is a gable roof.

DESIGN DATA

	Axial Load (kips)
Dead load, D	78
Live load, L	38
Roof live load, L_r	13
Balanced snow load, S	19

SOLUTION

The load combinations using strength design of IBC 1605.2 are summarized in Table 2.5 for this column. The load combinations in the table include only the applicable load effects from the design data.

Table 2.5 Summary of Load Combinations using Strength Design for Column in Example 2.1

Equation No.	Load Combination
16-1	$1.4D = 1.4 \times 78 = 109$ kips
16-2	$1.2D + 1.6L + 0.5(L_r \text{ or } S) = (1.2 \times 78) + (1.6 \times 38) + (0.5 \times 13) = 161$ kips $= (1.2 \times 78) + (1.6 \times 38) + (0.5 \times 19) = 164$ kips
16-3	$1.2D + 1.6(L_r \text{ or } S) + f_1 L = (1.2 \times 78) + (1.6 \times 13) + (0.5 \times 38) = 133$ kips $= (1.2 \times 78) + (1.6 \times 19) + (0.5 \times 38) = 143$ kips
16-4	$1.2D + f_1 L + 0.5(L_r \text{ or } S) = (1.2 \times 78) + (0.5 \times 38) + (0.5 \times 13) = 119$ kips $= (1.2 \times 78) + (0.5 \times 38) + (0.5 \times 19) = 122$ kips
16-5	$1.2D + f_1 L + f_2 S = (1.2 \times 78) + (0.5 \times 38) + (0.2 \times 19) = 116$ kips
16-6, 16-7	$0.9D = 0.9 \times 78 = 70$ kips

The constant f_1 is taken as 0.5, since the live load is less than 100 psf. The constant f_2 is taken as 0.2, since the gable roof can shed snow, unlike a sawtooth roof.

The largest axial load on the column is due to IBC Eq. 16-2 when the snow load is considered.

In this example, taking one or more of the variable loads (live, roof live or snow loads) in IBC Eqs. 16-1 through 16-5 equal to zero results in factored loads less than those shown in Table 2.5.

2.7.2 Example 2.2 – Column in Office Building, Strength Design Load Combinations for Axial Loads and Bending Moments

Determine the strength design load combinations for a column in a multistory office building using the nominal axial loads and maximum bending moments in the design data. The live load on the floors is less than 100 psf and the roof is essentially flat.

DESIGN DATA

	Axial Load (kips)	Bending Moment (ft-kips)
Dead load, D	78	15
Live load, L	38	5
Roof live load, L_r	13	0
Balanced snow load, S	19	0
Wind load, W	± 20	± 47

SOLUTION

The load combinations using strength design of IBC 1605.2 are summarized in Table 2.6 for this column. The load combinations in the table include only the applicable load effects from the design data. The constant f_1 is taken as 0.5, since the live load is less than 100 psf. The constant f_2 is taken as 0.2, since the flat roof can shed snow, unlike a sawtooth roof.

Since the wind loads cause the structure to sway to the right and to the left, load combinations must be investigated for both cases. This is accomplished by taking both "plus" and "minus" load effects of the wind.

In general, all of the load combinations in Table 2.6 must be investigated when designing the column. It is usually possible to anticipate which of the load combinations is the most critical. Taking one or more of the variable loads (live, roof live, snow or wind loads) in IBC Eqs. 16-1 through 16-5 equal to zero was not considered, since these load combinations typically do not govern the design of such members.

Table 2.6 Summary of Load Combinations using Strength Design for Column in Example 2.2

Equation No.	Load Combination	Axial Load (kips)	Bending Moment (ft-kips)
16-1	$1.4D$	109	21
16-2	$1.2D + 1.6L + 0.5L_r$	161	26
	$1.2D + 1.6L + 0.5S$	164	26
16-3	$1.2D + 1.6L_r + f_1 L$	133	21
	$1.2D + 1.6L_r + 0.8W$	130	56
	$1.2D + 1.6L_r - 0.8W$	98	-20
	$1.2D + 1.6S + f_1 L$	143	21
	$1.2D + 1.6S + 0.8W$	140	56
	$1.2D + 1.6S - 0.8W$	108	-20
16-4	$1.2D + 1.6W + f_1 L + 0.5L_r$	151	96
	$1.2D + 1.6W + f_1 L + 0.5S$	154	96
	$1.2D - 1.6W + f_1 L + 0.5L_r$	87	-55
	$1.2D - 1.6W + f_1 L + 0.5S$	90	-55
16-5	$1.2D + f_1 L + f_2 S$	116	21
16-6	$0.9D + 1.6W$	102	89
	$0.9D - 1.6W$	38	-62
16-7	$0.9D$	70	14

2.7.3 Example 2.3 – Beam in University Building, Strength Design Load Combinations for Shear Forces and Bending Moments

Determine the strength design load combinations for a beam in a university building using the nominal shear forces and bending moments in the design data. The occupancy of the building is classified as a place of public assembly.

DESIGN DATA

	Shear Force (kips)	Bending Moment (ft-kips)	
		Support	Midspan
Dead load, D	50	-250	170
Live load, L	15	-50	35
Wind load, W	± 10	± 100	---
Seismic, Q_E	± 5	± 50	---

The seismic design data are as follows:[15]

ρ = redundancy factor = 1.0

S_{DS} = design spectral response acceleration parameter at short periods = $0.5g$

SOLUTION

The load combinations using strength design of IBC 1605.2 are summarized in Table 2.7 for this beam. The load combinations in the table include only the applicable load effects from the design data.

The quantity Q_E is the effect of code-prescribed horizontal seismic forces on the beam determined from a structural analysis.

In accordance with ASCE/SEI 12.4.2, the seismic load effect E is defined as follows:

- For use in load combination 5 of ASCE/SEI 2.3.2, or, equivalently, in IBC Eq. 16-5:

$$E = E_h + E_v = \rho Q_E + 0.2 S_{DS} D = Q_E + 0.1D$$

- For use in load combination 7 of ASCE/SEI 2.3.2, or, equivalently, in IBC Eq. 16-7:

$$E = E_h - E_v = \rho Q_E - 0.2 S_{DS} D = Q_E - 0.1D$$

Substituting for E, IBC Eq. 16-5 becomes: $1.2D + 1.0E + f_1 L = 1.3D + L + Q_E$

Similarly, IBC Eq. 16-7 becomes: $0.9D + 1.0E = 0.8D + Q_E$

---

[15] More information on seismic design can be found in Chapter 6 of this publication.

Table 2.7 Summary of Load Combinations using Strength Design for the Beam in Example 2.3

Equation No.	Load Combination	Location	Bending Moment (ft-kips)	Shear Force (kips)
16-1	$1.4D$	Support	-350	70
		Midspan	238	---
16-2	$1.2D + 1.6L$	Support	-380	84
		Midspan	260	---
16-3	$1.2D + L$	Support	-350	75
		Midspan	239	---
	$1.2D + 0.8W$	Support	-380	68
		Midspan	204	---
16-4	$1.2D + L + 1.6W$	Support	-510	91
		Midspan	239	---
16-5	$1.3D + L + Q_E$	Support	-425	85
		Midspan	256	---
16-6	$0.9D - 1.6W$	Support	-65	29
		Midspan	153	---
16-7	$0.8D - Q_E$	Support	-150	35
		Midspan	136	---

Like wind loads, sidesway to the right and to the left must be investigated for seismic loads. In IBC Eq. 16-5, the maximum effect occurs when Q_E is added to the effects of the gravity loads. In IBC Eq. 16-7, Q_E is subtracted from the effect of the dead load, since maximum effects occur, in general, when minimum dead load and the effects from lateral loads counteract. The same reasoning is applied in IBC Eqs. 16-3, 16-4 and 16-6 for the wind effects W.

The constant f_1 is taken as 1.0, since the occupancy of the building is classified as a place of public assembly.

2.7.4 Example 2.4 – Beam in University Building, Basic Allowable Stress Design Load Combinations for Shear Forces and Bending Moments

Determine the basic allowable stress design load combinations of IBC 1605.3.1 for the beam described in Example 2.3.

SOLUTION

The basic load combinations using allowable stress design of IBC 1605.3.1 are summarized in Table 2.8 for this beam. The load combinations in the table include only the applicable load effects from the design data.

Table 2.8 Summary of Basic Load Combinations using Allowable Stress Design for Beam in Example 2.4

Equation No.	Load Combination	Location	Bending Moment (ft-kips)	Shear Force (kips)
16-8, 16-10	D	Support	-250	50
		Midspan	170	---
16-9	$D + L$	Support	-300	65
		Midspan	205	---
16-11	$D + 0.75L$	Support	-288	61
		Midspan	196	---
16-12	$D + W$	Support	-350	60
		Midspan	170	---
	$1.07D + 0.7Q_E$	Support	-303	57
		Midspan	182	---
16-13	$D + 0.75L + 0.75W$	Support	-363	69
		Midspan	196	---
	$1.05D + 0.75L + 0.525Q_E$	Support	-326	66
		Midspan	205	---
16-14	$0.6D - W$	Support	-50	20
		Midspan	102	---
16-15	$0.67D - 0.7Q_E$	Support	-133	30
		Midspan	114	---

In accordance with ASCE/SEI 12.4.2, the seismic load effect E is defined as follows:

- For use in load combinations 5 and 6 of ASCE/SEI 2.4.1 or, equivalently, in IBC Eqs. 16-12 and 16-13:

$$E = E_h + E_v = \rho Q_E + 0.2 S_{DS} D = Q_E + 0.1D$$

- For use in load combination 8 of ASCE/SEI 2.3.2 or, equivalently, in IBC Eq. 16-15:

$$E = E_h - E_v = \rho Q_E - 0.2 S_{DS} D = Q_E - 0.1D$$

Thus, substituting for E, IBC Eqs. 16-12 and 16-13 become, respectively,

$$D + 0.7E = 1.07D + 0.7Q_E$$

$$D + 0.75L + 0.525E = 1.05D + 0.75L + 0.525Q_E$$

Similarly, IBC Eq. 16-15 becomes: $0.6D + 0.7E = 0.67D + 0.7Q_E$

Like wind loads, sidesway to the right and to the left must be investigated for seismic loads. In IBC Eqs. 16-12 and 16-13, the maximum effect occurs when Q_E is added to the effects of the gravity loads. In IBC Eq. 16-15, Q_E is subtracted from the effect of the dead load, since maximum effects occur, in general, when minimum dead load and the effects from lateral loads counteract. The same reasoning is applied in IBC Eqs. 16-12, 16-13 and 16-14 for the wind effects W.

2.7.5 Example 2.5 – Beam in University Building, Alternative Basic Allowable Stress Design Load Combinations for Shear Forces and Bending Moments

Determine the alternative basic allowable stress design load combinations of IBC 1605.3.2 for the beam described in Example 2.3. Assume that the wind forces have been determined using the provisions of Chapter 6 of ASCE/SEI 7.

SOLUTION

The alternative basic load combinations using allowable stress design of IBC 1605.3.2 are summarized in Table 2.9 for this beam. The load combinations in the table include only the applicable load effects from the design data.

Table 2.9 Summary of Alternative Basic Load Combinations using Allowable Stress Design for Beam in Example 2.5

Equation No.	Load Combination	Location	Bending Moment (ft-kips)	Shear Force (kips)
16-16	$D + L$	Support	-300	65
		Midspan	205	---
16-17 16-18	$D + L + 1.3W$	Support	-430	78
		Midspan	205	---
	$0.67D + L - 1.3W$	Support	-88	36
		Midspan	149	---
16-19	$D + L + 0.65W$	Support	-365	72
		Midspan	205	---
16-20	$1.07D + L + (Q_E / 1.4)$	Support	-353	72
		Midspan	217	---
16-21	$0.83D - (Q_E / 1.4)$	Support	-172	38
		Midspan	141	---

The factor ω is taken as 1.3, since the wind forces have been determined by Chapter 6 of ASCE/SEI.

In accordance with ASCE/SEI 12.4.2, the seismic load effect E is defined as follows:

- For use in IBC Eq. 16-20:

$$E = E_h + E_v = \rho Q_E + 0.2 S_{DS} D = Q_E + 0.1D$$

- For use in IBC Eq. 16-21:

$$E = E_h - E_v = \rho Q_E - 0.2 S_{DS} D = Q_E - 0.1D$$

Thus, substituting for E, IBC Eq. 16-20 becomes: $D + L + E/1.4 = 1.07D + L + Q_E/1.4$

Similarly, IBC Eq. 16-21 becomes: $0.9D + E/1.4 = 0.83D + Q_E/1.4$

Like wind loads, sidesway to the right and to the left must be investigated for seismic loads. In IBC Eq. 16-20, the maximum effect occurs when Q_E is added to the effects of the gravity loads. In IBC Eq. 16-21, Q_E is subtracted from the effect of the dead load, since maximum effects occur, in general, when minimum dead load and the effects from lateral loads counteract.

In accordance with IBC 1605.3.2, two-thirds of the dead load is used in IBC Eqs. 16-17 and 16-18 to counter the maximum effects from the wind pressure.

2.7.6 Example 2.6 – Collector Beam in Residential Building, Load Combinations using Strength Design and Basic Load Combinations for Strength Design with Overstrength Factor for Axial Forces, Shear Forces, and Bending Moments

Determine the strength design load combinations and the basic combinations for strength design with overstrength factor for a simply supported collector beam in a residential building using the nominal axial loads, shear forces and bending moments in the design data. The live load on the floors is less than 100 psf.

DESIGN DATA			
	Axial Force (kips)	Shear Force (kips)	Bending Moment (ft-kips)
Dead load, D	0	56	703
Live load, L	0	19	235
Seismic, Q_E	± 50	0	0

The seismic design data are as follows:[16]

---

[16] More information on seismic design can be found in Chapter 6 of this publication.

Ω_o = system overstrength factor = 2.5

S_{DS} = design spectral response acceleration parameter at short periods = $1.0g$

Seismic design category: D

The axial seismic force Q_E corresponds to the portion of the diaphragm design force that is resisted by the collector beam. This force can be tensile or compressive.

SOLUTION

The governing load combination in IBC 1605.2.1 is as follows:

- IBC Eq. 16-2:

Bending moment: $1.2D + 1.6L = (1.2 \times 703) + (1.6 \times 235) = 1,220$ ft-kips

Shear force: $1.2D + 1.6L = (1.2 \times 56) + (1.6 \times 19) = 98$ kips

Since the beam is a collector beam in a building assigned to SDC D, the beam must be designed for the following basic combinations for strength design with overstrength factor (see IBC 1605.1 and 1605.2.1; ASCE/SEI 12.4.3.2):

- IBC Eq. 16-5: $(1.2 + 0.2S_{DS})D + \Omega_o Q_E + 0.5L$ [17]

Axial force: $\Omega_o Q_E = 2.5 \times 50 = 125$ kips tension or compression

Bending moment: $(1.2 + 0.2S_{DS})D + 0.5L = (1.4 \times 703) + (0.5 \times 235) = 1,102$ ft-kips

Shear force: $(1.2 + 0.2S_{DS})D + 0.5L = (1.4 \times 56) + (0.5 \times 19) = 88$ kips

- IBC Eq. 16-7: $(0.9 - 0.2S_{DS})D + \Omega_o Q_E$

Axial force: $\Omega_o Q_E = 2.5 \times 50 = 125$ kips tension or compression

Bending moment: $(0.9 - 0.2S_{DS})D = 0.7 \times 703 = 492$ ft-kips

Shear force: $(0.9 - 0.2S_{DS})D = 0.7 \times 56 = 39$ kips

---

[17] The load factor on L is permitted to equal 0.5 in accordance with Note 1 in ASCE 12.4.3.2.

The collector beam and its connections must be designed to resist the combined effects of (1) flexure and axial tension, (2) flexure and axial compression and (3) shear as set forth by the above load combinations.

2.7.7 Example 2.7 – Collector Beam in Residential Building, Load Combinations using Allowable Stress Design (Basic Load Combinations) and Basic Combinations for Allowable Stress Design with Overstrength Factor for Axial Forces, Shear Forces, and Bending Moments

Determine the load combinations using allowable stress design (basic load combinations) and the basic combinations for allowable stress design with overstrength factor for the beam in Example 2.6.

SOLUTION

The governing load combination in IBC 1605.3.1 is as follows:

- IBC Eq. 16-9:

    Bending moment: $D + L = 703 + 235 = 938$ ft-kips

    Shear force: $D + L = 56 + 19 = 75$ kips

Since the beam is a collector beam in a building assigned to SDC D, the beam must be designed for the following basic combinations for allowable stress design with overstrength factor (see IBC 1605.1 and 1605.3.1; ASCE/SEI 12.4.3.2):

- IBC Eq. 16-12: $(1.0 + 0.14 S_{DS})D + 0.7 \Omega_o Q_E$

    Axial force: $0.7 \Omega_o Q_E = 0.7 \times 2.5 \times 50 = 88$ kips tension or compression

    Bending moment: $(1.0 + 0.14 S_{DS})D = 1.14 \times 703 = 801$ ft-kips

    Shear force: $(1.0 + 0.14 S_{DS})D = 1.14 \times 56 = 64$ kips

- IBC Eq. 16-13: $(1.0 + 0.105 S_{DS})D + 0.525 \Omega_o Q_E + 0.75L$

    Axial force: $0.525 \Omega_o Q_E = 0.525 \times 2.5 \times 50 = 66$ kips tension or compression

    Bending moment: $1.105D + 0.75L = (1.105 \times 703) + (0.75 \times 235) = 953$ ft-kips

    Shear force: $1.105D + 0.75L = (1.105 \times 56) + (0.75 \times 19) = 76$ kips

- IBC Eq. 16-15: $(0.6 - 0.14S_{DS})D + 0.7\Omega_o Q_E$

    Axial force: $0.7\Omega_o Q_E = 0.7 \times 2.5 \times 50 = 88$ kips tension or compression

    Bending moment: $(0.6 - 0.14S_{DS})D = 0.46 \times 703 = 323$ ft-kips

    Shear force: $(0.6 - 0.14S_{DS})D = 0.46 \times 56 = 26$ kips

The collector beam and its connections must be designed to resist the combined effects of (1) flexure and axial tension, (2) flexure and axial compression and (3) shear as set forth by the above load combinations.[18]

2.7.8 Example 2.8 – Collector Beam in Residential Building, Load Combinations using Allowable Stress Design (Alternative Basic Load Combinations) and Basic Combinations for Allowable Stress Design with Overstrength Factor for Axial Forces, Shear Forces, and Bending Moments

Determine the load combinations using allowable stress design (alternative basic load combinations) and the basic combinations for allowable stress design with overstrength factor for the beam in Example 2.6.

SOLUTION

The governing load combination in IBC 1605.3.2 is as follows:

- IBC Eq. 16-16:

    Bending moment: $D + L = 703 + 235 = 938$ ft-kips

    Shear force: $D + L = 56 + 19 = 75$ kips

Since the beam is a collector beam in a building assigned to SDC D, the beam must be designed for the following basic combinations for allowable stress design with overstrength factor (see IBC 1605.1 and 1605.3.2; ASCE/SEI 12.4.3.2):

- IBC Eq. 16-20: $\left(1.0 + \dfrac{0.2S_{DS}}{1.4}\right)D + \dfrac{\Omega_o Q_E}{1.4} + L$

    Axial force: $\dfrac{\Omega_o Q_E}{1.4} = \dfrac{2.5 \times 50}{1.4} = 89$ kips tension or compression

---

[18] See ASCE/SEI 12.4.3.3 for allowable stress increase for load combinations with overstrength.

Bending moment: $\left[\left(1.0+\dfrac{0.2}{1.4}\right)\times 703\right]+235=1038$ ft-kips

Shear force: $\left[\left(1.0+\dfrac{0.2}{1.4}\right)\times 56\right]+19=83$ kips

- IBC Eq. 16-21: $\left(0.9-\dfrac{0.2S_{DS}}{1.4}\right)D+\dfrac{\Omega_o Q_E}{1.4}$

Axial force: $\dfrac{\Omega_o Q_E}{1.4}=\dfrac{2.5\times 50}{1.4}=89$ kips tension or compression

Bending moment: $\left(0.9-\dfrac{0.2}{1.4}\right)\times 703=532$ ft-kips

Shear force: $\left(0.9-\dfrac{0.2}{1.4}\right)\times 56=42$ kips

The collector beam and its connections must be designed to resist the combined effects of (1) flexure and axial tension, (2) flexure and axial compression and (3) shear as set forth by the above load combinations.[19]

2.7.9 Example 2.9 – Timber Pile in Residential Building, Basic Allowable Stress Design Load Combinations for Axial Forces

Determine the basic allowable stress design load combinations for a timber pile supporting a residential building using the nominal axial loads in the design data. The residential building is located in a Coastal A Zone.

DESIGN DATA

	Axial Force (kips)
Dead load, D	8
Live load, L	6
Roof live load, L_r	4
Wind, W	± 16
Flood, F_a	± 2

---

[19] See ASCE/SEI 12.4.3.3 for allowable stress increase for load combinations with overstrength.

SOLUTION

The basic load combinations using allowable stress design of IBC 1605.3.1 are summarized in Table 2.10 for this pile. The load combinations in the table include only the applicable load effects from the design data. Also, since flood loads F_a must be considered, the load combinations of ASCE/SEI 2.4.2 are used (IBC 1605.3.1.2). In particular, $1.5F_a$ is added to the other applicable loads in IBC Eqs. 16-12, 16-13 and 16-14.

Table 2.10 Summary of Basic Load Combinations using Allowable Stress Design (IBC 1605.3.1) for Timber Pile

Equation No.	Load Combination	Axial Load (kips)
16-8	D	8
16-9	$D + L$	14
16-10	$D + L_r$	12
16-11	$D + 0.75L + 0.75L_r$	16
16-12	$D + W + 1.5F_a$	27
16-12	$D - W - 1.5F_a$	-11
16-13	$D + 0.75W + 0.75L + 0.75L_r + 1.5F_a$	31
16-13	$D - 0.75W + 0.75L + 0.75L_r - 1.5F_a$	1
16-14	$0.6D + W$	21
16-14	$0.6D - W$	-11

The pile must be designed for the axial compression and tension forces in Table 2.10 in combination with bending moments caused by the wind and flood loads. Shear forces and deflection at the tip of the pile must also be checked. Finally, the embedment length of the pile must be sufficient to resist the maximum net tension force.[20]

---

[20] More information on flood loads can be found in Chapter 7 of this publication.

CHAPTER 3 DEAD, LIVE, AND RAIN LOADS

3.1 DEAD LOADS

Nominal dead loads D are determined in accordance with IBC 1606.[1] In general, design dead loads are the actual weights of construction materials and fixed service equipment that are attached to or supported by the building or structure. Various types of such loads are listed in IBC 1602.

Dead loads are considered to be permanent loads, i.e., loads in which variations over time are rare or of small magnitude. Variable loads, such as live loads and wind loads, are not permanent. It is important to know the distinction between permanent and variable loads when applying the provisions for load combinations.[2]

It is not uncommon that the weights of materials and service equipment (such as plumbing stacks and risers, HVAC equipment, elevators and elevator machinery, fire protection systems and similar fixed equipment) are not known during the design phase. Estimated material and/or equipment loads are often used in design. Typically, estimated dead loads are assumed to be greater than the actual dead loads so that the design is conservative. While such practice is acceptable when considering load combinations where the effects of gravity loads and lateral loads are additive, it is not acceptable when considering load combinations where gravity loads and lateral loads counteract. For example, it would be unconservative to design for uplift on a structure using an overestimated value of dead load.

ASCE/SEI Table C3-1 provides minimum design dead loads for various types of common construction components, including ceilings, roof and wall coverings, floor fill, floors and floor finishes, frame partitions and frame walls. Minimum densities for common construction materials are given in ASCE/SEI Table C3-2.

The weights in ASCE/SEI Tables C3-1 and C3-2 can be used as a guide when estimating dead loads. Actual weights of construction materials and equipment can be greater than tabulated values, so it is always prudent to verify weights with manufacturers or other similar resources prior to design. In cases where information on dead load is unavailable, values of dead loads used in design must be approved by the building official (IBC 1606.2).

---

[1] Nominal loads are those loads that are specified in Chapter 16 of the IBC.
[2] See IBC 1605, ASCE/SEI Chapter 2 and Chapter 2 of this publication for information on load combinations.

3.2 LIVE LOADS

3.2.1 General

Nominal live loads are determined in accordance with IBC 1607. Live loads are those loads produced by the use and occupancy of a building or structure and do not include construction loads, environmental loads (such as wind loads, snow loads, rain loads, earthquake loads and flood loads) or dead loads (IBC 1602).

In general, live loads are transient in nature and vary in magnitude over the life of a structure. Studies have shown that building live loads consist of both a sustained portion and a variable portion. The sustained portion is based on general day-to-day use of the facilities, and will generally vary during the life of the structure due to tenant modifications and changes in occupancy, for example. The variable portion of the live load is typically created by events such as remodeling, temporary storage and similar unusual events.

Nominal design values of uniformly distributed and concentrated live loads are given in IBC Table 1607.1 as a function of occupancy or use. The occupancy category listed in the table is not necessarily group specific.[3] As an example, an office building with a Business Group B classification may also have storage areas that may warrant live loads of 125 psf or 250 psf depending on the type of storage.

The design values in IBC Table 1607.1 are minimum values; actual design values can be determined to be larger than these values, but in no case shall the structure be designed for live loads that are less than the tabulated values. For occupancies that are not listed in the table, values of live load used in design must be approved by the building official. It is also important to note that the provisions do not require concurrent application of uniform and concentrated loads. Structural members are designed based on the maximum effects due to the application of either a uniform load or a concentrated load, and need not be designed for the effects of both loads applied at the same time.

Provisions for the weight of partitions are given in IBC 1607.5. Buildings where partitions can be relocated must include a live load of 15 psf if the nominal uniform floor live load is less than or equal to 80 psf. The weight of any built-in partitions that cannot be moved is considered a dead load in accordance with IBC 1602.

Minimum live loads for truck and bus garages are given in IBC 1607.6. IBC 1607.7 prescribes loads on handrails, guards, grab bars and vehicle barriers, and IBC 1607.8 contains provisions for impact loads that involve unusual vibration and impact forces, such as those from elevators and machinery.

ASCE/SEI Table 4-1 also contains minimum uniform and concentrated live loads. There are some differences between this table and IBC Table 1607.1. ASCE Tables C4-1 and C4-2 can also be used as a guide in establishing live loads for commonly encountered occupancies.

---

[3] Occupancy groups are defined in IBC Chapter 3.

3.2.2 Reduction in Live Loads

According to IBC 1607.9, the minimum nominal uniformly distributed live loads L_o of IBC Table 1607.1 and uniform live loads of special purpose roofs are permitted to be reduced by either the provisions of IBC 1607.9.1 or 1607.9.2. Both methods are discussed below. Roof live loads, other than special-purpose roofs, are not permitted to be reduced by these provisions; reduction of such roof live loads is covered in IBC 1607.11 and Section 3.2.4 of this publication.

General Method of Live Load Reduction

The general method in IBC 1607.9.1 of reducing uniform live loads other than uniform live loads at roofs is based on the provisions contained in ASCE/SEI 4.8. IBC Eq. 16-22 can be used to obtain a reduced live load L for members where $K_{LL}A_T \geq 400$ sq ft, subject to the limitations of IBC 1607.9.1.1 through 1607.9.1.5:

$$L = L_o \left(0.25 + \frac{15}{\sqrt{K_{LL}A_T}} \right)$$

$\geq 0.50L_o$ for members supporting one floor

$\geq 0.40L_o$ for members supporting two or more floors

where K_{LL} live load element factor given in IBC Table 1607.9.1 and A_T = tributary area in square feet.

The live load element factor K_{LL} converts the tributary area of a structural member A_T to an influence area, which is considered to be the adjacent floor area from which the member derives any of its load. For example, the influence area for an interior column is equal to the area of the four bays adjacent to the column, which is equal to four times the tributary area of the column. Thus, $K_{LL} = 4$ for an interior column. ASCE/SEI Figure C4 illustrates typical tributary and influence areas for a variety of elements. Figure 3.1 illustrates how the reduction multiplier $0.25 + 15/\sqrt{K_{LL}A_T}$ varies with respect to the influence area $K_{LL}A_T$. Included in the figure are the minimum influence area of 400 sq ft and the limits of 0.5 and 0.4, which are the maximum permitted reductions for members supporting one floor and two or more floors, respectively.

One-way Slabs. Live load reduction on one-way slabs is now permitted in the 2009 IBC provided the tributary area A_T does not exceed an area equal to the slab span times a width normal to the span of 1.5 times the slab span (i.e., an area with an aspect ratio of 1.5). The live load will be somewhat higher for a one-way slab with an aspect ratio of 1.5 than for a two-way slab with the same aspect ratio. This recognizes the benefits of higher redundancy that results from two-way action.

ASCE 4.8.5 has the same requirements for live load reduction on one-way slabs as that in IBC 1607.9.1.1.

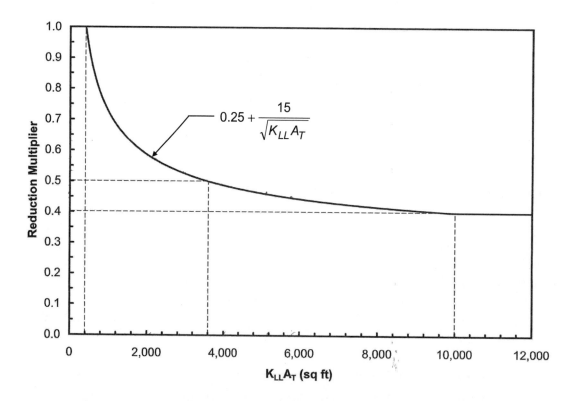

Figure 3.1 Reduction Multiplier for Live Load in accordance with IBC 1607.9.1

Heavy Live Loads. According to IBC 1607.9.1.2, live loads that are greater than 100 psf shall not be reduced except for the following: (1) Live loads for members supporting two or more floors are permitted to be reduced by a maximum of 20 percent, but L shall not be less than that calculated by IBC 1607.9.1, or (2) In occupancies other than storage, additional live load reduction is permitted if it can be shown by a registered design professional that such a reduction is warranted.

Passenger Vehicle Garages. The live load in passenger vehicle garages is not permitted to be reduced, except for members supporting two or more floors; in such cases, the maximum reduction is 20 percent, but L shall not be less than that calculated by IBC 1607.9.1 (IBC 1607.9.1.3).

Group A (Assembly) Occupancies. Due to the nature of assembly occupancies, there is a high probability that the entire floor is subjected to full uniform live load. Thus, IBC 1607.9.1.4 requires that live loads of 100 psf and live loads at areas where fixed seats are located shall not be reduced in such occupancies.

Roof Members. Live loads of 100 psf or less are not permitted to be reduced on roof members except as specified in IBC 1607.11.2 for flat, pitched and curved roofs and special-purpose roofs (IBC 1607.9.1.5).

A summary of the general method of live load reduction for floors in accordance with IBC 1607.9.1 is given in Figure 3.2.

LIVE LOADS 3-5

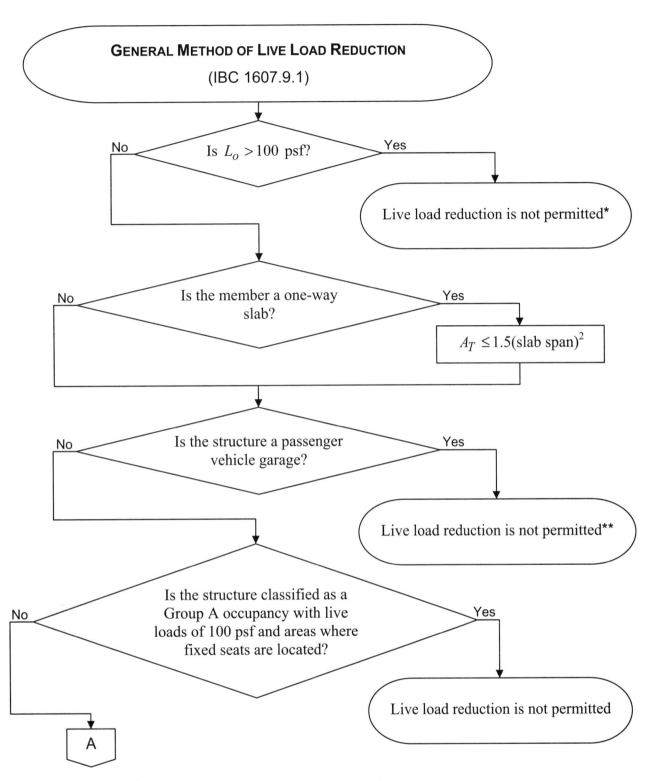

Figure 3.2 *General Method of Live Load Reduction in accordance with IBC 1607.9.1*

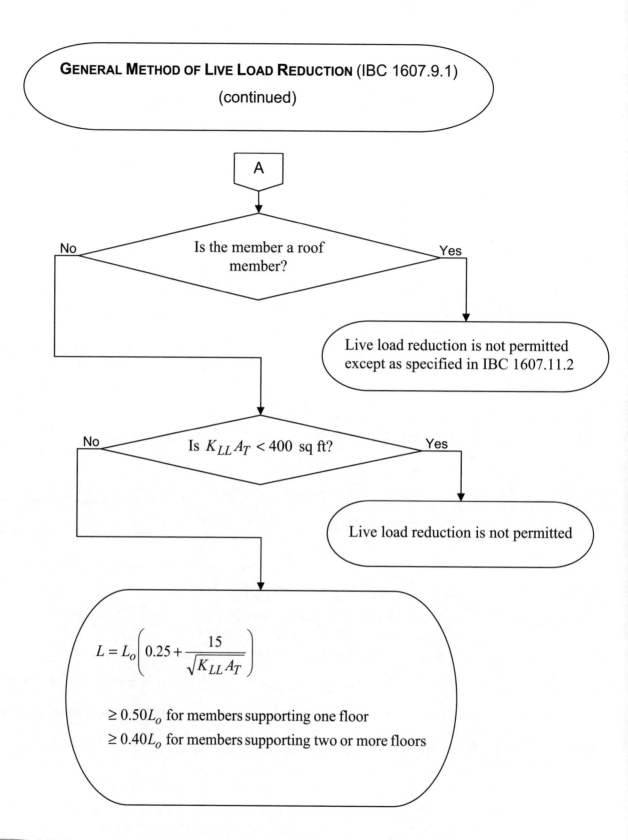

Figure 3.2 General Method of Live Load Reduction in accordance with IBC 1607.9.1 (continued)

Alternate Method of Floor Live Load Reduction

An alternate method of floor live load reduction, which is based on provisions in the 1997 *Uniform Building Code,* is given in IBC 1607.9.2. IBC Eq. 16-23 can be used to obtain a reduction factor R for members that support an area greater than or equal to 150 sq ft:

$$R = 0.08(A - 150)$$

$$\leq \text{ the smallest of } \begin{cases} 40 \text{ percent for horizontal members} \\ 60 \text{ percent for vertical members} \\ 23.1(1 + D/L_o) \end{cases}$$

where A = area of floor supported by a member in square feet and D = dead load per square foot of area supported.

The reduced live load L is then determined by

$$L = L_o \left(1 - \frac{R}{100}\right)$$

Similar to the general method of live load reduction, live loads are not permitted to be reduced in the following situations:

1. In Group A (assembly) occupancies.

2. Where the live load exceeds 100 psf except (a) for members supporting two or more floors in which case the live load may be reduced by a maximum of 20 percent or (b) in occupancies other than storage where it can be shown by a registered design professional that such a reduction is warranted.

3. In passenger vehicle garages except for members supporting two or more floors in which case the live load may be reduced by a maximum of 20 percent.

Reduction of live load on one-way slab systems is permitted by this method provided the area A is not taken greater than that prescribed in IBC 1607.9.2(5).

A summary of the alternate method of floor live load reduction in accordance with IBC 1607.9.2 is given in Figure 3.3.

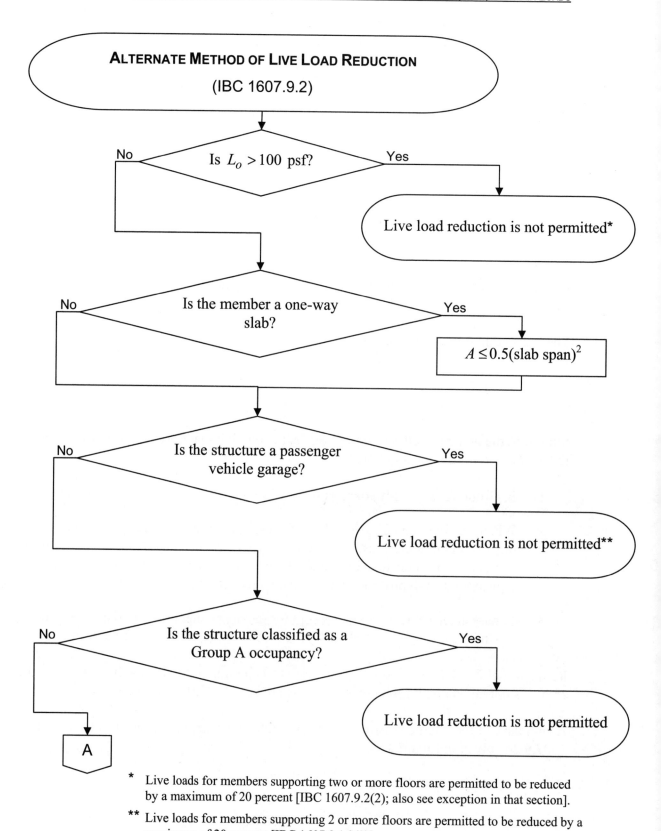

Figure 3.3 *Alternate Method of Live Load Reduction in accordance with IBC 1607.9.2*

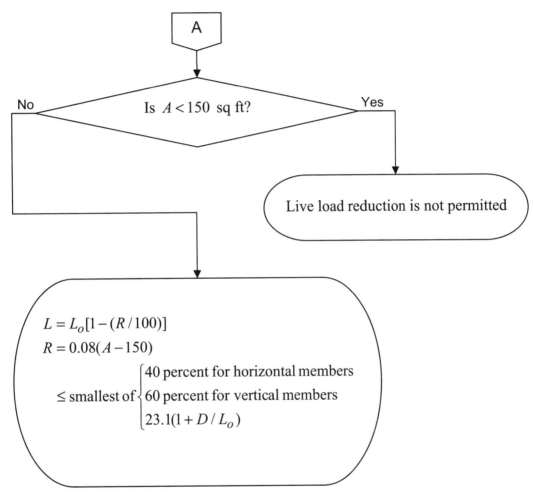

Figure 3.3 Alternate Method of Live Load Reduction in accordance with IBC 1607.9.2 (continued)

3.2.3 Distribution of Floor Loads

IBC 1607.10 requires that the effects of partial uniform live loading (or alternate span loading) be investigated when analyzing continuous floor members. Such loading produces greatest effects at different locations along the span. Reduced floor live loads may be used when performing this analysis.

Figure 3.4 illustrates four loading patterns that need to be investigated for a three-span continuous system subject to dead and live loads.

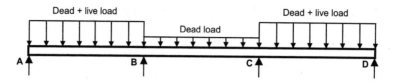

Loading pattern for maximum negative moment at support A or D
and maximum positive moment in span AB or CD

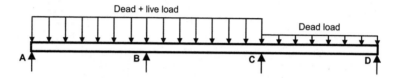

Loading pattern for maximum negative moment at support B

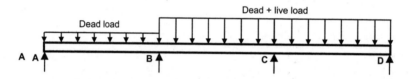

Loading pattern for maximum negative moment at support C

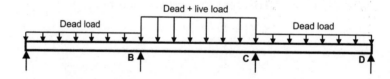

Loading pattern for maximum positive moment in span BC

Figure 3.4 Distribution of Floor Loads for a Three-span Continuous System in accordance with IBC 1607.10

3.2.4 Roof Loads

In general, roofs are to be designed to resist dead, live, wind, and where applicable, rain, snow and earthquake loads. A minimum roof live load of 20 psf is prescribed in IBC Table 1607.1 for typical roof structures, while larger live loads are required for roofs used as gardens or places of assembly.

IBC 1607.11.2 permits nominal roof live loads of 20 psf on flat, pitched and curved roofs to be reduced in accordance with Eq. 16-25:

$$L_r = L_o R_1 R_2, \quad 12 \leq L_r \leq 20$$

where L_r = reduced live load per square foot of horizontal roof projection

$$R_1 = \begin{cases} 1 & \text{for } A_t \leq 200 \text{ sq ft} \\ 1.2 - 0.001 A_t & \text{for } 200 \text{ sq ft} < A_t < 600 \text{ sq ft} \\ 0.6 & \text{for } A_t \geq 600 \text{ sq ft} \end{cases}$$

$$R_2 = \begin{cases} 1 & \text{for } F \leq 4 \\ 1.2 - 0.05 F & \text{for } 4 < F < 12 \\ 0.6 & \text{for } F > 12 \end{cases}$$

A_t = tributary area (span length multiplied by effective width) in square feet supported by a structural member

F = for a sloped roof, the number of inches of rise per foot; for an arch or dome, the rise-to-span ratio multiplied by 32

As seen from Eq. 16-25, roof live load reduction is based on tributary area of the member being considered and the slope of the roof. No live load reduction is permitted for members supporting less than or equal to 200 sq ft as well as for roof slopes less than or equal to 4:12. In no case is the reduced roof live load to be taken less than 12 psf.

Live load reduction of special purpose roofs (roofs used as promenade decks, roof gardens, roofs used as places of assembly, etc.) is permitted in accordance with the provisions of IBC 1607.9 for floors (IBC 1607.11.2.2). Live loads that are equal to or greater than 100 psf at areas of roofs that are classified as Group A (assembly) occupancies are not permitted to be reduced.

Landscaped roofs are to be designed for a minimum roof live load of 20 psf (IBC 1607.11.3). The weight of landscaping material is considered as dead load, considering the saturation level of the soil.

A minimum roof live load of 5 psf is required for awnings and canopies in accordance with IBC Table 1607.1 (IBC 1607.11.4). Such elements must also be designed for the combined effects of snow and wind loads in accordance with IBC 1605.

3.2.5 Crane Loads

A general description of the crane loads that must be considered is given in IBC 1607.12. In general, the support structure of the crane must be designed for the maximum wheel load, vertical impact and horizontal impact as a simultaneous load combination. Provisions on how to determine each of these component loads are given in IBC 1607.12.

3.2.6 Interior Walls and Partitions

Interior walls and partitions (including their finishing materials) greater than 6 ft in height are required to be designed for a horizontal load of 5 psf (IBC 1607.13). This requirement is intended to provide sufficient strength and durability of the wall framing and its finished construction when subjected to nominal impact loads, such as those from moving furniture or equipment, and from HVAC pressurization.

Requirements for fabric partitions are given in IBC 1607.13.1.

3.3 RAIN LOADS

IBC 1611 contains requirements for design rain loads. IBC Eq. 16-35 is used to determine the rain load R on an undeflected roof:

$$R = 5.2(d_s + d_h)$$

where d_s = depth of water on the undeflected roof up to the inlet of the secondary drainage system when the primary drainage system is blocked and d_h = additional depth of water on the undeflected roof above the inlet of the secondary drainage system at its design flow.[4]

The nominal rain load R represents the weight of accumulated rainwater on the roof, assuming that the primary roof drainage system is blocked. The primary roof drainage system is designed for the 100-year hourly rainfall rate indicated in IBC Figure 1611.1 and the area of roof that it drains; it can include, for example, roof drains, leaders, conductors and horizontal storm drains within the structure.[5]

When primary roof drainage is blocked, water will rise above the primary roof drain until it reaches the elevation of the roof edge, scuppers or other secondary drains. The depth of water above the primary drain at the design rainfall intensity is based on the flow rate of the secondary drainage system, which depends on the type of drainage system.[6]
Figure 3.5 illustrates the water depths d_s and d_h that are to be used in Eq. 16-35 for the case of a scupper acting as a secondary drain. Similarly, Figure 3.6 illustrates these water depths for a typical interior secondary drain.

---

[4] The constant in Eq. 16-35 is equal to the unit load per inch depth of water = 62.4/12 = 5.2 psf/in. An undeflected roof refers to the case where deflections from loads (including dead loads) are not considered when determining the amount of rainwater on the roof.

[5] IBC Figure 1611.1 provides the rainfall rates for a storm of 1-hour duration that has a 100-year return period. These rates are calculated by a statistical analysis of weather records. See Chapter 11 of the *International Plumbing Code®* (IPC®) for requirements on the design of roof drainage systems. Rainfall rates are given for various cities in the U.S. in Appendix B of the IPC. The rates are based on the maps in IPC Figure 1106.1, which have the same origin as the maps in the IBC.

[6] ASCE/SEI Table C8-1 gives flow rates in gallons per minute and corresponding hydraulic heads for various types of drainage systems. For example, a 6-inch wide by 4-inch high closed scupper with 3 inches of hydraulic head will discharge 90 gallons per minute.

RAIN LOADS

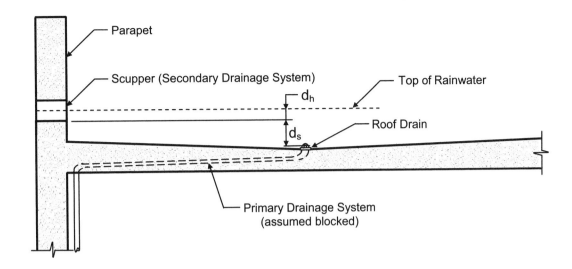

Figure 3.5 Example of Water Depths d_s and d_h in accordance with IBC 1611 for Typical Perimeter Scuppers

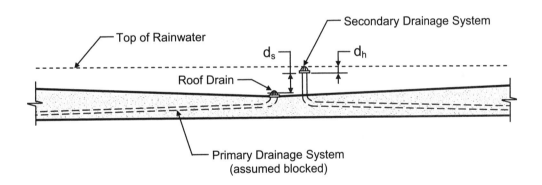

Figure 3.6 Example of Water Depths d_s and d_h in accordance with IBC 1611 for Typical Interior Drains

Where buildings are configured such that rainwater will not collect on the roof, no rain load is required in the design of the roof, and a secondary drainage system is not needed. What is important to note is that the provisions of IBC 1611 must be considered wherever the potential exists that water may accumulate on the roof.

3.4 EXAMPLES

The following examples illustrate the IBC requirements for live load reduction and rain loads.

3.4.1 Example 3.1 – Live Load Reduction, General Method of IBC 1607.9.1

The typical floor plan of a 10-story reinforced concrete office building is illustrated in Figure 3.7.

Determine reduced live loads for

    (1) column A3

    (2) column B3

    (3) column C1

    (4) column C4

    (5) column B6

    (6) two-way slab AB23

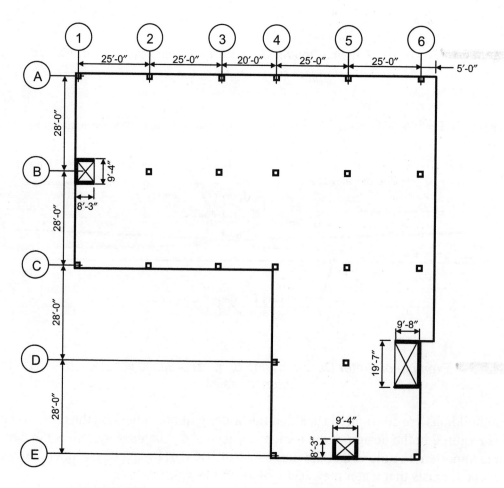

Figure 3.7 *Typical Floor Plan of 10-story Office Building*

The ninth floor is designated as a storage floor (125 psf) and all other floors are typical office floors with moveable partitions.

EXAMPLES

The roof is an ordinary flat roof (slope of 1/2 on 12), which is not used as a place of public assembly or for any special purposes. Assume that rainwater does not collect on the roof, and neglect snow loads.

Neglect lobby/corridor loads on the typical floors for this example.

SOLUTION

Nominal Loads

- Roof: 20 psf in accordance with IBC Table 1607.1, since the roof is an ordinary flat roof that is not used for public assembly or any other special purposes. (29)

- Ninth floor: storage load is given in the design criteria as 125 psf.

- Typical floor: 50 psf for office space in accordance with IBC Table 1607.1 and 15 psf for moveable partitions in accordance with IBC 1607.5, since the live load does not exceed 80 psf. The partition load is not reducible; only the minimum loads in IBC Table 1607.1 are permitted to be reduced (IBC 1607.9).

Part 1: Determine reduced live load for column A3

A summary of the reduced live loads is given in Table 3.1. Detailed calculations for various floor levels follow the table.

Table 3.1 Summary of Reduced Live Loads for Column A3

Story	Live Load (psf) N	Live Load (psf) R	$K_{LL}A_T$ (sq ft)	Reduction Multiplier	Reduced Live Load (psf)	N + R (kips)	Cumulative N + R (kips)
10	---	20	---*	---	17.8	5.6	5.6
9	125	---	---**	---	---	39.4	45.0
8	15	50	1,260	0.67	33.5	15.3	60.3
7	15	50	2,520	0.55	27.5	13.4	73.7
6	15	50	3,780	0.49	24.5	12.4	86.1
5	15	50	5,040	0.46	23.0	12.0	98.1
4	15	50	6,300	0.44	22.0	11.7	109.8
3	15	50	7,560	0.42	21.0	11.3	121.1
2	15	50	8,820	0.41	20.5	11.2	132.3
1	15	50	10,080	0.40	20.0	11.0	143.3

N = nonreducible live load, R = reducible live load
* Roof live load reduced in accordance with IBC 1607.11.2
** Live load > 100 psf is not permitted to be reduced (IBC 1607.9.1.2)

- Roof

  The reduced roof live load L_r is determined by Eq. 16-25:

$$L_r = L_o R_1 R_2 = 20 R_1 R_2$$

The tributary area A_t of column A3 $= (28/2) \times 22.5 = 315$ sq ft

Since 200 sq ft $< A_t <$ 600 sq ft, R_1 is determined by Eq. 16-27:

$$R_1 = 1.2 - 0.001 A_t = 1.2 - (0.001 \times 315) = 0.89$$

Since $F = 1/2 < 4$, $R_2 = 1$ (Eq. 16-29)

Thus, $L_r = 20 \times 0.89 \times 1 = 17.8$ psf

Axial load $= 17.8 \times 315 / 1,000 = 5.6$ kips

- Ninth floor

    Since the ninth floor is storage with a live load of 125 psf, which exceeds 100 psf, the live load is not permitted to be reduced (IBC 1607.9.1.2).

    Axial load $= 125 \times 315 / 1,000 = 39.4$ kips

- Typical floors

    Reducible nominal live load = 50 psf

    Since column A3 is an exterior column without a cantilever slab, the live load element factor $K_{LL} = 4$ (IBC Table 1607.9.1).[7]

    Reduced live load L is determined by Eq. 16-22:

$$L = L_o \left(0.25 + \frac{15}{\sqrt{K_{LL} A_T}} \right)$$

$\geq 0.50 L_o$ for members supporting one floor

$\geq 0.40 L_o$ for members supporting two or more floors

The reduction multiplier is equal to 0.40 where $K_{LL} A_T \geq 10,000$ sq ft (see Figure 3.1).

Axial load $= (L + 15) A_T = 315(L + 15)$

---

[7] K_{LL} = influence area/tributary area = 28(25 + 20)/315 = 4.

Part 2: Determine reduced live load for column B3

A summary of the reduced live loads is given in Table 3.2. Detailed calculations for various floor levels follow the table.

Table 3.2 Summary of Reduced Live Loads for Column B3

Story	Live Load (psf) N	Live Load (psf) R	$K_{LL}A_T$ (sq ft)	Reduction Multiplier	Reduced Live Load (psf)	N + R (kips)	Cumulative N + R (kips)
10	---	20	---*	---	12.0	7.6	7.6
9	125	---	---**	---	---	78.8	86.4
8	15	50	2,520	0.55	27.5	26.8	113.2
7	15	50	5,040	0.46	23.0	23.9	137.1
6	15	50	7,560	0.42	21.0	22.7	159.8
5	15	50	10,080	0.40	20.0	22.1	181.9
4	15	50	12,600	0.40	20.0	22.1	204.0
3	15	50	15,120	0.40	20.0	22.1	226.1
2	15	50	17,640	0.40	20.0	22.1	248.2
1	15	50	20,160	0.40	20.0	22.1	270.3

N = nonreducible live load, R = reducible live load
* Roof live load reduced in accordance with IBC 1607.11.2
** Live load > 100 psf is not permitted to be reduced (IBC 1607.9.1.2)

- Roof

    The reduced roof live load L_r is determined by Eq. 16-25:

$$L_r = L_o R_1 R_2 = 20 R_1 R_2$$

    The tributary area A_t of column B3 $= 28 \times 22.5 = 630$ sq ft

    Since $A_t > 600$ sq ft, R_1 is determined by Eq. 16-28: $R_1 = 0.6$

    Since $F = 1/2 < 4$, $R_2 = 1$ (Eq. 16-29)

    Thus, $L_r = 20 \times 0.6 \times 1 = 12.0$ psf

    Axial load $= 12.0 \times 630 / 1,000 = 7.6$ kips

- Ninth floor

    Since the ninth floor is storage with a live load of 125 psf, which exceeds 100 psf, the live load is not permitted to be reduced (IBC 1607.9.1.2).

Axial load $= 125 \times 630 / 1{,}000 = 78.8$ kips

- Typical floors

Reducible nominal live load = 50 psf

Since column B3 is an interior column, the live load element factor $K_{LL} = 4$ (IBC Table 1607.9.1).[8]

Reduced live load L is determined by Eq. 16-22:

$$L = L_o \left(0.25 + \frac{15}{\sqrt{K_{LL} A_T}} \right)$$

$\geq 0.50 L_o$ for members supporting one floor

$\geq 0.40 L_o$ for members supporting two or more floors

The reduction multiplier is equal to 0.40 where $K_{LL} A_T \geq 10{,}000$ sq ft (see Figure 3.1).

Axial load $= (L + 15) A_T = 630(L + 15)$

Part 3: Determine reduced live load for column C1

A summary of the reduced live loads is given in Table 3.3. Detailed calculations for various floor levels follow the table.

Table 3.3 Summary of Reduced Live Loads for Column C1

Story	Live Load (psf) N	Live Load (psf) R	$K_{LL} A_T$ (sq ft)	Reduction Multiplier	Reduced Live Load (psf)	N + R (kips)	Cumulative N + R (kips)
10	---	20	---*	---	20.0	3.5	3.5
9	125	---	---**	---	---	21.9	25.4
8	15	50	700	0.82	41.0	9.8	35.2
7	15	50	1,400	0.65	32.5	8.3	43.5
6	15	50	2,100	0.58	29.0	7.7	51.2
5	15	50	2,800	0.53	26.5	7.3	58.5
4	15	50	3,500	0.50	25.0	7.0	65.5
3	15	50	4,200	0.48	24.0	6.8	72.3
2	15	50	4,900	0.46	23.0	6.7	79.0
1	15	50	5,600	0.45	22.5	6.6	85.6

N = nonreducible live load, R = reducible live load
* Roof live load reduced in accordance with IBC 1607.11.2
** Live load > 100 psf is not permitted to be reduced (IBC 1607.9.1.2)

---

[8] K_{LL} = influence area/tributary area = $2[(28 \times 25) + (28 \times 20)]/630 = 4$.

EXAMPLES   3-19

- Roof

    The reduced roof live load L_r is determined by Eq. 16-25:

    $$L_r = L_o R_1 R_2 = 20 R_1 R_2$$

    The tributary area A_t of column C1 $= 28 \times 25 / 4 = 175$ sq ft

    Since $A_t < 200$ sq ft, R_1 is determined by Eq. 16-26: $R_1 = 1$

    Since $F = 1/2 < 4$, $R_2 = 1$ (Eq. 16-29)

    Thus, $L_r = 20 \times 1 \times 1 = 20.0$ psf

    Axial load $= 20.0 \times 175 / 1,000 = 3.5$ kips

- Ninth floor

    Since the ninth floor is storage with a live load of 125 psf, which exceeds 100 psf, the live load is not permitted to be reduced (IBC 1607.9.1.2).

    Axial load $= 125 \times 175 / 1,000 = 21.9$ kips

- Typical floors

    Reducible nominal live load = 50 psf

    Since column C1 is an exterior column without a cantilever slab, the live load element factor $K_{LL} = 4$ (IBC Table 1607.9.1).[9]

    Reduced live load L is determined by Eq. 16-22:

    $$L = L_o \left(0.25 + \frac{15}{\sqrt{K_{LL} A_T}} \right)$$

    $\geq 0.50 L_o$ for members supporting one floor
    $\geq 0.40 L_o$ for members supporting two or more floors

    Axial load $= (L + 15) A_T = 175(L + 15)$

---

[9] K_{LL} = influence area/tributary area = $(28 \times 25)/175 = 4$.

Part 4: Determine reduced live load for column C4

A summary of the reduced live loads is given in Table 3.4. Detailed calculations for various floor levels follow the table.

Table 3.4 Summary of Reduced Live Loads for Column C4

Story	Live Load (psf) N	Live Load (psf) R	$K_{LL}A_T$ (sq ft)	Reduction Multiplier	Reduced Live Load (psf)	N + R (kips)	Cumulative N + R (kips)
10	---	20	---*	---	14.2	7.0	7.0
9	125	---	---**	---	---	61.3	68.3
8	15	50	1,960	0.59	29.5	21.8	90.1
7	15	50	3,920	0.49	24.5	19.4	109.5
6	15	50	5,880	0.45	22.5	18.4	127.9
5	15	50	7,840	0.42	21.0	17.6	145.5
4	15	50	9,800	0.40	20.0	17.2	162.7
3	15	50	11,760	0.40	20.0	17.2	179.9
2	15	50	13,720	0.40	20.0	17.2	197.1
1	15	50	15,680	0.40	20.0	17.2	214.3

N = nonreducible live load, R = reducible live load
* Roof live load reduced in accordance with IBC 1607.11.2
** Live load > 100 psf is not permitted to be reduced (IBC 1607.9.1.2)

- Roof

    The reduced roof live load L_r is determined by Eq. 16-25:

    $L_r = L_o R_1 R_2 = 20 R_1 R_2$

    The tributary area A_t of column C4 $= (28 \times 20)/4 + (28 \times 25)/2 = 490$ sq ft

    Since 200 sq ft $< A_t <$ 600 sq ft, R_1 is determined by Eq. 16-27:

    $R_1 = 1.2 - 0.001 A_t = 1.2 - (0.001 \times 490) = 0.71$

    Since $F = 1/2 < 4$, $R_2 = 1$ (Eq. 16-29)

    Thus, $L_r = 20.0 \times 0.71 \times 1 = 14.2$ psf

    Axial load $= 14.2 \times 490 / 1,000 = 7.0$ kips

- Ninth floor

    Since the ninth floor is storage with a live load of 125 psf, which exceeds 100 psf, the live load is not permitted to be reduced (IBC 1607.9.1.2).

    Axial load $= 125 \times 490 / 1{,}000 = 61.3$ kips

- Typical floors

    Reducible nominal live load = 50 psf

    Since column C1 is an exterior column without a cantilever slab, the live load element factor $K_{LL} = 4$ (IBC Table 1607.9.1).[10]

    Reduced live load L is determined by Eq. 16-22:

$$L = L_o \left(0.25 + \frac{15}{\sqrt{K_{LL} A_T}} \right)$$

$\geq 0.50 L_o$ for members supporting one floor

$\geq 0.40 L_o$ for members supporting two or more floors

The reduction multiplier is equal to 0.40 where $K_{LL} A_T \geq 10{,}000$ sq ft.

Axial load $= (L + 15) A_T = 490(L + 15)$

Part 5: *Determine reduced live load for column B6*

A summary of the reduced live loads is given in Table 3.5. Detailed calculations for various floor levels follow.

- Roof

    The reduced roof live load L_r is determined by Eq. 16-25:

$$L_r = L_o R_1 R_2 = 20 R_1 R_2$$

The tributary area A_t of column B6 $= 28 \times (25/2 + 5) = 490$ sq ft

Since 200 sq ft $< A_t <$ 600 sq ft, R_1 is determined by Eq. 16-27:

$$R_1 = 1.2 - 0.001 A_t = 1.2 - (0.001 \times 490) = 0.71$$

---

[10] K_{LL} = influence area/tributary area = $28[20 + (2 \times 25)]/490 = 4$.

Since $F = 1/2 < 4$, $R_2 = 1$ (Eq. 16-29)

Thus, $L_r = 20 \times 0.71 \times 1 = 14.2$ psf

Axial load $= 14.2 \times 490/1{,}000 = 7.0$ kips

Table 3.5 Summary of Reduced Live Loads for Column B6

Story	Live Load (psf) N	Live Load (psf) R	$K_{LL}A_T$ (sq ft)	Reduction Multiplier	Reduced Live Load (psf)	N + R (kips)	Cumulative N + R (kips)
10	---	20	---*	---	14.2	7.0	7.0
9	125	---	---**	---	---	61.3	68.3
8	15	50	1,470	0.64	32.0	23.0	91.3
7	15	50	2,940	0.53	26.5	20.3	111.6
6	15	50	4,410	0.48	24.0	19.1	130.7
5	15	50	5,880	0.45	22.5	18.4	149.1
4	15	50	7,350	0.42	21.0	17.6	166.7
3	15	50	8,820	0.41	20.5	17.4	184.1
2	15	50	10,290	0.40	20.0	17.2	201.3
1	15	50	11,760	0.40	20.0	17.2	218.5

N = nonreducible live load, R = reducible live load
* Roof live load reduced in accordance with IBC 1607.11.2
** Live load > 100 psf is not permitted to be reduced (IBC 1607.9.1.2)

- Ninth floor

    Since the ninth floor is storage with a live load of 125 psf, which exceeds 100 psf, the live load is not permitted to be reduced (IBC 1607.9.1.2).

    Axial load $= 125 \times 490/1{,}000 = 61.3$ kips

- Typical floors

    Reducible nominal live load = 50 psf

    Column B6 is an exterior column with a cantilever slab; thus, the live load element factor $K_{LL} = 3$ (IBC Table 1607.9.1).[11]

    Reduced live load L is determined by Eq. 16-22:

---

[11] Actual influence area/tributary area $= 2[(28 \times 25) + (5 \times 28)]/490 = 3.4$. IBC Table 1607.9.1 requires $K_{LL} = 3$, which is slightly conservative.

$$L = L_o \left(0.25 + \frac{15}{\sqrt{K_{LL} A_T}} \right)$$

$\geq 0.50 L_o$ for members supporting one floor

$\geq 0.40 L_o$ for members supporting two or more floors

The reduction multiplier is equal to 0.40 where $K_{LL} A_T \geq 10{,}000$ sq ft (see Figure 3.1).

Axial load $= (L+15) A_T = 490(L+15)$

Part 6: Determine reduced live load for two-way slab AB23

- Roof

The reduced roof live load L_r is determined by Eq. 16-25:

$$L_r = L_o R_1 R_2 = 20 R_1 R_2$$

The tributary area A_t of this slab $= 25 \times 28 = 700$ sq ft

Since $A_t > 600$ sq ft, R_1 is determined by Eq. 16-28: $R_1 = 0.6$

Since $F = 1/2 < 4$, $R_2 = 1$ (Eq. 16-29)

Thus, $L_r = 20 \times 0.6 \times 1 = 14.0$ psf

- Ninth floor

Since the ninth floor is storage with a live load of 125 psf, which exceeds 100 psf, the live load is not permitted to be reduced (IBC 1607.9.1.2).

Live load = 125 psf

- Typical floors

Reducible nominal live load = 50 psf

According to IBC Table 1607.9.1, $K_{LL} = 1$ for a two-way slab.

Reduced live load L is determined by Eq. 16-22:

$$L = L_o\left(0.25 + \frac{15}{\sqrt{K_{LL}A_T}}\right) = L_o\left(0.25 + \frac{15}{\sqrt{700}}\right) = 0.82 L_o$$

$= 41.0$ psf

$> 0.50 L_o$ for members supporting one floor

Total live load $= 41 + 15 = 56$ psf

3.4.2   Example 3.2 – Live Load Reduction, Alternate Method of IBC 1607.9.2

Determine the reduced live loads for the elements in Example 3.1 using the alternate floor live load reduction of IBC 1607.9.2. Assume a nominal dead-to-live load ratio of 2.

SOLUTION

Part 1: Determine reduced live load for column A3

A summary of the reduced live loads is given in Table 3.6. Detailed calculations for various floor levels follow the table.

Table 3.6 Summary of Reduced Live Loads for Column A3

Story	Live Load (psf) N	Live Load (psf) R	A (sq ft)	Reduction Factor, R (%)	Reduced Live Load (psf)	N + R (kips)	Cumulative N + R (kips)
10	---	20	---*	---	17.8	5.6	5.6
9	125	---	---**	---	---	39.4	45.0
8	15	50	315	13	43.5	18.4	63.4
7	15	50	630	38	31.0	14.5	77.9
6	15	50	945	60	20.0	11.0	88.9
5	15	50	1,260	60	20.0	11.0	99.9
4	15	50	1,575	60	20.0	11.0	110.9
3	15	50	1,890	60	20.0	11.0	121.9
2	15	50	2,205	60	20.0	11.0	132.9
1	15	50	2,520	60	20.0	11.0	143.9

N = nonreducible live load, R = reducible live load
* Roof live load reduced in accordance with IBC 1607.11.2
** Live load > 100 psf is not permitted to be reduced [IBC 1607.9.2(2)]

- Roof

The reduced roof live load L_r is determined by Eq. 16-25:

$$L_r = L_o R_1 R_2 = 20 R_1 R_2$$

The tributary area A_t of column A3 $= (28/2) \times 22.5 = 315$ sq ft

Since 200 sq ft $< A_t <$ 600 sq ft, R_1 is determined by Eq. 16-27:

$R_1 = 1.2 - 0.001 A_t = 1.2 - (0.001 \times 315) = 0.89$

Since $F = 1/2 < 4$, $R_2 = 1$ (Eq. 16-29)

Thus, $L_r = 20 \times 0.89 \times 1 = 17.8$ psf

Axial load $= 17.8 \times 315 / 1,000 = 5.6$ kips

- Ninth floor

Since the ninth floor is storage with a live load of 125 psf, which exceeds 100 psf, the live load is not permitted to be reduced [IBC 1607.9.2(2)].

Axial load $= 125 \times 315 / 1,000 = 39.4$ kips

- Typical floors

Reducible nominal live load = 50 psf

Reduction factor R is given by Eq. 16-23:

$R = 0.08(A - 150)$

$\le$ the smallest of $\begin{cases} 60 \text{ percent for vertical members (governs)} \\ 23.1(1 + D/L_o) = 23.1(1 + 2) = 69 \text{ percent} \end{cases}$

Axial load $= [L_o(1 - 0.01R) + 15]A = 315[50(1 - 0.01R) + 15]$

Part 2: Determine reduced live load for column B3

A summary of the reduced live loads is given in Table 3.7. Detailed calculations for various floor levels follow.

- Roof

The reduced roof live load L_r is determined by Eq. 16-25:

$L_r = L_o R_1 R_2 = 20 R_1 R_2$

The tributary area A_t of column B3 $= 28 \times 22.5 = 630$ sq ft

Since $A_t > 600$ sq ft, R_1 is determined by Eq. 16-28: $R_1 = 0.6$

Since $F = 1/2 < 4$, $R_2 = 1$ (Eq. 16-29)

Thus, $L_r = 20 \times 0.6 \times 1 = 12.0$ psf

Axial load $= 12.0 \times 630/1,000 = 7.6$ kips

Table 3.7 Summary of Reduced Live Loads for Column B3

Story	Live Load (psf) N	Live Load (psf) R	A (sq ft)	Reduction Factor, R (%)	Reduced Live Load (psf)	N + R (kips)	Cumulative N + R (kips)
10	---	20	---*	---	12.0	7.6	7.6
9	125	---	---**	---	---	78.8	86.4
8	15	50	630	38	31.0	29.0	115.4
7	15	50	1,260	60	20.0	22.1	137.5
6	15	50	1,890	60	20.0	22.1	159.6
5	15	50	2,520	60	20.0	22.1	181.7
4	15	50	3,150	60	20.0	22.1	203.8
3	15	50	3,780	60	20.0	22.1	225.9
2	15	50	4,410	60	20.0	22.1	248.0
1	15	50	5,040	60	20.0	22.1	270.1

N = nonreducible live load, R = reducible live load
* Roof live load reduced in accordance with IBC 1607.11.2
** Live load > 100 psf is not permitted to be reduced [IBC 1607.9.2(2)]

- Ninth floor

Since the ninth floor is storage with a live load of 125 psf, which exceeds 100 psf, the live load is not permitted to be reduced [IBC 1607.9.2(2)].

Axial load $= 125 \times 630/1,000 = 78.8$ kips

- Typical floors

Reducible nominal live load = 50 psf

Reduction factor R is given by Eq. 16-23:

$R = 0.08(A - 150)$

EXAMPLES 3-27

$$\leq \text{the smallest of} \begin{cases} 60 \text{ percent for vertical members (governs)} \\ 23.1(1+D/L_o) = 23.1(1+2) = 69 \text{ percent} \end{cases}$$

Axial load $= [L_o(1-0.01R)+15]A = 630[50(1-0.01R)+15]$

Part 3: Determine reduced live load for column C1

A summary of the reduced live loads is given in Table 3.8. Detailed calculations for various floor levels follow the table.

Table 3.8 Summary of Reduced Live Loads for Column C1

Story	Live Load (psf) N	Live Load (psf) R	A (sq ft)	Reduction Factor, R (%)	Reduced Live Load (psf)	N + R (kips)	Cumulative N + R (kips)
10	---	20	---*	---	20.0	3.5	3.5
9	125	---	---**	---	---	21.9	25.4
8	15	50	175	2	49.0	11.2	36.6
7	15	50	350	16	42.0	10.0	46.6
6	15	50	525	30	35.0	8.8	55.4
5	15	50	700	44	28.0	7.5	62.9
4	15	50	875	58	21.0	6.3	69.2
3	15	50	1,050	60	20.0	6.1	75.3
2	15	50	1,225	60	20.0	6.1	81.4
1	15	50	1,400	60	20.0	6.1	87.5

N = nonreducible live load, R = reducible live load
* Roof live load reduced in accordance with IBC 1607.11.2
** Live load > 100 psf is not permitted to be reduced [IBC 1607.9.2(2)]

- Roof

The reduced roof live load L_r is determined by Eq. 16-25:

$$L_r = L_o R_1 R_2 = 20 R_1 R_2$$

The tributary area A_t of column C1 $= 28 \times 25 / 4 = 175$ sq ft

Since $A_t < 200$ sq ft, R_1 is determined by Eq. 16-26: $R_1 = 1$

Since $F = 1/2 < 4$, $R_2 = 1$ (Eq. 16-29)

Thus, $L_r = 20 \times 1 \times 1 = 20.0$ psf

Axial load = $20.0 \times 175/1{,}000 = 3.5$ kips

- Ninth floor

Since the ninth floor is storage with a live load of 125 psf, which exceeds 100 psf, the live load is not permitted to be reduced [IBC 1607.9.2(2)].

Axial load = $125 \times 175/1{,}000 = 21.9$ kips

- Typical floors

Reducible nominal live load = 50 psf

Reduction factor R is given by Eq. 16-23:

$$R = 0.08(A - 150)$$

$$\leq \text{the smallest of } \begin{cases} 60 \text{ percent for vertical members (governs)} \\ 23.1(1 + D/L_o) = 23.1(1+2) = 69 \text{ percent} \end{cases}$$

Axial load = $[L_o(1 - 0.01R) + 15]A = 175[50(1 - 0.01R) + 15]$

Part 4: Determine reduced live load for column C4

A summary of the reduced live loads is given in Table 3.9. Detailed calculations for various floor levels follow the table.

Table 3.9 Summary of Reduced Live Loads for Column C4

Story	Live Load (psf)		A (sq ft)	Reduction Factor, R (%)	Reduced Live Load (psf)	N + R (kips)	Cumulative N + R (kips)
	N	R					
10	---	20	---*	---	14.2	7.0	7.0
9	125	---	---**	---	---	61.3	68.3
8	15	50	490	27	36.5	25.2	93.5
7	15	50	980	60	20.0	17.2	110.7
6	15	50	1,470	60	20.0	17.2	127.9
5	15	50	1,960	60	20.0	17.2	145.1
4	15	50	2,450	60	20.0	17.2	162.3
3	15	50	2,940	60	20.0	17.2	179.5
2	15	50	3,430	60	20.0	17.2	196.7
1	15	50	3,920	60	20.0	17.2	213.9

N = nonreducible live load, R = reducible live load
* Roof live load reduced in accordance with IBC 1607.11.2
** Live load > 100 psf is not permitted to be reduced [IBC 1607.9.2(2)]

- Roof

  The reduced roof live load L_r is determined by Eq. 16-25:

  $$L_r = L_o R_1 R_2 = 20 R_1 R_2$$

  The tributary area A_t of column C4 $= (28 \times 20)/4 + (28 \times 25)/2 = 490$ sq ft

  Since 200 sq ft $< A_t < 600$ sq ft, R_1 is determined by Eq. 16-27:

  $$R_1 = 1.2 - 0.001 A_t = 1.2 - (0.001 \times 490) = 0.71$$

  Since $F = 1/2 < 4$, $R_2 = 1$ (Eq. 16-29)

  Thus, $L_r = 20.0 \times 0.71 \times 1 = 14.2$ psf

  Axial load $= 14.2 \times 490 / 1,000 = 7.0$ kips

- Ninth floor

  Since the ninth floor is storage with a live load of 125 psf, which exceeds 100 psf, the live load is not permitted to be reduced [IBC 1607.9.2(2)].

  Axial load $= 125 \times 490 / 1,000 = 61.3$ kips

- Typical floors

  Reducible nominal live load $= 50$ psf

  Reduction factor R is given by Eq. 16-23:

  $$R = 0.08(A - 150)$$

  $$\leq \text{the smallest of } \begin{cases} 60 \text{ percent for vertical members (governs)} \\ 23.1(1 + D/L_o) = 23.1(1 + 2) = 69 \text{ percent} \end{cases}$$

  Axial load $= [L_o(1 - 0.01R) + 15]A = 490[50(1 - 0.01R) + 15]$

Part 5: Determine reduced live load for column B6

A summary of the reduced live loads is given in Table 3.10. Detailed calculations for various floor levels follow the table.

Table 3.10 Summary of Reduced Live Loads for Column B6

Story	Live Load (psf) N	Live Load (psf) R	A (sq ft)	Reduction Factor, R (%)	Reduced Live Load (psf)	N + R (kips)	Cumulative N + R (kips)
10	---	20	---*	---	14.2	7.0	7.0
9	125	---	---**	---	---	61.3	68.3
8	15	50	490	27	36.5	25.2	93.5
7	15	50	980	60	20.0	17.2	110.7
6	15	50	1,470	60	20.0	17.2	127.9
5	15	50	1,960	60	20.0	17.2	145.1
4	15	50	2,450	60	20.0	17.2	162.3
3	15	50	2,940	60	20.0	17.2	179.5
2	15	50	3,430	60	20.0	17.2	196.7
1	15	50	3,920	60	20.0	17.2	213.9

N = nonreducible live load, R = reducible live load
* Roof live load reduced in accordance with IBC 1607.11.2
** Live load > 100 psf is not permitted to be reduced [IBC 1607.9.2(2)]

- Roof

The reduced roof live load L_r is determined by Eq. 16-25:

$$L_r = L_o R_1 R_2 = 20 R_1 R_2$$

The tributary area A_t of column B6 $= 28 \times (25/2 + 5) = 490$ sq ft

Since 200 sq ft $< A_t <$ 600 sq ft, R_1 is determined by Eq. 16-27:

$$R_1 = 1.2 - 0.001 A_t = 1.2 - (0.001 \times 490) = 0.71$$

Since $F = 1/2 < 4$, $R_2 = 1$ (Eq. 16-29)

Thus, $L_r = 20 \times 0.71 \times 1 = 14.2$ psf

Axial load $= 14.2 \times 490 / 1,000 = 7.0$ kips

- Ninth floor

Since the ninth floor is storage with a live load of 125 psf, which exceeds 100 psf, the live load is not permitted to be reduced [IBC 1607.9.2(2)].

Axial load $= 125 \times 490 / 1,000 = 61.3$ kips

- Typical floors

    Reducible nominal live load = 50 psf

    Reduction factor R is given by Eq. 16-23:

    $R = 0.08(A - 150)$

    $\leq$ the smallest of $\begin{cases} 60 \text{ percent for vertical members (governs)} \\ 23.1(1 + D/L_o) = 23.1(1+2) = 69 \text{ percent} \end{cases}$

    Axial load = $[L_o(1 - 0.01R) + 15]A = 490[50(1 - 0.01R) + 15]$

Part 6: Determine reduced live load for two-way slab AB23

- Roof

    The reduced roof live load L_r is determined by Eq. 16-25:

    $L_r = L_o R_1 R_2 = 20 R_1 R_2$

    The tributary area A_t of this slab $= 25 \times 28 = 700$ sq ft

    Since $A_t > 600$ sq ft, R_1 is determined by Eq. 16-28: $R_1 = 0.6$

    Since $F = 1/2 < 4$, $R_2 = 1$ (Eq. 16-29)

    Thus, $L_r = 20 \times 0.6 \times 1 = 14.0$ psf

- Ninth floor

    Since the ninth floor is storage with a live load of 125 psf, which exceeds 100 psf, the live load is not permitted to be reduced [IBC 1607.9.2(2)].

    Live load = 125 psf

- Typical floors

    Reducible nominal live load = 50 psf

    Reduction factor R is given by Eq. 16-23:

    $R = 0.08(A - 150) = 0.08(700 - 150) = 44$ percent

$$> \text{ the smallest of } \begin{cases} 40 \text{ percent for horizontal members (governs)} \\ 23.1(1 + D/L_o) = 23.1(1+2) = 69 \text{ percent} \end{cases}$$

Reduced live load $L = L_o(1 - R) = 50(1 - 0.4) = 30$ psf

Total live load = 30 + 15 = 45 psf

Note: the reduced live load on the shear walls in this example, as well as in Example 3.1, can be determined using the same procedure as for columns. For example, the shear walls located at E5 can be collectively considered to be an edge column without a cantilever slab.

3.4.3 Example 3.3 – Live Load Reduction on a Girder

Determine the reduced live load on a typical interior girder of the warehouse shown in Figure 3.8. The roof is an ordinary flat roof.

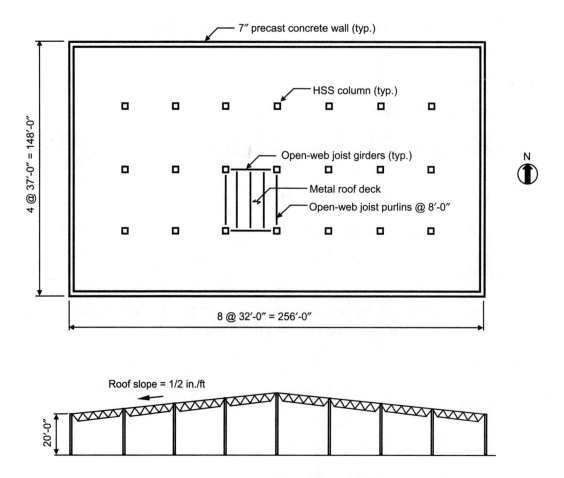

Figure 3.8 Plan and Elevation of Warehouse Building

SOLUTION

Nominal roof live load: 20 psf in accordance with IBC Table 1607.1, since the roof is an ordinary flat roof that is not used for public assembly or any other special purposes.

The reduced roof live load L_r is determined by Eq. 16-25:

$$L_r = L_o R_1 R_2 = 20 R_1 R_2$$

The tributary area A_t of this girder $= 32 \times 37 = 1,184$ sq ft

Since $A_t > 600$ sq ft, R_1 is determined by Eq. 16-28: $R_1 = 0.6$

Since $F = 1/2 < 4$, $R_2 = 1$ (Eq. 16-29)

Thus, $L_r = 20 \times 0.6 \times 1 = 14.0$ psf

The code-prescribed snow and wind loads on the roof of this warehouse are given in Chapters 4 and 5 of this publication, respectively.

3.4.4 Example 3.4 – Rain Load, IBC 1611

Determine the rain load R on a roof located in Madison, Wisconsin, similar to the one depicted in Figure 3.5 given the following design data:

- Tributary area of primary roof drain = 6,200 sq ft
- Closed scupper size: 6 in. wide (b) by 4 in. high (h)
- Vertical distance from primary roof drain to inlet of scupper (static head distance d_s) = 6 in.
- Rainfall rate = 3.0 in./hr (IBC Figure 1611.1)

SOLUTION

To determine the rain load R, the hydraulic head d_h must be determined, based on the required flow rate.

Required flow rate Q = tributary area of roof drain × rainfall rate = $6,200 \times 3/12$ = 1,550 cu ft/hr = 25.83 cu ft/min = 193.2 gpm[12]

---

[12] 1 gallon = 0.1337 cu ft

The hydraulic head d_h is determined by the following equation, which is applicable for closed scuppers where the free surface of the water is above the top of the scupper:[13]

$$Q = 2.9b(d_h^{1.5} - h_1^{1.5}) = 2.9b[(h+h_1)^{1.5} - h_1^{1.5}]$$

where b = width of the scupper, h = depth of the scupper, and $h_1 = d_h - h$ = distance from the free surface of the water to the top of the scupper.

For a flow rate of 193.2 gpm, $h_1 = 1.5$ in. and $d_h = 4 + 1.5 = 5.5$ in.[14]

The rain load R is determined by Eq. 16-35:

$$R = 5.2(d_s + d_h) = 5.2(6 + 5.5) = 59.8 \text{ psf}$$

---

[13] Equations for flow rate for various types of roof drains can be found in various references, including "Roof Loads for New Construction," *FM Global Property Loss Prevention Data Sheet 1-54*, Factory Mutual Insurance Company, 2006. See also Appendix Table C8-1 of ASCE/SEI 7 for flow rates and hydraulic heads of various drainage systems.

[14] By interpolating the values in Table C8-1 for a flow rate of 193.2 gpm, the hydraulic head d_h is equal to 5.6 inches.

CHAPTER 4　　　　　　　　　　　　　　　　　　　　　　*SNOW LOADS*

4.1　INTRODUCTION

IBC 1608.1 requires that design snow loads on buildings and structures be determined by the provisions of Chapter 7 of ASCE/SEI 7-05. These provisions are based on over 40 years of ground snow load data.

The ground snow load p_g is obtained from ASCE/SEI Figure 7-1 or IBC Figure 1608.2 for the conterminous U.S. and from ASCE/SEI Table 7-1 or IBC Table 1608.2 for locations in Alaska. The snow loads on the maps have a 2-percent annual probability of being exceeded (i.e., a 50-year mean recurrence interval). Table C7-1 in the commentary of ASCE/SEI 7 contains ground snow loads at 204 national weather service locations where load measurements are made.[1]

In some areas of the U.S., the ground snow load is too variable to allow mapping. Such regions are noted on the maps as "CS," which indicates that a site-specific case study is required. More information on site-specific case studies can be found in C7.2. The maps also provide ground snow loads in mountainous areas based on elevation. Numbers in parentheses represent the upper elevation limits in feet for the ground snow load values that are given below the elevation. Where a building is located at an elevation greater than that shown on the maps, a site-specific case study must be conducted to establish the ground snow load.

Once a ground snow load has been established, a flat roof snow load p_f is determined by Eq. 7-1 (7.3). This snow load is used for flat roofs (roof slope less than or equal to 5 degrees) and is a function of roof exposure, roof thermal condition and occupancy of the structure. Minimum values of p_f are given for low-slope roofs in 7.3.[2]

Design snow loads for all structures are based on the sloped roof snow load p_s, which is determined by modifying the flat roof snow load p_f by a roof slope factor C_s (Eq. 7-2). The factor C_s depends on the slope and temperature of the roof, the presence or absence of obstructions and the degree of slipperiness of the roof surface. A list of roof materials that are considered to be slippery and those that are not is given in 7.4. Figure 7-2 contains graphs of C_s for various conditions, and equations for C_s are given in C7.4.

---

[1] According to Note a in Table C7-1, it is not appropriate to use only the site-specific information in this table to determine design snow loads. See C7.2 for more information.
[2] Low-slope roofs are defined in 7.3.4. The minimum roof snow load is a separate load case, and it is not to be combined with drifting, sliding or other types of snow loading.

According to 7.4.3, $C_s = 0$ for portions of curved roofs that have a slope exceeding 70 degrees, i.e, $p_s = 0$. Balanced snow loads[3] for curved roofs are determined from the loading diagrams in Figure 7-3 with C_s determined from the appropriate curve in Figure 7-2. Multiple folded plate, sawtooth and barrel vault roofs are to be designed using $C_s = 1$, i.e., $p_s = p_f$ (7.4.4). These types of roofs collect additional snow in their valleys by wind drifting and snow sliding, so no reduction in snow load based on roof slope is applied.

The partial loading provisions of 7.5 must be satisfied for continuous roof framing systems and all other roof systems where removal of snow load on one span (by wind or thermal effects, for example) causes an increase in stress or deflection in an adjacent span. For simplicity, only the three load cases given in Figure 7-4 need to be investigated; comprehensive alternate span (or, checkerboard) loading analyses are not required. Partial loading provisions need not be considered for structural members that span perpendicular to the ridgeline of gable roofs with slopes greater than the larger of 2.38 degrees and $(70/W) + 0.5$ where W is the horizontal distance from the eave to the ridge in feet. Also, the minimum roof load requirements of 7.3.4 are not applicable in the partial load provisions.

Unbalanced load occurs on sloped roofs from wind and sunlight. Wind tends to reduce the snow load on the windward portion and increase the snow load on the leeward portion. This is unlike partial loading where snow is removed on one portion of the roof and is not added to another portion. Provisions for unbalanced snow loads are given in 7.6.1 for hip and gable roofs, in 7.6.2 for curved roofs, in 7.6.3 for multiple folded plate, sawtooth and barrel vault roofs, and in 7.6.4 for dome roofs. Figures 7-3, 7-5 and 7-6 illustrate balanced and unbalanced snow loads for curved roofs, hip and gable roofs, and sawtooth roofs, respectively.

Section 7.7 contains provisions for snow drifts that can occur on lower roofs of a building due to (1) wind depositing snow from higher portions of the same building or an adjacent building or terrain feature (such as a hill) to a lower roof and (2) wind depositing snow from the windward portion of a lower roof to the portion of a lower roof adjacent to a taller part of the building. These two types of drifts, which are called leeward and windward drifts, respectively, are illustrated in Figure 7-7. Loads from drifting snow are superimposed on the balanced snow load, as shown in Figure 7-8. Drift loads on sides of roof projections (including rooftop equipment) and at parapet walls are determined by the provisions of 7.8, which are based on the drift requirements of 7.7.1.

The load caused by snow sliding off a sloped roof onto a lower roof is determined by the provisions of 7.9. Such loads are superimposed on the balanced snow load of the lower roof.

---

[3] A balanced snow load is defined in the snow load provisions as the sloped roof snow load p_s determined by Eq. 7-2. This load is assumed to act on the horizontal projection of the entire roof surface.

INTRODUCTION

A rain-on-snow surcharge load of 5 psf is to be added on all roofs that meet the conditions of 7.10. This surcharge load applies only to the balanced load case, and need not be used in combination with drift, sliding, unbalanced or partial loads.

Provisions for ponding instability and progressive deflection of roofs with a slope less than ¼ inch per foot are given in 7.11 and 8.4. Requirements for increased snow loads on existing roofs due to additions and alterations are covered in 7.12.

The following general procedure, which is based on that given in C7.0, can be used to determine design snow loads in accordance with Chapter 7 of ASCE/SEI 7-05:

1. Determine ground snow load p_g (7.2).

2. Determine flat roof snow load p_f by Eq. 7-1 (7.3).

3. Determine sloped roof snow load p_s by Eq. 7-2 (7.4).

4. Consider partial loading (7.5).

5. Consider unbalanced snow loads (7.6).

6. Consider snow drifts on lower roofs (7.7) and roof projections (7.8).

7. Consider sliding snow (7.9).

8. Consider rain-on-snow loads (7.10).

9. Consider ponding instability (7.11).

10. Consider existing roofs (7.12).

It is possible that snow loads in excess of the design values computed by Chapter 7 may occur on a building or structure. The snow load to dead load ratio of a roof structure is an important consideration when evaluating the implications of excess loads. Section C7.0 provides additional information on this topic.

Section C7.13 gives information on wind tunnel tests and other experimental and computational methods that have been employed to establish design snow loads for roof geometries and complicated sites not addressed in the provisions.

Section 4.2 of this document contains flowcharts for determining design snow loads and load cases, based on the design procedure outlined above.

Section 4.3 contains completely worked-out examples that illustrate the design requirements for snow.

4.2 FLOWCHARTS

A summary of the flowcharts provided in this chapter is given in Table 4.1.

Table 4.1 Summary of Flowcharts Provided in Chapter 4

Flowchart	Title
Flowchart 4.1	Flat Roof Snow Load, p_f
Flowchart 4.2	Roof Slope Factor, C_s
Flowchart 4.3	Sloped Roof Snow Load, p_s
Flowchart 4.4	Unbalanced Roof Snow Loads – Hip and Gable Roofs
Flowchart 4.5	Unbalanced Roof Snow Loads – Curved and Dome Roofs
Flowchart 4.6	Unbalanced Roof Snow Loads – Multiple Folded Plate, Sawtooth, and Barrel Vault Roofs
Flowchart 4.7	Drifts on Lower Roof of a Structure

FLOWCHARTS

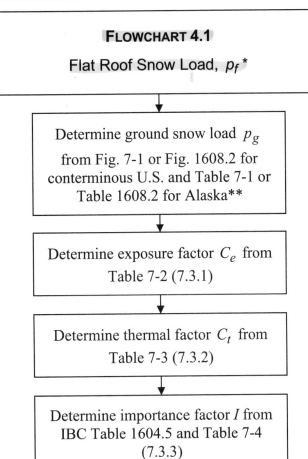

FLOWCHART 4.1
Flat Roof Snow Load, p_f*

- Determine ground snow load p_g from Fig. 7-1 or Fig. 1608.2 for conterminous U.S. and Table 7-1 or Table 1608.2 for Alaska**
- Determine exposure factor C_e from Table 7-2 (7.3.1)
- Determine thermal factor C_t from Table 7-3 (7.3.2)
- Determine importance factor I from IBC Table 1604.5 and Table 7-4 (7.3.3)
- Determine flat roof snow load $p_f = 0.7 C_e C_t I p_g$ by Eq. 7-1†

* A flat roof is defined as a roof with a slope that is less than or equal to 5 degrees.

** "CS" in the maps signifies areas where a site-specific study must be conducted to determine p_g. Numbers in parentheses represent the upper elevation limit in feet for the ground snow load values given below. Site-specific studies are required at elevations not covered in the maps.

† Minimum values of p_f are specified in 7.3 for low-slope roofs, which are defined in 7.3.4.

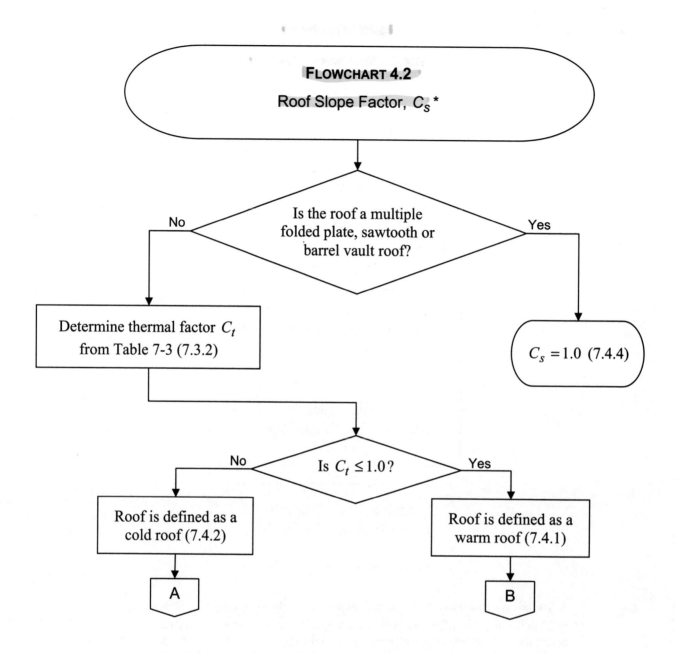

* Portions of curved roofs having a slope exceeding 70 degrees shall be considered free of snow load, i.e., $C_s = 0$ (7.4.3).

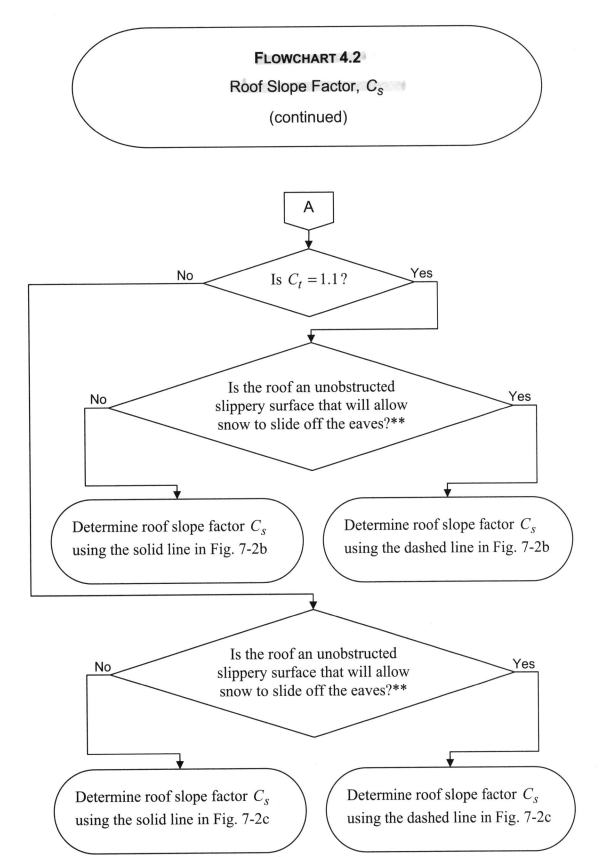

** See 7.4 for definitions of unobstructed and slippery surfaces.

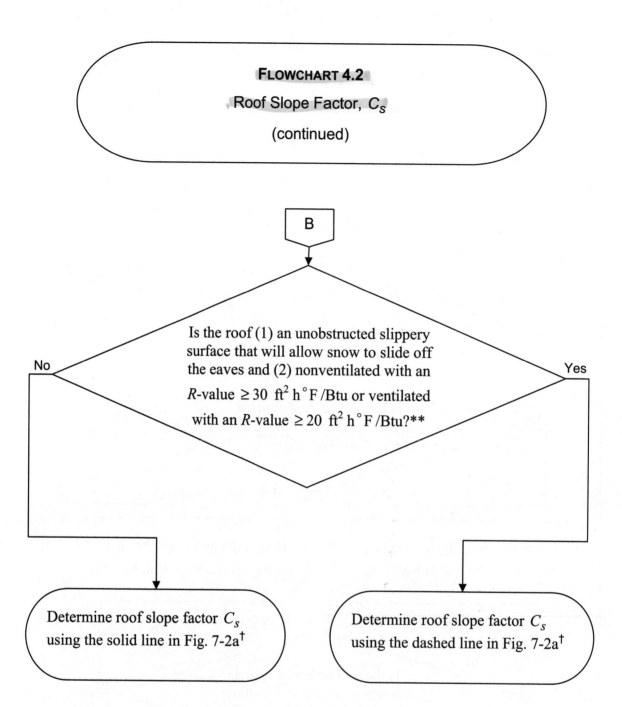

** See 7.4 for definitions of unobstructed and slippery surfaces. An R-value for a roof is defined as its thermal resistance.

† See 7.4.5 for an additional uniformly distributed load that is to be applied on overhanging portions of warm roofs due to formation of ice dams and icicles along eaves.

FLOWCHART 4.3

Sloped Roof Snow Load, p_s

↓

Determine flat snow load p_f from Flowchart 4.1

↓

Determine roof slope factor C_s from Flowchart 4.2

↓

Determine sloped roof snow load $p_s = C_s p_f$ by Eq. 7-2

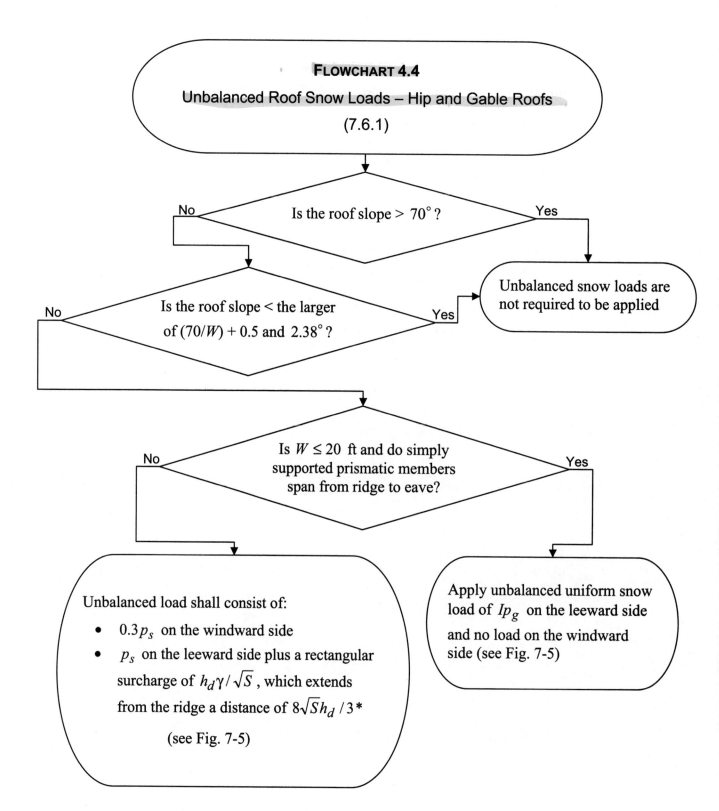

* h_d is the drift height from Fig. 7-9 with W substituted for ℓ_u, γ = snow density determined by Eq. 7-3, and S = roof slope run for a rise of one

FLOWCHARTS

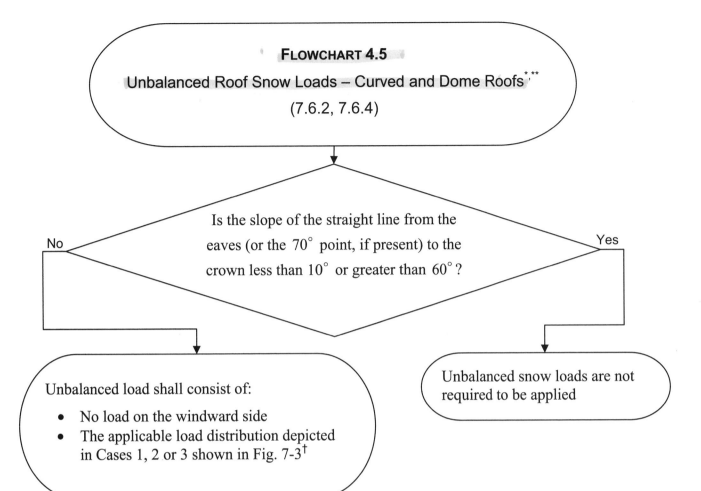

FLOWCHART 4.5

Unbalanced Roof Snow Loads – Curved and Dome Roofs[*][**]

(7.6.2, 7.6.4)

Is the slope of the straight line from the eaves (or the 70° point, if present) to the crown less than 10° or greater than 60°?

- **No** → Unbalanced load shall consist of:
  - No load on the windward side
  - The applicable load distribution depicted in Cases 1, 2 or 3 shown in Fig. 7-3[†]

- **Yes** → Unbalanced snow loads are not required to be applied

---

[*] Portions of curved roofs having a slope > 70° shall be considered free of snow.

[**] See 7.6.4 for provisions related to dome roofs.

[†] See 7.6.2 where ground or another roof abuts a Case 2 or Case 3 curved roof at or within 3 ft of its eaves.

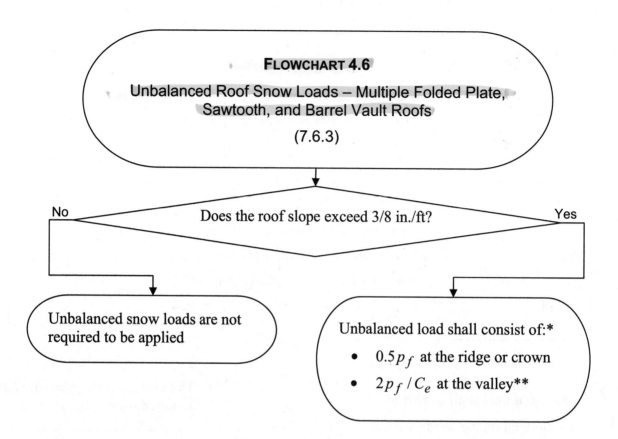

* Figure 7-6 illustrates balanced and unbalanced snow loads for a sawtooth roof.

** Snow surface above the valley shall not be at an elevation higher than the snow above the ridge. Snow depths shall be determined by dividing the snow load by the snow density given by Eq. 7-3.

FLOWCHARTS   4-13

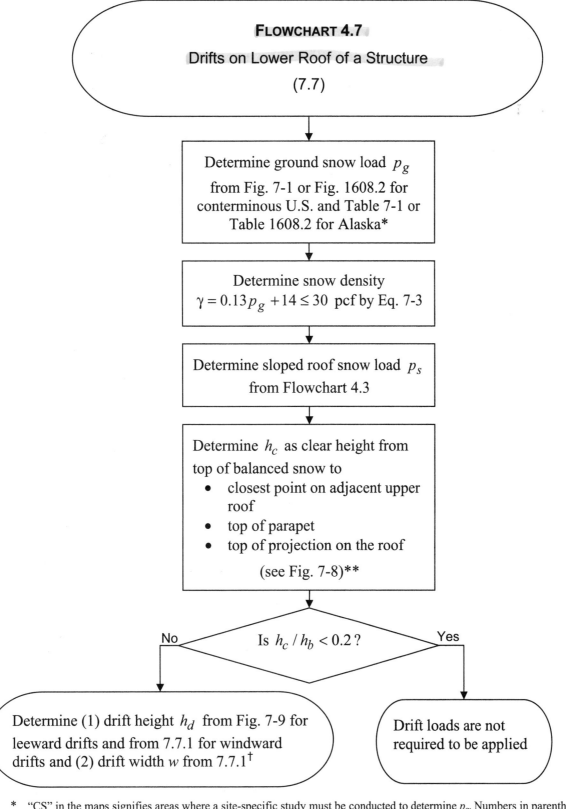

*   "CS" in the maps signifies areas where a site-specific study must be conducted to determine p_g. Numbers in parentheses represent the upper elevation limit in feet for the ground snow load values given below. Site-specific studies are required at elevations not covered in the maps.

** Height of balanced snow $h_b = p_s/\gamma$ or p_f/γ (7.7.1)

†   See 7.7.2 for drift loads caused by adjacent structures and terrain features. See 7.8 for drift loads on roof projections and parapet walls.

4.3 EXAMPLES

The following sections contain examples that illustrate the snow load design provisions of Chapter 7 in ASCE/SEI 7-05.

4.3.1 Example 4.1 – Warehouse Building, Roof Slope of 1/2 on 12

Determine the design snow loads for the one-story warehouse illustrated in Figure 4.1.

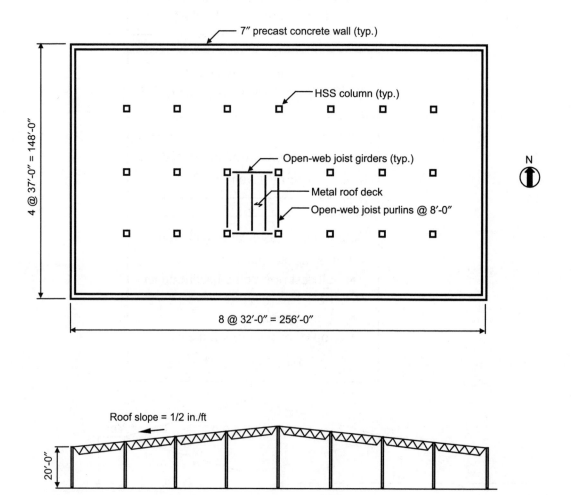

Figure 4.1 Plan and Elevation of Warehouse Building

EXAMPLES

DESIGN DATA

Location:	St. Louis, MO
Terrain category:	C (open terrain with scattered obstructions less than 30 ft in height)
Occupancy:	Warehouse use. Less than 300 people congregate in one area and the building is not used to store hazardous or toxic materials
Thermal condition:	Structure is kept just above freezing
Roof exposure condition:	Partially exposed
Roof surface:	Rubber membrane
Roof framing:	All members are simply supported

SOLUTION

1. Determine ground snow load p_g.

    From Figure 7-1 or Figure 1608.2, the ground snow load is equal to 20 psf for St. Louis, MO.

2. Determine flat roof snow load p_f by Eq. 7-1.

    Use Flowchart 4.1 to determine p_f.

    a. Determine exposure factor C_e from Table 7-2.

    From the design data, the terrain category is C and the roof exposure is partially exposed. Therefore, $C_e = 1.0$ from Table 7-2.

    b. Determine thermal factor C_t from Table 7-3.

    From the design data, the structure is kept just above freezing during the winter, so $C_t = 1.1$ from Table 7-3.

    c. Determine the importance factor I from Table 7-4.

    From IBC Table 1604.5, the Occupancy Category is II, based on the occupancy given in the design data. Thus, $I = 1.0$ from Table 7-4.

    Therefore,

$$p_f = 0.7 C_e C_t I p_g$$

$$= 0.7 \times 1.0 \times 1.1 \times 1.0 \times 20 = 15.4 \text{ psf}$$

Check if the minimum snow load requirements are applicable:

Minimum values of p_f in accordance with 7.3 apply to hip and gable roofs with slopes less than the larger of 2.38 degrees (1/2 on 12) (governs) and $(70/W) + 0.5 = (70/128) + 0.5 = 1.05$ degrees. Since the roof slope in this example is equal to 2.38 degrees, minimum roof snow loads do not apply.

3. Determine sloped roof snow load p_s by Eq. 7-1.

   Use Flowchart 4.2 to determine roof slope factor C_s.

   a. Determine thermal factor C_t from Table 7-3.

      From item 2 above, thermal factor $C_t = 1.1$.

   b. Determine if the roof is warm or cold.

      Since $C_t = 1.1$, the roof is defined as a cold roof in accordance with 7.4.2.

   c. Determine if the roof is unobstructed or not and if the roof is slippery or not.

      There are no obstructions on the roof that inhibit the snow from sliding off the eaves.[4] Also, the roof surface is a rubber membrane. According to 7.4, rubber membranes are considered to be slippery surfaces.

   Since this roof is unobstructed and slippery, use the dashed line in Figure 7-2b to determine C_s:

   For a roof slope of 2.38 degrees, $C_s = 1.0$.

   Therefore, $p_s = C_s p_f = 1.0 \times 15.4 = 15.4$ psf. This is the balanced snow load for this roof.

---

[4] In general, large vent pipes, snow guards, parapet walls and large rooftop equipment are a few common examples of obstructions that could prevent snow from sliding off the roof. Ice dams and icicles along eaves can also possibly inhibit snow from sliding off of two types of warm roofs, which are described in 7.4.5.

4. Consider partial loading.

   Since all of the members are simply supported, partial loading is not considered (7.5).

5. Consider unbalanced snow loads.

   Flowchart 4.4 is used to determine if unbalanced loads on this gable roof need to be considered or not.

   Unbalanced snow loads must considered for this roof, since the slope is greater than or equal to the larger of $(70/W) + 0.5 = 1.05$ degrees and 2.38 degrees (governs).

   Since $W = 128$ ft > 20 ft, the unbalanced load consists of the following (see Figure 7-5):

   - Windward side: $0.3 p_s = 0.3 \times 15.4 = 4.6$ psf
   - Leeward side: $p_s = 15.4$ psf along the entire leeward length plus a uniform pressure of $h_d \gamma / \sqrt{S} = (3.6 \times 16.6)/\sqrt{24} = 12.2$ psf, which extends from the ridge a distance of $8 h_d \sqrt{S}/3 = (8 \times 3.6 \times \sqrt{24})/3 = 47.0$ ft where

   h_d = drift length from Figure 7-9 with $W = 128$ ft substituted for ℓ_u
    = $0.43(W)^{1/3}(p_g + 10)^{1/4} - 1.5 = 3.6$ ft
   γ = snow density (Eq. 7-3)
    = $0.13 p_g + 14 = 16.6$ pcf < 30 pcf
   S = roof slope run for a rise of one = 24

6. Consider snow drifts on lower roofs and roof projections.

   Not applicable.

7. Consider sliding snow.

   Not applicable.

8. Consider rain-on-snow loads.

   In accordance with 7.10, a rain-on-snow surcharge of 5 psf is required for locations where the ground snow load p_g is 20 psf or less (but not zero) with roof slopes less than $W/50$.

In this example, $p_g = 20$ psf and $W/50 = 128/50 = 2.56$ degrees, which is greater than the roof slope of 2.38 degrees. Thus, an additional 5 psf must be added to the balanced load of 15.4 psf.[5]

9. Consider ponding instability.

   Since the roof slope in this example is greater than 1/4 in./ft, progressive roof deflection and ponding instability from rain-on-snow or from snow meltwater need not be investigated (7.11 and 8.4).

10. Consider existing roofs.

    Not applicable.

The balanced and unbalanced snow loads are depicted in Figure 4.2.

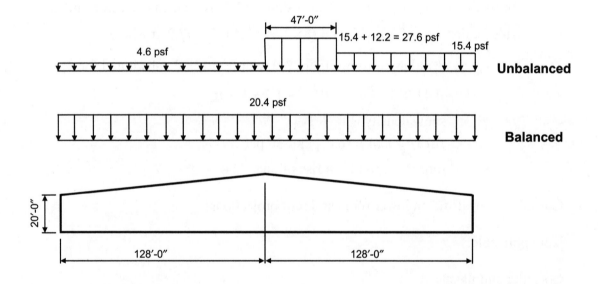

Figure 4.2 Balanced and Unbalanced Snow Loads for Warehouse Building

---

[5] The rain-on-snow load applies only to the balanced load case and need not be used in combination with drift, sliding, unbalanced or partial loads (7.10).

4.3.2 Example 4.2 – Warehouse Building, Roof Slope of 1/4 on 12

For the one-story warehouse depicted in Figure 4.1, determine the design snow loads for a roof slope of 1/4 on 12. Use the same design data given in Example 4.1.

SOLUTION

---

1. Determine ground snow load p_g.

   From Figure 7-1 or Figure 1608.2, the ground snow load is equal to 20 psf for St. Louis, MO.

2. Determine flat roof snow load p_f by Eq. 7-1.

   It was determined in item 2 of Example 4.1 that $p_f = 15.4$ psf.

   Check if the minimum snow load requirements are applicable:

   Minimum values of p_f in accordance with 7.3 apply to hip and gable roofs with slopes less than the larger of 2.38 degrees (1/2 on 12) (governs) and $(70/W) + 0.5 = (70/128) + 0.5 = 1.05$ degrees. Since the roof slope in this example is equal to 1.19 degrees, minimum flat roof snow loads apply.

   In accordance with 7.3, minimum flat roof snow load = $Ip_g = 1.0 \times 20 = 20$ psf, since p_g is equal to 20 psf or less.

3. Determine sloped roof snow load p_s by Eq. 7-1.

   It was determined in item 3 of Example 4.1 that $C_s = 1.0$ for a roof slope of 2.38 degrees. Using the dashed line in Figure 7-2b, $C_s = 1.0$ for a roof slope of 1.19 degrees as well.

   Therefore, $p_s = 1.0 \times 15.4 = 15.4$ psf.

4. Consider partial loading.

   Since all of the members are simply supported, partial loading is not considered (7.5).

5. Consider unbalanced snow loads.

   Flowchart 4.4 is used to determine if unbalanced loads on this gable roof need to be considered or not.

Unbalanced snow loads need not be considered for this roof, since the slope is less than the larger of $(70/W) + 0.5 = 1.05$ degrees and 2.38 degrees (governs).

6. Consider snow drifts on lower roofs and roof projections.

    Not applicable.

7. Consider sliding snow.

    Not applicable.

8. Consider rain-on-snow loads.

    In accordance with 7.10, a rain-on-snow surcharge of 5 psf is required for locations where the ground snow load p_g is 20 psf or less (but not zero) with roof slopes less than $W/50$.

    In this example, $p_g = 20$ psf and $W/50 = 128/50 = 2.56$ degrees, which is greater than the roof slope of 1.19 degrees. Thus, an additional 5 psf must be added to the sloped roof snow load of 15.4 psf.[6]

9. Consider ponding instability.

    Since the roof slope in this example is not less than 1/4 in./ft, progressive roof deflection and ponding instability from rain-on-snow or from snow meltwater need not be investigated (7.11 and 8.4).

10. Consider existing roofs.

    Not applicable.

In this example, the uniform load of $15.4 + 5 = 20.4$ psf (balanced plus rain-on-snow) governs, since it is greater than the minimum roof snow load of 20 psf. The 20.4 psf snow load is uniformly distributed over the entire length of the roof, as depicted in Figure 4.2. This is the only load that needs to be considered in this example.

4.3.3 Example 4.3 – Warehouse Building (Roof Slope of 1/2 on 12) and Adjoining Office Building (Roof Slope of 1/2 on 12)

A new one-story office building is to be constructed adjacent to the existing one-story warehouse in Example 4.1 (see Figure 4.3). Determine the design snow loads on the roof of the office building.[7] Both structures have a roof slope of 1/2 on 12.

---

[6] The rain-on-snow load applies only to the balanced load case and need not be used in combination with drift, sliding, unbalanced or partial loads (7.10).

[7] A summary of the design snow loads for the warehouse is given in Figure 4.2 in Example 4.1.

EXAMPLES

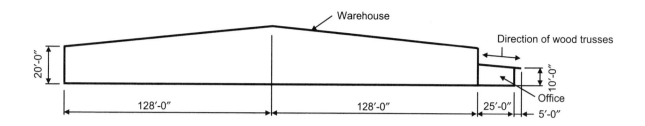

Figure 4.3 Elevation of Warehouse and Office Buildings

DESIGN DATA FOR OFFICE BUILDING

Location:	St. Louis, MO
Terrain category:	C (open terrain with scattered obstructions less than 30 ft in height)
Occupancy:	Business (less than 300 people congregate in one area)
Thermal condition:	Heated with unventilated roof (R-value less than 30 ft²h°F/Btu)
Roof exposure condition:	Partially exposed (due in part to the presence of the adjacent taller warehouse building)
Roof surface:	Asphalt shingles
Roof framing:	Wood trusses spaced 25 ft on center that overhang a masonry wall and wood purlins spaced 5 ft on center that frame between the trusses (see Figure 4.3)

SOLUTION

1. Determine ground snow load p_g.

   From Figure 7-1 or Figure 1608.2, the ground snow load is equal to 20 psf for St. Louis, MO.

2. Determine flat roof snow load p_f by Eq. 7-1.

   Use Flowchart 4.1 to determine p_f.

   a. Determine exposure factor C_e from Table 7-2.

      From the design data, the terrain category is C and the roof exposure is partially exposed. Therefore, $C_e = 1.0$ from Table 7-2.

b. Determine thermal factor C_t from Table 7-3.

From the design data, the structure is heated with an unventilated roof, so $C_t = 1.0$ from Table 7-3.

c. Determine the importance factor I from Table 7-4.

From IBC Table 1604.5, the Occupancy Category is II, based on the occupancy given in the design data. Thus, $I = 1.0$ from Table 7-4.

Therefore,

$$p_f = 0.7 C_e C_t I p_g$$

$$= 0.7 \times 1.0 \times 1.0 \times 1.0 \times 20 = 14.0 \text{ psf}$$

Check if the minimum snow load requirements are applicable:

Minimum values of p_f in accordance with 7.3 apply to monoslope roofs with slopes less than 15 degrees. Since the roof slope in this example is equal to 2.38 degrees, minimum roof snow loads apply.

In accordance with 7.3, minimum roof snow load = $I p_g$ 1.0 × 20 = 20 psf, since p_g is equal to 20 psf or less.

3. Determine sloped roof snow load p_s by Eq. 7-1.

Use Flowchart 4.2 to determine roof slope factor C_s.

a. Determine thermal factor C_t from Table 7-3.

From item 2 above, thermal factor $C_t = 1.0$.

b. Determine if the roof is warm or cold.

Since $C_t = 1.0$, the roof is defined as a warm roof in accordance with 7.4.1.

c. Determine if the roof is unobstructed or not and if the roof is slippery or not.

In accordance with the design data, the roof surface is asphalt shingles. According to 7.4, asphalt shingles are not considered to be slippery.

Also, since the roof is unventilated with an R-value less than 30 ft^{2}h°F/Btu, it is possible for an ice dam to form at the eave, which can prevent the snow from sliding off of the roof (7.4.5). This is considered to be an obstruction.

Thus, use the solid line in Figure 7-2a to determine C_s:

For a roof slope of 2.38 degrees, $C_s = 1.0$.

Therefore, $p_s = C_s p_f = 1.0 \times 14.0 = 14.0$ psf.

In accordance with 7.4.5, a uniformly distributed load of $2p_f = 2 \times 14.0 = 28.0$ psf must be applied on the 5-ft overhanging portion of the roof to account for ice dams. Only the dead load is to be present when this uniformly distributed load is applied.

4. Consider partial loading.

   It is assumed that the roof purlins are connected to the wood trusses by metal hangers, which are essentially simple supports, so partial loads do not have to be considered for the roof purlins. Therefore, with a spacing of 5 ft, the uniform snow load on a purlin is equal to $20.0 \times 5.0 = 100$ plf (minimum snow load governs).

   The roof trusses are continuous over the masonry wall; thus, partial loading must be considered (7.5). The balanced snow load to be used in partial loading cases is that determined by Eq. 7-2, which is equal to 14.0 psf. With a spacing of 25 ft, the balanced and partial loads on a typical roof truss are

   Balanced load = $14.0 \times 25.0 = 350$ plf

   Partial load = one-half of balanced load = 175 plf

   Shown in Figure 4.4 are the balanced and partial load cases that must be considered for the roof trusses in this example, including the ice dam load on the overhang, which was determined in item 3 above. Note that the minimum snow load of 20 psf is not applicable in the partial load cases and in the ice dam load case.

5. Consider unbalanced snow loads.

   Not applicable.

6. Consider snow drifts on lower roofs and roof projections.

   Use Flowchart 4.7 to determine the leeward and windward drifts that form on the lower (office) roof.

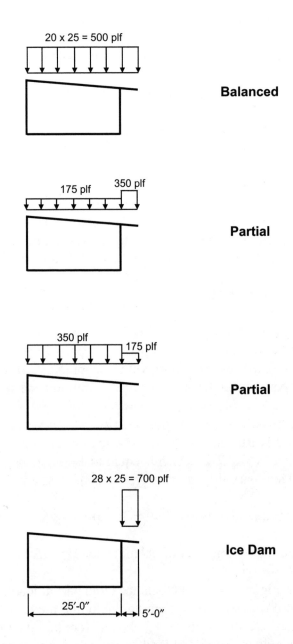

Figure 4.4 Balanced and Partial Load Cases for Roof Trusses

a. Determine ground snow load p_g.

From item 1 above, the ground snow load is equal to 20 psf for St. Louis, MO.

b. Determine snow density γ by Eq. 7-3.

$$\gamma = 0.13 p_g + 14 = (0.13 \times 20) + 14 = 16.6 \text{ pcf} < 30 \text{ pcf}$$

c. Determine sloped roof snow load p_s from Flowchart 4.3.

   From item 3 above, $p_s = 14.0$ psf.

d. Determine clear height h_c.

   In this example, the clear height h_c is from the top of the balanced snow to the top of the warehouse eave (see Figure 7-8). The height of the balanced snow $h_b = p_s/\gamma = 14.0/16.6 = 0.8$ ft.

   Thus, $h_c = (10 - 25 \tan 2.38°) - 0.8 = 8.2$ ft

e. Determine if drift loads are required or not.

   Drift loads are not required where $h_c / h_b < 0.2$ (7.7.1). In this example, $h_c / h_b = 8.2/0.8 > 0.2$, so drift loads must be considered.

f. Determine drift load.

   Both leeward and windward drift heights h_d must be determined by the provisions of 7.7.1. The larger of these two heights is used to determine the drift load.

   - Leeward drift

     A leeward drift occurs when snow from the warehouse roof is deposited by wind to the office roof (wind from left to right in Figure 4.3).

     For leeward drifts, the drift height h_d is determined from Figure 7-9 using the length of the upper roof ℓ_u. In this example, $\ell_u = 256$ ft and the ground snow load $p_g = 20$ psf. Using the equation in Figure 7-9:

     $$h_d = 0.43(\ell_u)^{1/3}(p_g + 10)^{1/4} - 1.5$$
     $$= 0.43(256)^{1/3}(20 + 10)^{1/4} - 1.5 = 4.9 \text{ ft}$$

   - Windward drift

     A windward drift occurs when snow from the office roof is deposited adjacent to the wall of the warehouse building (wind from right to left in Figure 4.3).

     For windward drifts, the drift height h_d is 75 percent of that determined from Figure 7-9 using the length of the lower roof for ℓ_u:

$$h_d = 0.75[0.43(\ell_u)^{1/3}(p_g+10)^{1/4} - 1.5]$$

$$= 0.75[0.43(30)^{1/3}(20+10)^{1/4} - 1.5] = 1.2 \text{ ft}$$

Thus, the leeward drift controls and $h_d = 4.9$ ft.

Since $h_d = 4.9$ ft $< h_c = 8.2$ ft, the drift width $w = 4h_d = 4 \times 4.9 = 19.6$ ft.

The maximum surcharge drift load $p_d = h_d \gamma = 4.9 \times 16.6 = 81.3$ psf.

The total load at the step is the balanced load on the office roof plus the drift surcharge = 14.0 + 81.3 = 95.3 psf, which is illustrated in Figure 4.5.

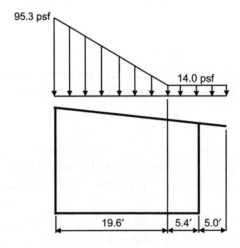

Figure 4.5 Balanced and Drift Loads on Office Roof

The snow loads on the purlins and trusses are obtained by multiplying the loads depicted in Figure 4.5 by the respective tributary widths. As expected, the purlins closest to the warehouse have the largest loads.

If the office and warehouse were separated, the drift load on the office roof would be reduced by the factor $(20 - s)/20$ where s is the separation distance in feet (7.7.2). For example, if the buildings were separated by 5 ft, the modified drift height is equal to $[(20 - 5)/20] \times 4.9 = 3.7$ ft and the maximum surcharge load at the step is equal to $3.7 \times 16.6 = 61.4$ psf. Also, the drift width is equal to $4 \times 3.7 = 14.8$ ft. Drift loads on lower roofs are not considered for structures separated by a distance of 20 ft or more.

7. Consider sliding snow.

The provisions of 7.9 are used to determine if a load due to snow sliding off of the warehouse roof on to the office roof must be considered.

Load caused by snow sliding must be considered, since the warehouse roof is slippery with a slope greater than 1/4 on 12. This load is in addition to the balanced load acting on the lower roof.

The total sliding load per unit length of eave is equal to $0.4\,p_f W$ where W is the horizontal distance from the eave to the ridge of the warehouse roof. In this example, $W = 128$ ft and the sliding load $= 0.4 \times 14.0 \times 128 = 717$ plf. This load is to be uniformly distributed over a distance of 15 ft from the warehouse eave.[8] Thus, the sliding load is equal to $717/15 = 47.8$ psf.

The total load over the 15-ft width is equal to the balanced load plus the sliding load $= 14.0 + 47.8 = 61.8$ psf. The total depth of snow for the total load is equal to $61.8/16.6 = 3.7$ ft, which is less than the distance from the warehouse eave to the top of the office roof at the interface. Thus, sliding snow is not blocked and the full load can be developed over the 15-ft length.[9]

Depicted in Figure 4.6 is the load case including sliding snow. The total balanced and sliding snow load is less than the total balanced and drift snow load (see Figure 4.5).

8. Consider rain-on-snow loads.

In accordance with 7.10, a rain-on-snow surcharge of 5 psf is required for locations where the ground snow load p_g is 20 psf or less (but not zero) with roof slopes less than $W/50$.

In this example, $p_g = 20$ psf and $W/50 = 25/50 = 0.5$ degrees, which is less than the roof slope of 2.38 degrees. Thus, rain-on-snow loads are not considered.

---

[8] If the width of the lower roof is less than 15 ft, the sliding load is to be reduced proportionally (7.9). For example, if the width of the office building in this example was 12 ft, the reduced sliding load $= (12/15) \times 717 = 574$ plf. This load would be applied uniformly over the 12-ft width.

[9] If the calculated total snow depth on the lower roof exceeds the distance from the upper roof eave to the top of the lower roof, sliding snow is blocked and a fraction of the sliding snow is forced to remain on the upper roof. In such cases, the total load on the lower roof near the upper roof eave is equal to the density of the snow multiplied by the distance from the upper roof eave to the top of the lower roof. This load is uniformly distributed over a distance of 15 ft or the width of the lower roof, whichever is less.

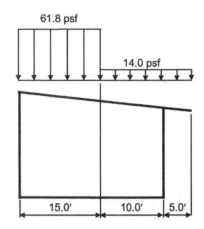

Figure 4.6 Balanced and Sliding Snow Loads on Office Roof

9. Consider ponding instability.

   Since the roof slope in this example is greater than 1/4 in./ft, progressive roof deflection and ponding instability from rain-on-snow or from snow meltwater need not be investigated (7.11 and 8.4).

10. Consider existing roofs.

    Not applicable.

The loads depicted in Figures 4.4, 4.5 and 4.6 must be considered when designing the purlins and trusses.

4.3.4 Example 4.4 – Six-Story Hotel with Parapet Walls

Determine the design snow loads for the six-story hotel depicted in Figure 4.7. Parapet walls are on all four sides of the building and the roof is nominally flat except for localized areas around roof drains that are sloped to facilitate drainage.

DESIGN DATA

Ground snow load, p_g:	40 psf
Terrain category:	B (urban area with numerous closely spaced obstructions having the size of single-family dwellings or larger)
Occupancy:	Residential (less than 300 people congregate in one area)
Thermal condition:	Cold, ventilated roof (R-value between the ventilated space and the heated space exceeds 25 $ft^2h°F/Btu$)
Roof exposure condition:	Fully exposed
Roof surface:	Concrete slab with waterproofing

EXAMPLES   4-29

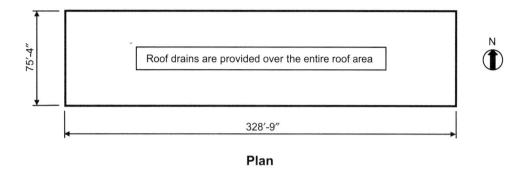

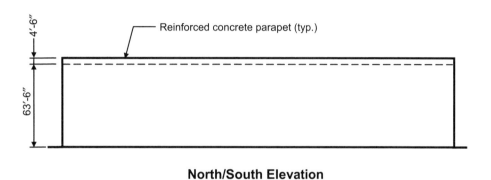

North/South Elevation

Figure 4.7 Plan and Elevation of Six-story Hotel with Parapet Walls

---
SOLUTION
---

1. Determine ground snow load p_g.

   From the design data, the ground snow load p_g is equal to 40 psf.

2. Determine flat roof snow load p_f by Eq. 7-1.

   Use Flowchart 4.1 to determine p_f.

   a. Determine exposure factor C_e from Table 7-2.

      From the design data, the terrain category is B and the roof exposure is fully exposed. Therefore, $C_e = 0.9$ from Table 7-2.

b. Determine thermal factor C_t from Table 7-3.

From the design data, the roof is cold and ventilated with an *R*-value between the ventilated space and the heated space that exceeds 25 ft²h°F/Btu, so $C_t = 1.1$ from Table 7-3.

c. Determine the importance factor *I* from Table 7-4.

From IBC Table 1604.5, the Occupancy Category is II, based on the occupancy given in the design data. Thus, $I = 1.0$ from Table 7-4.

Therefore,

$$p_f = 0.7 C_e C_t I p_g$$

$$= 0.7 \times 0.9 \times 1.1 \times 1.0 \times 40 = 27.7 \text{ psf}$$

3. Consider drift loading at parapet walls.

   According to 7.8, drift loads at parapet walls and other roof projections are determined using the provisions of 7.7.1.

   Windward drifts occur at parapet walls, and Flowchart 4.7 is used to determine the windward drift load.

   a. Determine snow density γ by Eq. 7-3.

   $$\gamma = 0.13 p_g + 14 = (0.13 \times 40) + 14 = 19.2 \text{ pcf} < 30 \text{ pcf}$$

   b. Determine clear height h_c.

   The clear height h_c is from the top of the balanced snow to the top of the parapet wall. For a flat roof, the height of the balanced snow is determined as follows: $h_b = p_f / \gamma = 27.7/19.2 = 1.4$ ft.

   Thus, $h_c = 4.5 - 1.4 = 3.1$ ft

   c. Determine if drift loads are required or not.

   Drift loads are not required where $h_c / h_b < 0.2$ (7.7.1). In this example, $h_c / h_b = 3.1/1.4 > 0.2$, so drift loads must be considered.

d. Determine drift load.

Windward drift height h_d must be determined by the provisions of 7.7.1 using three-quarters of the drift height h_d from Figure 7-9 with ℓ_u equal to the length of the roof upwind of the parapet wall (7.8). Wind in both the north-south and east-west directions must be examined.

- Wind in north-south direction

    The equation in Figure 7-9 yields the following for the drift height h_d based on a ground snow load $p_g = 40$ psf and an upwind fetch $\ell_u = 75.33$ ft:

    $$h_d = 0.75[0.43(\ell_u)^{1/3}(p_g + 10)^{1/4} - 1.5]$$
    $$= 0.75[0.43(75.33)^{1/3}(40 + 10)^{1/4} - 1.5] = 2.5 \text{ ft}$$

    Since $h_d = 2.5$ ft $< h_c = 3.1$ ft, the drift width $w = 4h_d = 4 \times 2.5 = 10.0$ ft.

    The maximum surcharge drift load $p_d = h_d \gamma = 2.5 \times 19.2 = 48.0$ psf.

    The total load at the face of the parapet wall is the balanced load plus the drift surcharge = 27.7 + 48.0 = 75.7 psf.

- Wind in east-west direction

    The equation in Figure 7-9 yields the following for the drift height h_d based on an upwind fetch $\ell_u = 328.75$ ft:

    $$h_d = 0.75[0.43(\ell_u)^{1/3}(p_g + 10)^{1/4} - 1.5]$$
    $$= 0.75[0.43(328.75)^{1/3}(40 + 10)^{1/4} - 1.5] = 4.8 \text{ ft}$$

    Since $h_d = 4.8$ ft $> h_c = 3.1$ ft, the drift height is limited to 3.1 ft and the drift width $w = 4h_d^2/h_c = (4 \times 4.8^2)/3.1 = 29.7$ ft $> 8h_c = 8 \times 3.1 = 24.8$ ft. Therefore, use $w = 24.8$ ft.

    The total load at the face of the parapet wall is $4.5 \times 19.2 = 86.4$ psf.

Balanced and drift loads at the parapet walls in both directions are shown in Figure 4.8.

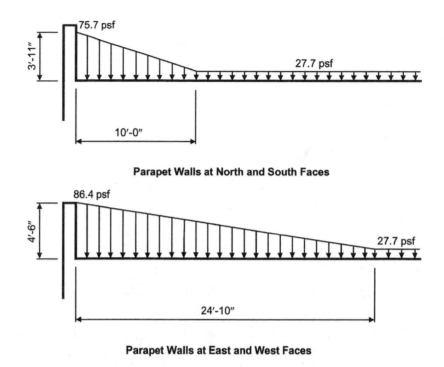

Figure 4.8 Balanced and Drift Snow Loads at Parapet Walls

Since the ground snow load p_g exceeds 20 psf, the minimum snow load is $20I = 20 \times 1.0 = 20$ psf (7.3), which is less than the flat roof snow load $p_f = 27.7$ psf, and, in accordance with 7.10, a rain-on-snow surcharge load is not considered.

The only other load cases that need to be considered are the partial load cases of 7.5. For illustration purposes, assume that in the N-S direction the framing consists of a 3-span moment frame (cast-in-place concrete columns and beams) with 25 ft-2 in. exterior spans and a 25 ft-0 in. interior span. Balanced and partial loading diagrams for the concrete beams are illustrated in Figure 4.9. Partial loads are determined in accordance with 7.5 and Figure 7-4.

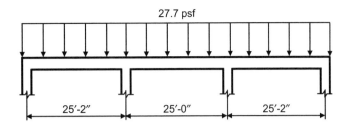

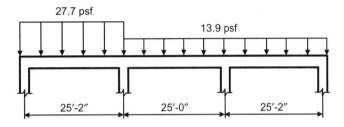

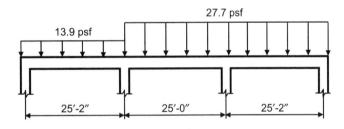

Figure 4.9 Balanced and Partial Loading Diagrams for the Concrete Beams Spanning in the N-S Direction

4.3.5   Example 4.5 – Six-Story Hotel with Rooftop Unit

For the six-story hotel in Example 4.4, determine the drift loads at the rooftop unit depicted in Figure 4.10. Use the same design data as in Example 4.4 and assume that the roof has no parapets.

SOLUTION

The following were determined in Example 4.4 and are used in this example:

Sloped roof snow load $p_s = 27.7$ psf

Snow density $\gamma = 19.2$ pcf

Height of the balanced snow $h_b = p_s / \gamma = 27.7/19.2 = 1.4$ ft

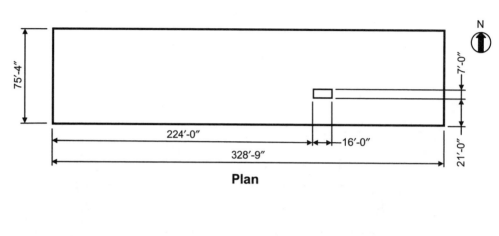

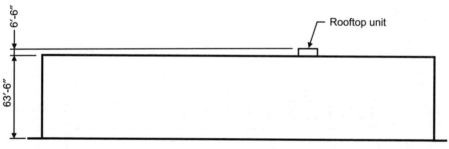

North/South Elevation

Figure 4.10 Plan and Elevation of Six-story Hotel with Rooftop Unit

The clear height to the top of the rooftop unit $h_c = 6.5 - 1.4 = 5.1$ ft

Drift loads are not required where $h_c / h_b < 0.2$ (7.7.1). In this example, $h_c / h_b = 5.1/1.4 > 0.2$, so drift loads must be considered.

Since the plan dimension of the rooftop unit in the N-S direction is less than 15 ft, a drift load is not required to be applied to those sides for wind in the E-W direction (7.8). Drift loads must be considered for the other sides of the rooftop unit, since those sides are greater than 15 ft.

For a N-S wind, the larger of the upwind fetches is $75.33 - 21.0 - 3.5 = 50.83$ ft. For simplicity, this fetch is used for drift on both sides of the rooftop unit.

The equation in Figure 7-9 yields the following for the drift height h_d based on a ground snow load $p_g = 40$ psf and an upwind fetch $\ell_u = 50.83$ ft:

$$h_d = 0.75[0.43(\ell_u)^{1/3}(p_g+10)^{1/4} - 1.5]$$
$$= 0.75[0.43(50.83)^{1/3}(40+10)^{1/4} - 1.5] = 2.1 \text{ ft}$$

Since $h_d = 2.1$ ft $< h_c = 5.1$ ft, the drift width $w = 4h_d = 4 \times 2.1 = 8.4$ ft.

The maximum surcharge drift load $p_d = h_d \gamma = 2.1 \times 19.2 = 40.3$ psf.

The total load at the face of the parapet wall is the balanced load plus the drift surcharge = 27.7 + 40.3 = 68.0 psf.

Drift loads at the rooftop unit are illustrated in Figure 4.11.

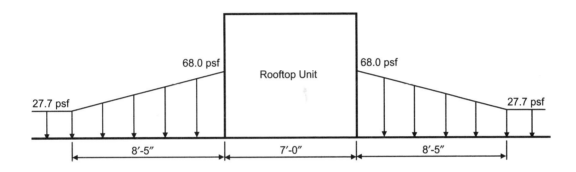

Figure 4.11 Balanced and Drift Loads at Rooftop Unit

4.3.6 Example 4.6 – Agricultural Building

Determine the design snow loads for the agricultural building depicted in Figure 4.12.

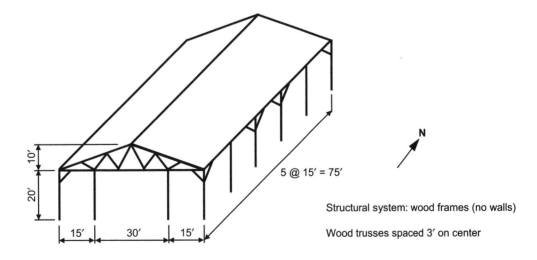

Figure 4.12 Agricultural Building

DESIGN DATA

Ground snow load, p_g:	30 psf
Terrain category:	C (open terrain with scattered obstructions having heights less than 30 ft)
Occupancy:	Utility and miscellaneous occupancy
Thermal condition:	Unheated structure
Roof exposure condition:	Sheltered
Roof surface:	Wood shingles

SOLUTION

1. Determine ground snow load p_g.

   From the design data, the ground snow load p_g is equal to 30 psf.

2. Determine flat roof snow load p_f by Eq. 7-1.

   Use Flowchart 4.1 to determine p_f.

   a. Determine exposure factor C_e from Table 7-2.

      From the design data, the terrain category is B and the roof exposure is sheltered. Therefore, $C_e = 1.2$ from Table 7-2.

   b. Determine thermal factor C_t from Table 7-3.

      From the design data, the structure is unheated, so $C_t = 1.2$ from Table 7-3.

   c. Determine the importance factor I from Table 7-4.

      From IBC Table 1604.5, the Occupancy Category is I for an agricultural facility. Thus, $I = 0.8$ from Table 7-4.

   Therefore,

   $$p_f = 0.7 C_e C_t I p_g$$

   $$= 0.7 \times 1.2 \times 1.2 \times 0.8 \times 30 = 24.2 \text{ psf}$$

   Check if the minimum snow load requirements are applicable:

Minimum values of p_f in accordance with 7.3 apply to hip and gable roofs with slopes less than the larger of 2.38 degrees (1/2 on 12) and $(70/W) + 0.5 = (70/30) + 0.5 = 2.83$ degrees (governs). Since the roof slope in this example is equal to 18.4 degrees, minimum roof snow loads do not apply.

3. Determine sloped roof snow load p_s by Eq. 7-1.

   Use Flowchart 4.2 to determine roof slope factor C_s.

   a. Determine thermal factor C_t from Table 7-3.

      From item 2 above, thermal factor $C_t = 1.2$.

   b. Determine if the roof is warm or cold.

      Since $C_t = 1.2$, the roof is defined as a cold roof in accordance with 7.4.2.

   c. Determine if the roof is unobstructed or not and if the roof is slippery or not.

      There are no obstructions on the roof that inhibit the snow from sliding off the eaves.[10] Also, the roof surface has wood shingles. According to 7.4, wood shingles are not considered to be slippery.

   Since this roof is unobstructed and not slippery, use the solid line in Figure 7-2c to determine C_s:

   For a roof slope of 18.4 degrees, $C_s = 1.0$.

   Therefore, $p_s = C_s p_f = 1.0 \times 24.2 = 24.2$ psf. This is the balanced snow load for this roof.

4. Consider partial loading.

   Partial loads need not be applied to structural members that span perpendicular to the ridgeline in gable roofs with slopes greater than the larger of 2.38 degrees (1/2 on 12) and $(70/W) + 0.5 = (70/30) + 0.5 = 2.83$ degrees (governs).

---

[10] In general, large vent pipes, snow guards, parapet walls and large rooftop equipment are a few common examples of obstructions that could prevent snow from sliding off the roof. Ice dams and icicles along eaves can also possibly inhibit snow from sliding off of two types of warm roofs, which are described in 7.4.5.

Since the roof slope is greater than 2.83 degrees, partial loading is not considered.[11]

5. Consider unbalanced snow loads.

   Flowchart 4.4 is used to determine if unbalanced loads on this gable roof need to be considered or not.

   Unbalanced snow loads must considered for this roof, since the slope is greater than or equal to the larger of $(70/W) + 0.5 = 2.83$ degrees (governs) and 2.38 degrees.

   Since $W = 30$ ft > 20 ft, the unbalanced load consists of the following (see Figure 7-5):

   - Windward side: $0.3 p_s = 0.3 \times 24.2 = 7.3$ psf
   - Leeward side: $p_s = 24.2$ psf along the entire leeward length plus a uniform pressure of $h_d \gamma / \sqrt{S} = (1.9 \times 17.9)/\sqrt{3} = 19.6$ psf, which extends from the ridge a distance of $8 h_d \sqrt{S}/3 = (8 \times 1.9 \times \sqrt{3})/3 = 8.8$ ft where

   $$\begin{aligned}
   h_d &= \text{drift length from Figure 7-9 with } W = 30 \text{ ft substituted for } \ell_u \\
   &= 0.43(W)^{1/3}(p_g + 10)^{1/4} - 1.5 = 1.9 \text{ ft} \\
   \gamma &= \text{snow density (Eq. 7-3)} \\
   &= 0.13 p_g + 14 = 17.9 \text{ pcf} < 30 \text{ pcf} \\
   S &= \text{roof slope run for a rise of one} = 3
   \end{aligned}$$

6. Consider snow drifts on lower roofs and roof projections.

   Not applicable.

7. Consider sliding snow.

   Not applicable.

8. Consider rain-on-snow loads.

   In accordance with 7.10, a rain-on-snow surcharge of 5 psf is required for locations where the ground snow load p_g is 20 psf or less (but not zero) with roof slopes less than $W/50$.

---

[11] Partial loads on individual members of roof trusses such as those illustrated in Figure 4.12 are generally not considered.

In this example, $p_g = 30$ psf so an additional 5 psf need not be added to the balanced load of 24.2 psf.[12]

9. Consider ponding instability.

   Since the roof slope in this example is greater than 1/4 in./ft, progressive roof deflections and ponding instability from rain-on-snow or from snow meltwater need not be investigated (7.11 and 8.4).

10. Consider existing roofs.

    Not applicable.

The balanced and unbalanced snow loads are depicted in Figure 4.13.

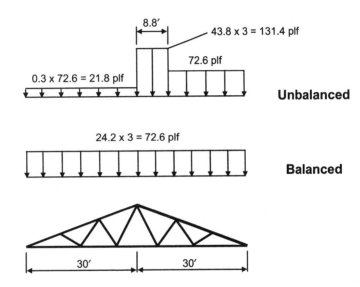

Figure 4.13 Balanced and Unbalanced Snow Loads for Agricultural Building

4.3.7 Example 4.7 – University Facility with Sawtooth Roof

Determine the design snow loads for the university facility depicted in Figure 4.14.

---

[12] The rain-on-snow load applies only to the balanced load case and need not be used in combination with drift, sliding, unbalanced or partial loads (7.10).

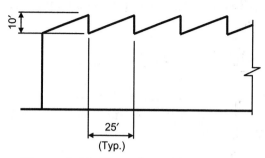

Figure 4.14 Elevation of University Facility

DESIGN DATA

Ground snow load, p_g: 25 psf

Terrain category: C (open terrain with scattered obstructions having heights less than 30 ft)

Occupancy: Educational with an occupant load greater than 500

Thermal condition: Cold, ventilated roof (R-value between the ventilated space and the heated space exceeds 25 ft^{2}h°F/Btu)

Roof exposure condition: Partially exposed

Roof surface: Glass

SOLUTION

1. Determine ground snow load p_g.

   From the design data, the ground snow load p_g is equal to 25 psf.

2. Determine flat roof snow load p_f by Eq. 7-1.

   Use Flowchart 4.1 to determine p_f.

   a. Determine exposure factor C_e from Table 7-2.

   From the design data, the terrain category is C and the roof exposure is partially exposed. Therefore, $C_e = 1.0$ from Table 7-2.

   b. Determine thermal factor C_t from Table 7-3.

   From the design data, the roof is cold and ventilated roof with an R-value between the ventilated space and the heated space that exceeds 25 ft^{2}h°F/Btu. Thus, $C_t = 1.1$ from Table 7-3.

c. Determine the importance factor I from Table 7-4.

From IBC Table 1604.5, the Occupancy Category is III for this college educational facility that has an occupant load greater than 500 people. Thus, $I = 1.1$ from Table 7-4.

Therefore,

$$p_f = 0.7 C_e C_t I p_g$$

$$= 0.7 \times 1.0 \times 1.1 \times 1.1 \times 25 = 21.2 \text{ psf}$$

3. Determine sloped roof snow load p_s from Eq. 7-1.

Use Flowchart 4.2 to determine roof slope factor C_s.

In accordance with 7.4.4, $C_s = 1.0$ for sawtooth roofs.

Thus, $p_s = p_f = 21.2$ psf.

4. Consider unbalanced snow loads.

Flowchart 4.6 is used to determine if unbalanced loads on this sawtooth roof need to be considered or not.

Unbalanced snow loads must be considered, since the slope is greater than 1.79 degrees (7.6.3).

In accordance with 7.6.3, the load at the ridge or crown is equal to $0.5 p_f = 0.5 \times 21.2 = 10.6$ psf. At the valley, the load is $2 p_f / C_e = 2 \times 21.2 / 1.0 = 42.4$ psf.

The load at the valley is limited by the space that is available for snow accumulation. The unit weight of the snow is determined by Eq. 7-3:

$$\gamma = 0.13 p_g + 14 = (0.13 \times 25) + 14 = 17.3 \text{ pcf} < 30 \text{ pcf}$$

The maximum permissible load is equal to the load at the ridge plus the load corresponding to 10 ft of snow: $10.6 + (10 \times 17.3) = 183.6$ psf.

Since the unbalanced load of 42.4 psf at the valley is less than 183.6 psf, the load at the valley is not reduced.

Balanced and unbalanced snow loads are illustrated in Figure 4.15.

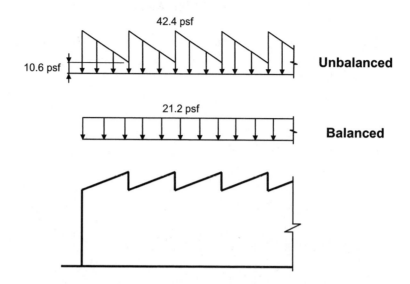

Figure 4.15 Balanced and Unbalanced Snow Loads for University Facility

4.3.8 Example 4.8 – Public Utility Facility with Curved Roof

Determine the design snow loads for the public utility facility depicted in Figure 4.16. The facility is required to remain operational during an emergency.

DESIGN DATA	
Ground snow load, p_g:	60 psf
Terrain category:	D (flat unobstructed area near water)
Occupancy:	Essential facility
Thermal condition:	Unheated structure
Roof exposure condition:	Fully exposed
Roof surface:	Rubber membrane

SOLUTION

1. Determine ground snow load p_g.

   From the design data, the ground snow load p_g is equal to 60 psf.

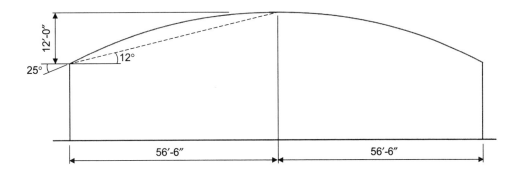

Figure 4.16 Elevation of Public Utility Facility

2. Determine flat roof snow load p_f by Eq. 7-1.

   Use Flowchart 4.1 to determine p_f.

   a. Determine exposure factor C_e from Table 7-2.

      From the design data, the terrain category is D and the roof exposure is fully exposed. Therefore, $C_e = 0.8$ from Table 7-2.

   b. Determine thermal factor C_t from Table 7-3.

      From the design data, the structure is unheated. Thus, $C_t = 1.2$ from Table 7-3.

   c. Determine the importance factor I from Table 7-4.

      From IBC Table 1604.5, the Occupancy Category is IV for this essential facility. Thus, $I = 1.2$ from Table 7-4.

Therefore,

$p_f = 0.7 C_e C_t I p_g$

$= 0.7 \times 0.8 \times 1.2 \times 1.2 \times 60 = 48.4$ psf

Check if the minimum snow load requirements are applicable:

Minimum values of p_f in accordance with 7.3 apply to curved roofs where the vertical angle from the eaves to the crown is less than 10 degrees. Since that slope in this example is equal to 12 degrees, minimum roof snow loads do not apply.

3. Determine sloped roof snow load p_s from Eq. 7-1.

   Use Flowchart 4.2 to determine roof slope factor C_s.

   a. Determine thermal factor C_t from Table 7-3.

      From item 2 above, thermal factor $C_t = 1.2$.

   b. Determine if the roof is warm or cold.

      Since $C_t = 1.2$, the roof is defined as a cold roof in accordance with 7.4.2.

   c. Determine if the roof is unobstructed or not and if the roof is slippery or not.

      There are no obstructions on the roof that inhibit the snow from sliding off the eaves.[13] Also, the roof surface is a rubber membrane. According to 7.4, rubber membranes are considered to be slippery.

   Since this roof is unobstructed and slippery, use the dashed line in Figure 7-2c to determine C_s.

   For the tangent slope of 25 degrees at the eave, the roof slope factor C_s is determined by the equation in C7.4 for cold roofs with $C_t = 1.2$:

   $$C_s = 1.0 - \frac{(\text{slope} - 15°)}{55°} = 1.0 - \frac{(25° - 15°)}{55°} = 0.82$$

   Therefore, $p_s = C_s p_f = 0.82 \times 48.4 = 39.7$ psf, which is the balanced snow load at the eaves.

   Away from the eaves, the roof slope factor C_s is equal to 1.0 where the tangent roof slope is less than or equal to 15 degrees (see dashed line in Figure 7-2c). This occurs at distances of approximately 20.7 ft from the eaves at both ends of the roof. Therefore, in the center portion of the roof, $p_s = C_s p_f = 1.0 \times 48.4 = 48.4$ psf.

   The balanced snow load is depicted in Figure 4.17, which is based on Case 1 in Figure 7-3 for slope at eaves less than 30 degrees.

---

[13] In general, large vent pipes, snow guards, parapet walls and large rooftop equipment are a few common examples of obstructions that could prevent snow from sliding off the roof. Ice dams and icicles along eaves can also possibly inhibit snow from sliding off of two types of warm roofs, which are described in 7.4.5.

4. Consider unbalanced snow loads.

   Flowchart 4.5 is used to determine if unbalanced loads on this curved roof need to be considered or not.

   Since the slope of the straight line from the eaves to the crown is greater than 10 degrees and is less than 60 degrees, unbalanced snow loads must be considered (7.6.2).

   Unbalanced loads for this roof are given in Case 1 of Figure 7-3. No snow loads are applied on the windward side. On the leeward side, the snow load is equal to $0.5\,p_f = 0.5 \times 48.4 = 24.2$ psf at the crown and $2\,p_f C_s / C_e = 2 \times 48.4 \times 0.82 / 0.8 = 99.2$ psf at the eaves where C_s is based on the slope at the eaves. The unbalanced snow loads are shown in Figure 4.17.

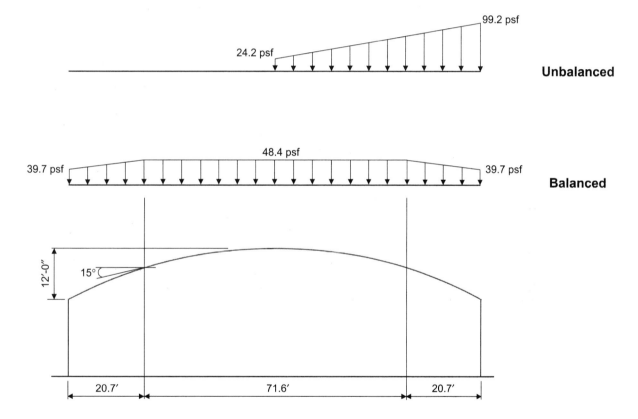

Figure 4.17 Balanced and Unbalanced Snow Loads for Public Utility Facility

CHAPTER 5   *WIND LOADS*

5.1 INTRODUCTION

In accordance with IBC 1609.1.1, wind loads on buildings and structures are to be determined by the provisions of Chapter 6 of ASCE/SEI 7-05 or by the alternate all-heights method of IBC 1609.6. Exceptions are also given in this section, which permit wind forces to be determined on certain types of structures using industry standards other than ASCE/SEI 7. Wind tunnel tests that conform to the provisions of ASCE/SEI 6.6 are also permitted, provided the limitations of IBC 1609.1.1.2 are satisfied.

The basic wind speed, the exposure category and the type of opening protection required may be determined in accordance with the provisions of IBC 1609 or ASCE/SEI 7-05.

Figure 1609 in the IBC and Figure 6-1 in ASCE/SEI 7-05 are identical and provide basic wind speeds based on 3-second gusts at 33 ft above ground for Exposure C. The design wind speeds on these maps do not include effects of tornadoes. Since some referenced standards contain criteria or applications based on fastest-mile wind speed, which was the wind speed utilized in earlier editions of ASCE/SEI 7 and the legacy codes, IBC 1609.3.1 provides an equation and a table that can be used to convert from one wind speed to the other.

Chapter 6 of ASCE/SEI 7-05 contains three methods to determine design wind pressures or loads:

- Method 1 – Simplified Procedure
- Method 2 – Analytical Procedure
- Method 3 – Wind Tunnel Procedure

The design requirements of Method 1 can be found in 6.4. This method is based on the low-rise buildings procedure in Method 2, and can be used to determine wind pressures on the main wind-force-resisting system (MWFRS) of a building, provided the conditions of 6.4.1.1 are met. Although there are eight conditions that need to be satisfied, a large number of typical low-rise buildings meet these criteria. Definitions for enclosed, low-rise, regularly shaped, and simple diaphragm buildings are given in 6.2, along with other important definitions (these types of buildings are listed under the conditions in 6.4.1.1).

Method 1 can also be used to determine wind pressures on components and cladding (C&C) of a building, provided the five conditions in 6.4.1.2 are satisfied. If a building satisfies only those criteria in 6.4.1.2 for C&C, the design wind pressures on the MWFRS must be determined by Method 2 or 3.

Wind pressures are tabulated in Figures 6-2 and 6-3 as a function of the basic wind speed for a specific set of conditions: Occupancy Category II buildings with a mean roof height of 30 ft that is located on primarily flat ground in Exposure B. Adjustments are made to these tabulated pressures based on actual building height, exposure, occupancy and topography at the site. The adjusted pressures are applied normal to projected areas of the building in accordance with Figures 6-2 and 6-3.

Method 2 provides wind pressures and forces for the design of MWFRSs and C&C of enclosed and partially enclosed rigid and flexible buildings, open buildings and other structures, including freestanding walls, signs, rooftop equipment and other structures (6.5).

In general, this procedure entails the determination of velocity pressures (which are a function of exposure, height, topographic effects, wind directionality, wind velocity and building occupancy), gust effect factors, external and internal pressure coefficients and force coefficients. The Analytical Procedure can be used to determine pressures and forces on a wide range of buildings and structures, provided

1. The building or structure is regularly-shaped, i.e., the building has no unusual geometrical irregularity in spatial form (both vertical and horizontal).

2. The building or structure responds to wind primarily in the same direction as that of the wind, i.e., it does not have response characteristics that make it subject to across-wind loading, vortex shedding or any other dynamic load effects, which are common in tall, slender buildings and structures and cylindrical buildings and structures.

3. The building or structure is located on a site where there are no channeling effects or buffeting in the wake of upwind obstructions.

The Wind Tunnel Procedure (Method 3) in 6.6 can be utilized for any building or structure in lieu of Methods 1 or 2, and must be used where the conditions of Methods 1 or 2 are not satisfied. Requirements for proper wind tunnel testing are given in 6.6.2. As noted previously, the limitations of IBC 1609.1.1.2 must be satisfied where this procedure is utilized. The provisions in IBC 1609.1.1.2.1 and 1609.1.1.2.2 prescribe lower limits on the magnitude of the base overturning moments on the main wind-force-resisting system and the pressures for components and cladding, respectively.

The alternate all-heights method in IBC 1609.6, which is based on Method 2 of ASCE/SEI 7-05, can be used to determine wind pressures on regularly shaped buildings and structures that meet the five conditions listed in IBC 1609.6.1. In this method, terms in the design pressure equation of Method 2 are combined to produce pressure coefficients C_{net}, which are provided in IBC Table 1609.6.2(2) for surfaces in main wind-force-resisting systems and components and cladding. Net wind pressures calculated by IBC Eq. 16-34 are applied simultaneously on, and in a direction normal to, all wall and roof surfaces.

Provisions for minimum design wind loading are given in 6.1.4. Figure C6-1 in the commentary of ASCE/SEI 7 illustrates the minimum wind pressure that must be applied horizontally on the entire vertical projection of a building or structure for the design of the MWFRS. This 10-psf pressure is to be applied as a separate load case in addition to the other load cases specified in Chapter 6. For C&C, a minimum net pressure of 10 psf acting in either direction normal to the surface is required.

The same minimum pressures given in 6.1.4 of ASCE/SEI 7 are in IBC 1609.6.3 for the alternate all-heights method.

Section 5.2 of this document contains various flowcharts for determining wind loads. Included are two flowcharts that present information on when the various methods can be utilized and eight flowcharts that provide step-by-step procedures on how to determine design wind pressures and forces on buildings and other structures.

Section 5.3 contains completely worked-out design examples that illustrate the design requirements for wind.

5.2 FLOWCHARTS

A summary of the flowcharts provided in this chapter is given in Table 5.1. Included is a description of the content of each flowchart.

Table 5.1 Summary of Flowcharts Provided in Chapter 5

Flowchart	Title	Description
ASCE/SEI 5.2.1 Allowed Procedures		
Flowchart 5.1	Allowed Procedures – MWFRS	Summarizes procedures that are allowed in determining design wind pressures on MWFRSs.
Flowchart 5.2	Allowed Procedures – C&C	Summarizes procedures that are allowed in determining design wind pressures on C&C.
ASCE/SEI 5.2.2 Method 1 – Simplified Procedure		
Flowchart 5.3	MWFRS – Net Design Wind Pressure, p_s	Provides step-by-step procedure on how to determine the net design wind pressures p_s on MWFRSs.
Flowchart 5.4	C&C – Net Design Wind Pressure, p_{net}	Provides step-by-step procedure on how to determine the net design wind pressures p_{net} on C&C.

(continued)

Table 5.1 Summary of Flowcharts Provided in Chapter 5 (continued)

Flowchart	Title	Description
ASCE/SEI 5.2.3 Method 2 – Analytical Procedure		
Flowchart 5.5	Velocity Pressures, q_z and q_h	Outlines the procedure for determining velocity pressures that are used in Method 2.
Flowchart 5.6	Gust Effect Factors, G and G_f	Outlines methods for determining gust effect factors that are used in Method 2.
Flowchart 5.7	Buildings, MWFRS	Provides step-by-step procedures on how to determine design wind pressures on MWFRSs of enclosed, partially enclosed and open buildings.
Flowchart 5.8	Buildings, C&C	Provides step-by-step procedures on how to determine design wind pressures on C&C of enclosed, partially enclosed and open buildings.
Flowchart 5.9	Structures Other than Buildings	Provides step-by-step procedures on how to determine design wind forces on solid freestanding walls, solid signs, rooftop structures and equipment, and other structures that are not buildings.
IBC 1609.6 Alternate All-heights Method		
Flowchart 5.10	Net wind pressure, p_{net}	Provides step-by-step procedure on how to determine design wind pressures on MWFRSs and C&C.

MWFRS = Main wind-force-resisting system
C&C = Components and cladding

5.2.1 Allowed Procedures

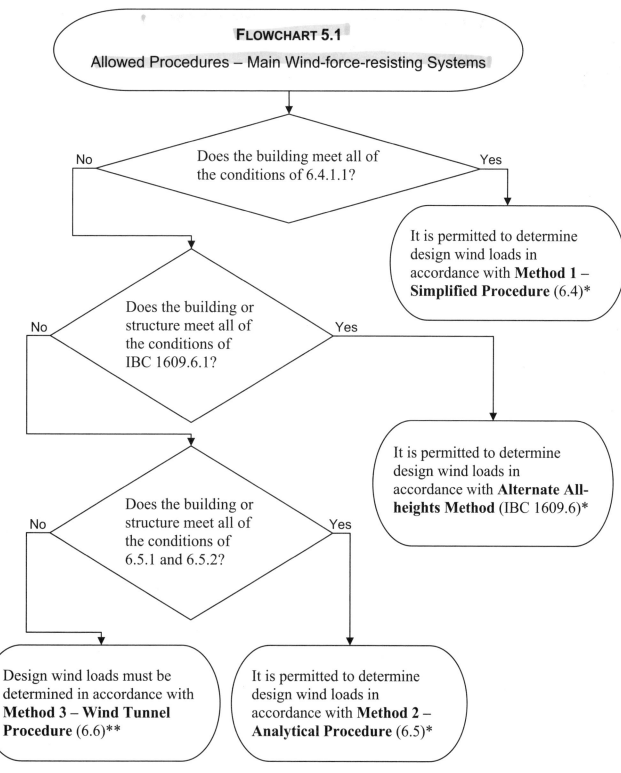

* Wind tunnel testing (ASCE/SEI 6.6.1) is permitted for any building or structure subject to the limitations of IBC 1609.1.1.2.
** Limitations of IBC 1609.1.1.2 must also be satisfied.

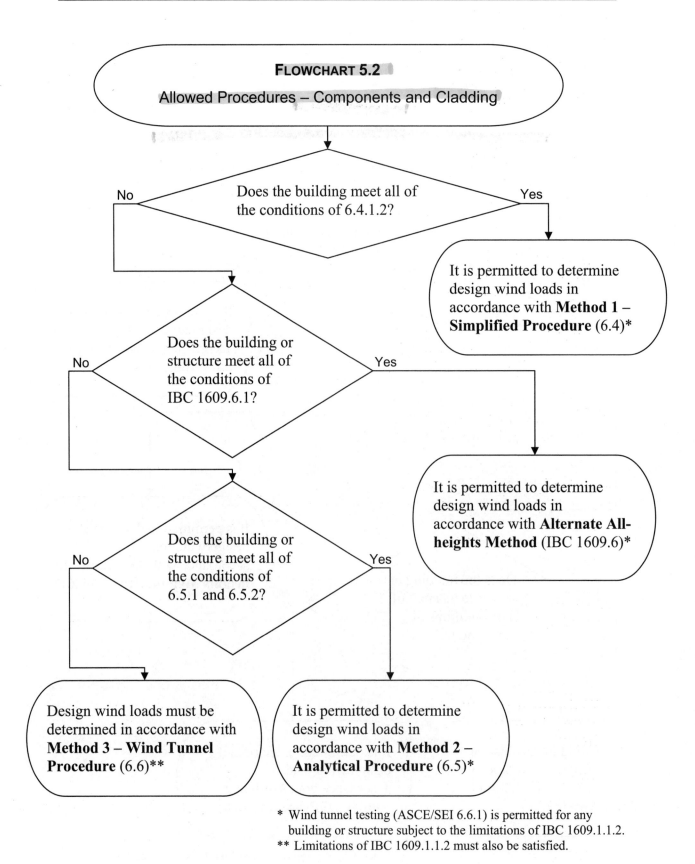

* Wind tunnel testing (ASCE/SEI 6.6.1) is permitted for any building or structure subject to the limitations of IBC 1609.1.1.2.
** Limitations of IBC 1609.1.1.2 must also be satisfied.

5.2.2 Method 1 – Simplified Procedure

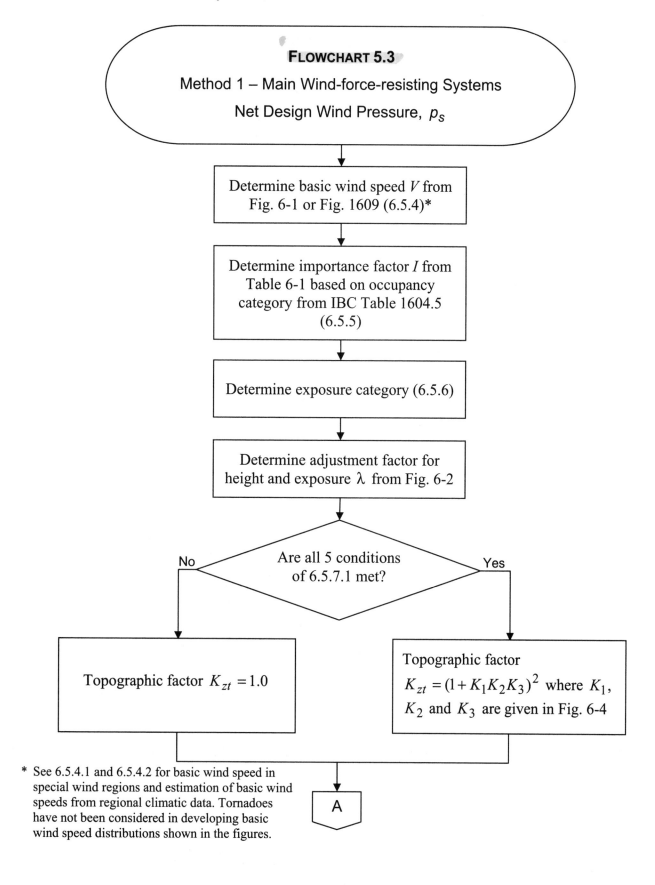

* See 6.5.4.1 and 6.5.4.2 for basic wind speed in special wind regions and estimation of basic wind speeds from regional climatic data. Tornadoes have not been considered in developing basic wind speed distributions shown in the figures.

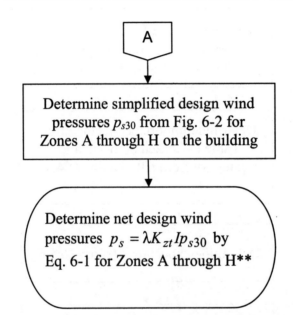

FLOWCHART 5.3

Method 1 – Main Wind-force-resisting Systems
Net Design Wind Pressure, p_s (continued)

** Notes:
1. For horizontal pressure zones, p_s is the sum of the windward and leeward net (sum of internal and external) pressures on vertical projection of Zones A, B, C, and D. For vertical pressure zones, p_s is the net (sum of internal and external) pressure on horizontal projection of Zones E, F, G and H.
2. The load patterns shown in Fig. 6-2 shall be applied to each corner of the building in turn as the reference corner. See other notes in Fig. 6-2.
3. Load effects of the design wind pressures determined by Eq. 6-1 shall not be less than those from the minimum load case of 6.1.4.1. It is assumed that the pressures p_s for Zones A, B, C and D are equal to +10 psf and the pressures for Zones E, F, G and H are equal to 0 psf.

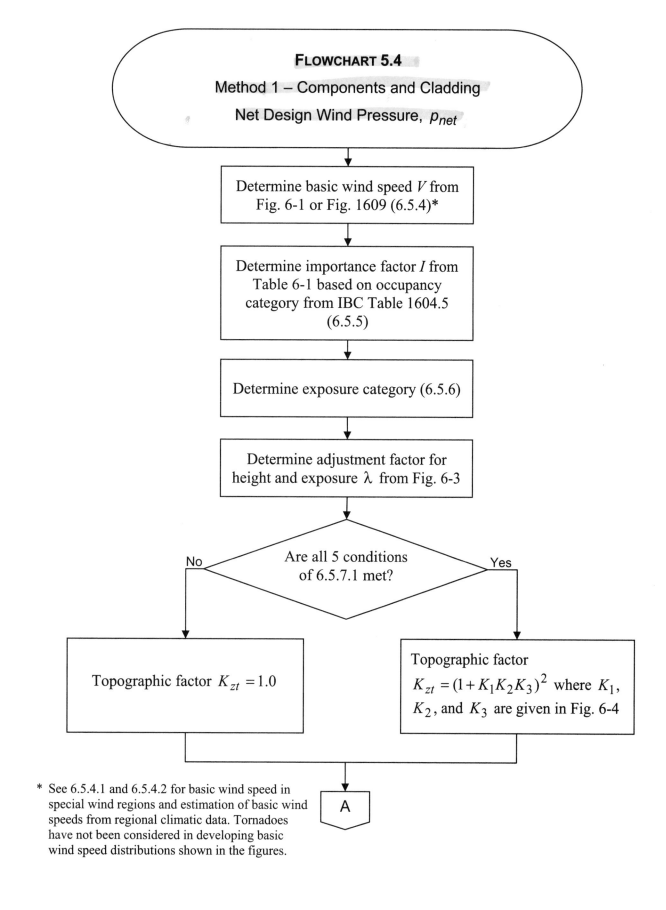

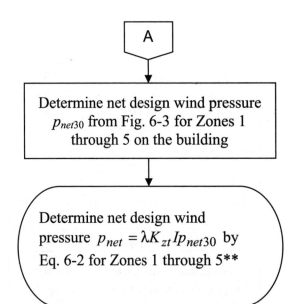

** Notes:
1. p_{net} represents the net pressures (sum of internal and external) to be applied normal to each building surface as shown in Fig. 6-3. See notes in Fig. 6-3.
2. The positive design wind pressures p_{net} determined by Eq. 6-2 shall not be less than +10 psf and the negative design wind pressures p_{net} determined by Eq. 6-2 shall not be less than −10 psf.

5.2.3 Method 2 – Analytical Procedure

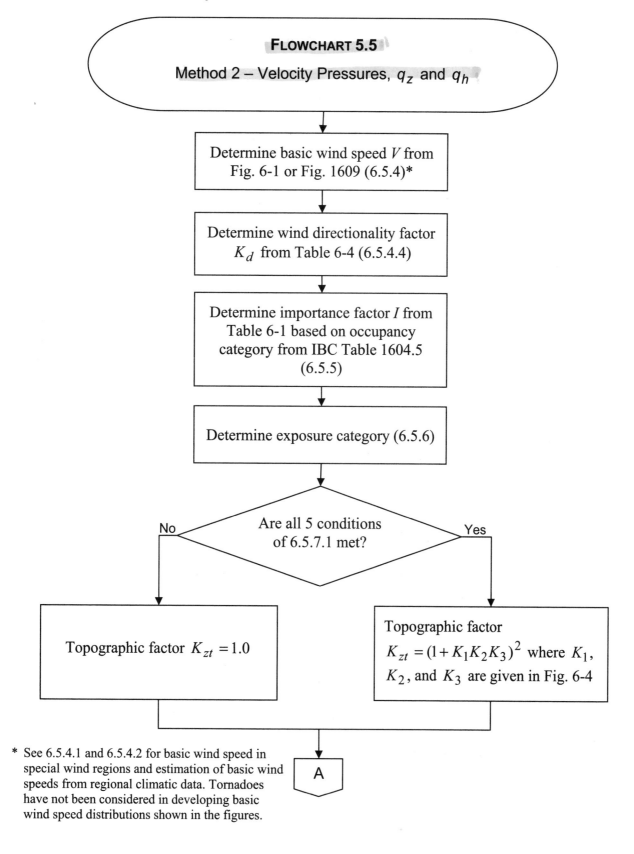

FLOWCHART 5.5

Method 2 – Velocity Pressures, q_z and q_h

Determine basic wind speed V from Fig. 6-1 or Fig. 1609 (6.5.4)*

Determine wind directionality factor K_d from Table 6-4 (6.5.4.4)

Determine importance factor I from Table 6-1 based on occupancy category from IBC Table 1604.5 (6.5.5)

Determine exposure category (6.5.6)

Are all 5 conditions of 6.5.7.1 met?

No: Topographic factor $K_{zt} = 1.0$

Yes: Topographic factor $K_{zt} = (1 + K_1 K_2 K_3)^2$ where K_1, K_2, and K_3 are given in Fig. 6-4

A

* See 6.5.4.1 and 6.5.4.2 for basic wind speed in special wind regions and estimation of basic wind speeds from regional climatic data. Tornadoes have not been considered in developing basic wind speed distributions shown in the figures.

FLOWCHART 5.5

Method 2 – Velocity Pressures, q_z and q_h

(continued)

A

Determine velocity pressure exposure coefficients K_z and K_h from Table 6-3 (6.5.6.6)

Determine velocity pressure at height z and h by Eq. 6-15:

$$q_z = 0.00256 K_z K_{zt} K_d V^2 I$$

$$q_h = 0.00256 K_h K_{zt} K_d V^2 I \ **$$

** Notes:
1. q_z = velocity pressure evaluated at height z
2. q_h = velocity pressure evaluated at mean roof height h
3. The numerical constant of 0.00256 should be used except where sufficient weather data are available to justify a different value (see C6.5.10)

FLOWCHARTS 5-13

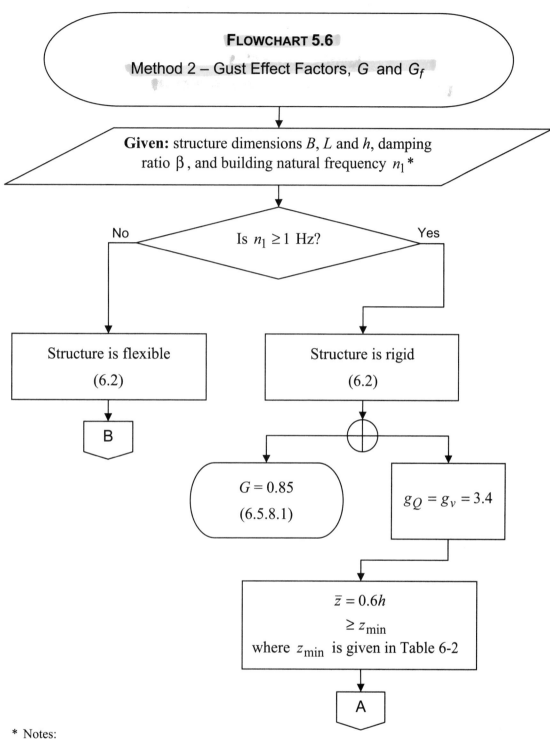

* Notes:
    1. Information on structural damping can be found in C6.5.8.
    2. n_1 can be determined from a rational analysis or estimated from approximate equations given in C6.5.8.

FLOWCHART 5.6
Method 2 – Gust Effect Factors, G and G_f
(continued)

A

$$I_{\bar{z}} = c\left(\frac{33}{\bar{z}}\right)^{1/6} \quad \text{Eq. 6-5}$$

where c is given in Table 6-2

$$L_{\bar{z}} = \ell\left(\frac{\bar{z}}{33}\right)^{\bar{\epsilon}} \quad \text{Eq. 6-7}$$

where ℓ and $\bar{\epsilon}$ are given in Table 6-2

$$Q = \sqrt{\frac{1}{1+0.63\left(\frac{B+h}{L_{\bar{z}}}\right)^{0.63}}} \quad \text{Eq. 6-6}$$

$$G = 0.925\left(\frac{1+1.7g_Q I_{\bar{z}} Q}{1+1.7g_v I_{\bar{z}}}\right) \quad \text{Eq. 6-4}$$

FLOWCHART 5.6
Method 2 – Gust Effect Factors, G and G_f
(continued)

B

$g_Q = g_v = 3.4$

$$g_R = \sqrt{2\ln(3{,}600n_1)} + \frac{0.577}{\sqrt{2\ln(3{,}600n_1)}} \quad \text{Eq. 6-9}$$

$\bar{z} = 0.6h$
$\geq z_{\min}$

where $z_{\min}$ is given in Table 6-2

$$I_{\bar{z}} = c\left(\frac{33}{\bar{z}}\right)^{1/6} \quad \text{Eq. 6-5}$$

where c is given in Table 6-2

C

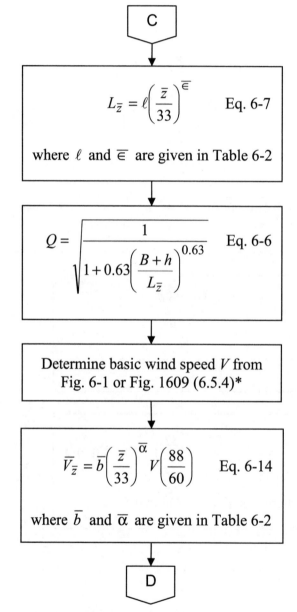

FLOWCHART 5.6

Method 2 – Gust Effect Factors, G and G_f

(continued)

C

$$L_{\bar{z}} = \ell \left(\frac{\bar{z}}{33}\right)^{\bar{\epsilon}} \quad \text{Eq. 6-7}$$

where ℓ and $\bar{\epsilon}$ are given in Table 6-2

$$Q = \sqrt{\frac{1}{1 + 0.63\left(\frac{B+h}{L_{\bar{z}}}\right)^{0.63}}} \quad \text{Eq. 6-6}$$

Determine basic wind speed V from Fig. 6-1 or Fig. 1609 (6.5.4)*

$$\bar{V}_{\bar{z}} = \bar{b}\left(\frac{\bar{z}}{33}\right)^{\bar{\alpha}} V\left(\frac{88}{60}\right) \quad \text{Eq. 6-14}$$

where $\bar{b}$ and $\bar{\alpha}$ are given in Table 6-2

D

* See 6.5.4.1 and 6.5.4.2 for basic wind speed in special wind regions and estimation of basic wind speeds from regional climatic data. Tornadoes have not been considered in developing basic wind speed distributions shown in the figures.

FLOWCHART 5.6

Method 2 – Gust Effect Factors, G and G_f

(continued)

D

$$N_1 = \frac{n_1 L_{\bar{z}}}{\bar{V}_{\bar{z}}} \qquad \text{Eq. 6-12}$$

$$R_n = \frac{7.47 N_1}{(1+10.3 N_1)^{5/3}} \qquad \text{Eq. 6-11}$$

$$R_h = \frac{1}{\eta} - \frac{1}{2\eta^2}\left(1 - e^{-2\eta}\right) \text{ for } \eta > 0$$

$$R_h = 1 \text{ for } \eta = 0$$

where $\eta = 4.6 n_1 h / \bar{V}_{\bar{z}}$

E

FLOWCHART 5.6

Method 2 – Gust Effect Factors, G and G_f

(continued)

E

$$R_B = \frac{1}{\eta} - \frac{1}{2\eta^2}\left(1 - e^{-2\eta}\right) \text{ for } \eta > 0$$

$$R_B = 1 \text{ for } \eta = 0$$

where $\eta = 4.6 n_1 B / \overline{V}_{\bar{z}}$

$$R_L = \frac{1}{\eta} - \frac{1}{2\eta^2}\left(1 - e^{-2\eta}\right) \text{ for } \eta > 0$$

$$R_L = 1 \text{ for } \eta = 0$$

where $\eta = 15.4 n_1 L / \overline{V}_{\bar{z}}$

$$R = \sqrt{\frac{1}{\beta} R_n R_h R_B (0.53 + 0.47 R_L)} \quad \text{Eq. 6-10}$$

$$G_f = 0.925 \left(\frac{1 + 1.7 I_{\bar{z}} \sqrt{g_Q^2 Q^2 + g_R^2 R^2}}{1 + 1.7 g_v I_{\bar{z}}} \right) \quad \text{Eq. 6-8}$$

FLOWCHARTS 5-19

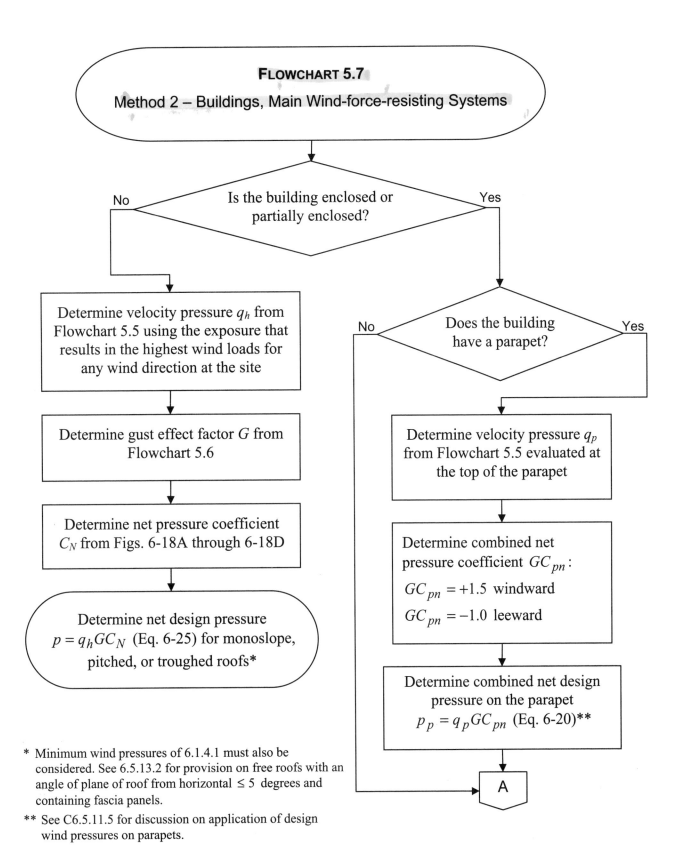

FLOWCHART 5.7 — Method 2 – Buildings, Main Wind-force-resisting Systems

* Minimum wind pressures of 6.1.4.1 must also be considered. See 6.5.13.2 for provision on free roofs with an angle of plane of roof from horizontal ≤ 5 degrees and containing fascia panels.

** See C6.5.11.5 for discussion on application of design wind pressures on parapets.

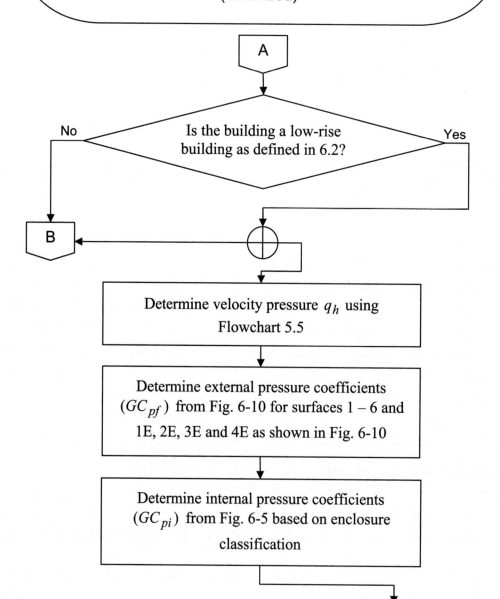

FLOWCHART 5.7

Method 2 – Buildings, Main Wind-force-resisting Systems

(continued)

Determine whether the building is rigid or flexible and the corresponding gust effect factor G or G_f from Flowchart 5.6

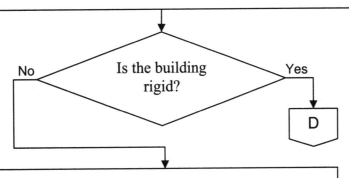

Determine velocity pressure q_z for windward walls along the height of the building and q_h for leeward walls, side walls and roof using Flowchart 5.5

Determine pressure coefficients C_p for the walls and roof from Fig. 6-6 or 6-8

Determine q_i for the walls and roof using Flowchart 5.5‡

‡ $q_i = q_h$ or $q_i = q_z$ depending on enclosure classification (see 6.5.12.2.1). q_i may conservatively be evaluated at height h ($q_i = q_h$) where applicable.

FLOWCHART 5.7
Method 2 – Buildings, Main Wind-force-resisting Systems
(continued)

C

Determine internal pressure coefficients (GC_{pi}) from Fig. 6-5 based on enclosure classification

Determine design wind pressures by Eq. 6-19:
- Windward walls: $p_z = q_z G_f C_p - q_h(GC_{pi})$
- Leeward walls, side walls, and roofs: $p_h = q_h G_f C_p - q_h(GC_{pi})$ [+]

[+] Notes:
1. See 6.5.12.3 and Fig. 6-9 for the load cases that must be considered.
2. Minimum wind pressures of 6.1.4.1 must also be considered.

FLOWCHART 5.7

Method 2 – Buildings, Main Wind-force-resisting Systems
(continued)

Determine velocity pressure q_z for windward walls along the height of the building and q_h for leeward walls, side walls and roof using Flowchart 5.5

↓

Determine pressure coefficients C_p for the walls and roof from Fig. 6-6 or 6-8

↓

Determine q_i for the walls and roof using Flowchart 5.5[‡‡]

↓

Determine internal pressure coefficients (GC_{pi}) from Fig. 6-5 based on enclosure classification

↓

Determine design wind pressures by Eq. 6-17:
- Windward walls: $p_z = q_z GC_p - q_h(GC_{pi})$
- Leeward walls, side walls, and roofs: $p_h = q_h GC_p - q_h(GC_{pi})$ [++]

[‡‡] $q_i = q_h$ or $q_i = q_z$ depending on enclosure classification (see 6.5.12.2.1). qi may conservatively be evaluated at height h ($q_i = q_h$) where applicable.

[++] Notes:
1. See 6.5.12.3 and Fig. 6-9 for the load cases that must be considered.
2. Minimum wind pressures of 6.1.4.1 must also be considered.

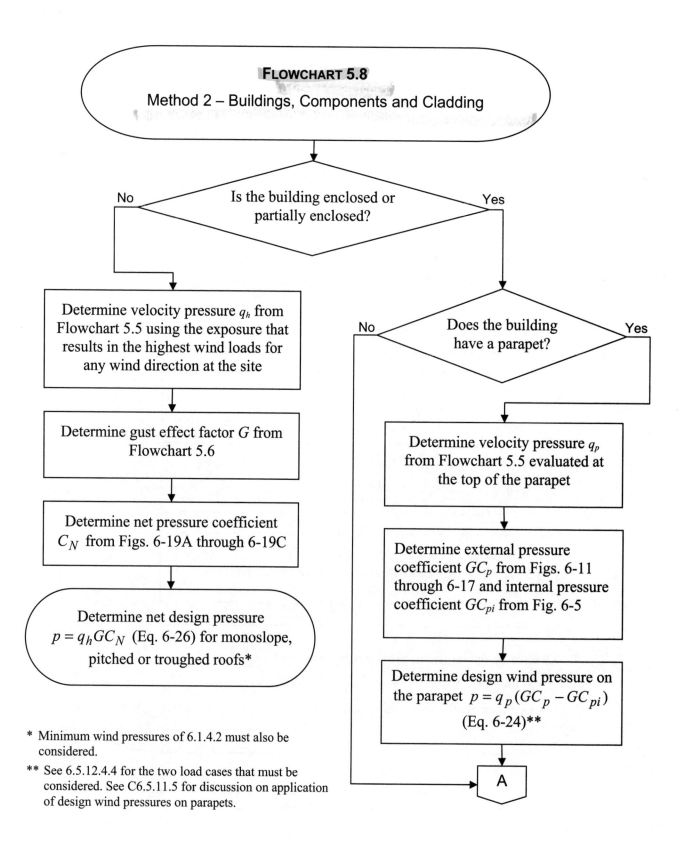

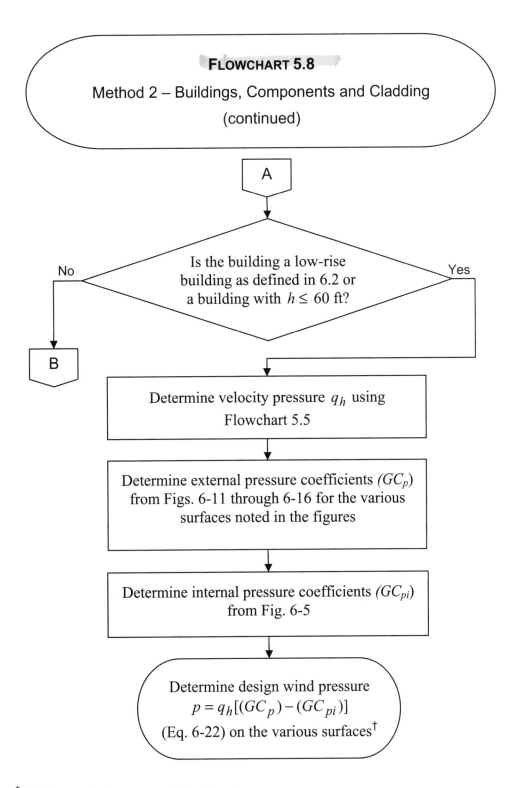

† Minimum wind pressures of 6.1.4.2 must also be considered.

FLOWCHART 5.8

Method 2 – Buildings, Components and Cladding
(continued)

B

↓

Determine velocity pressure q_z for windward walls along the height of the building and q_h for leeward walls, side walls and roof using Flowchart 5.5

↓

Determine external pressure coefficients (GC_p) for the walls and roof from Fig. 6-17

↓

Determine q_i for the walls and roof using Flowchart 5.5‡‡

↓

Determine internal pressure coefficients (GC_{pi}) from Fig. 6-5 based on enclosure classification

↓

Determine design wind pressures by Eq. 6-23:
- Windward walls: $p_z = q_z(GC_p) - q_h(GC_{pi})$
- Leeward walls, side walls, and roofs: $p_h = q_h(GC_p) - q_h(GC_{pi})$ ++

---

‡‡ $q_i = q_h$ or $q_i = q_z$ depending on enclosure classification (see 6.5.12.4.2).
q_i may conservatively be evaluated at height h ($q_i = q_h$) where applicable.

++ Notes:
1. Minimum wind pressures of 6.1.4.2 must also be considered.
2. An alternate method for C&C in buildings with 60 ft < h < 90 ft is given in 6.5.12.4.3.

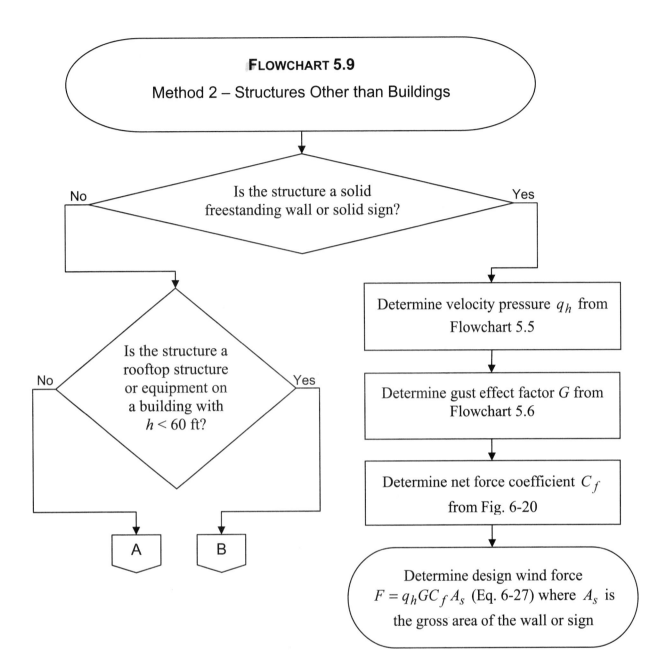

FLOWCHART 5.9

Method 2 – Structures Other than Buildings
(continued)

A

Determine velocity pressure q_z evaluated at height z of the centroid of area A_f based on the exposure defined in 6.5.6.3 using Flowchart 5.5

Determine gust effect factor G from Flowchart 5.6

Determine force coefficient C_f from Figs. 6-21 through 6-23

Determine design wind force $F = q_z G C_f A_f$ (Eq. 6-28) where A_f is the projected area normal to the wind except where C_f is specified for the actual surface area

FLOWCHART 5.9

Method 2 – Structures Other than Buildings
(continued)

B

Determine velocity pressure q_z evaluated at height z of the centroid of area A_f based on the exposure defined in 6.5.6.3 using Flowchart 5.5

Determine gust effect factor G from Flowchart 5.6

Determine force coefficient C_f from Figs. 6-21 through 6-23

Is $A_f < 0.1Bh$?

- Yes: Determine design wind force by increasing force obtained by Eq. 6-28 by 1.9: $F = 1.9 q_z G C_f A_f$
- No: Is $A_f \geq Bh$?
  - Yes: Determine design wind force by Eq. 6-28: $F = q_z G C_f A_f$
  - No: Determine design wind force by increasing force obtained by Eq. 6-28 by a factor that decreases linearly from 1.9 to 1.0 as A_f increases from $0.1Bh$ to Bh

5.2.4 Alternate All-heights Method

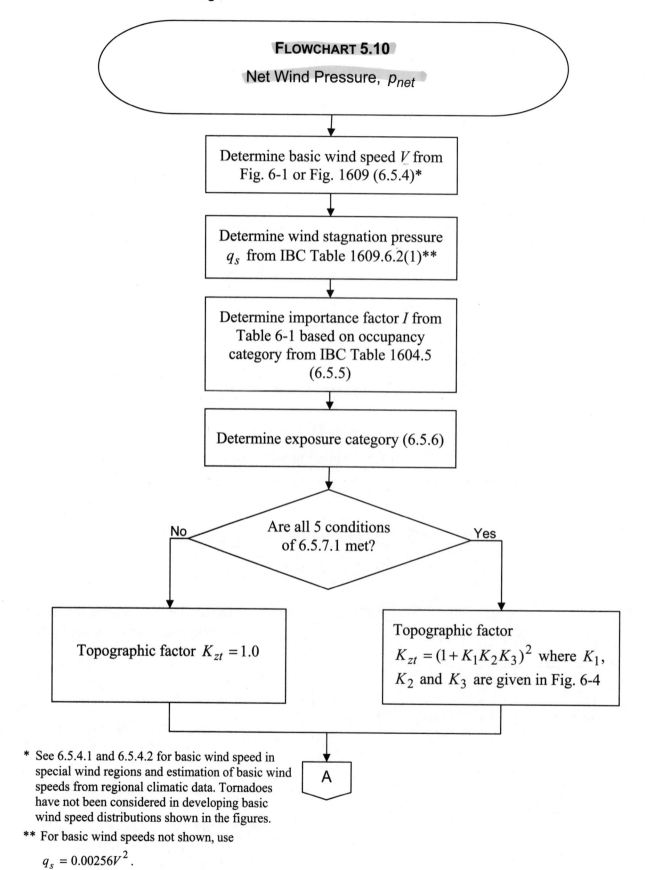

* See 6.5.4.1 and 6.5.4.2 for basic wind speed in special wind regions and estimation of basic wind speeds from regional climatic data. Tornadoes have not been considered in developing basic wind speed distributions shown in the figures.

** For basic wind speeds not shown, use

$q_s = 0.00256V^2$.

FLOWCHARTS 5-31

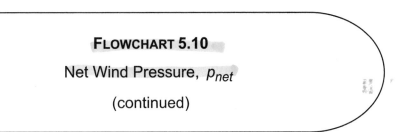

FLOWCHART 5.10

Net Wind Pressure, p_{net}

(continued)

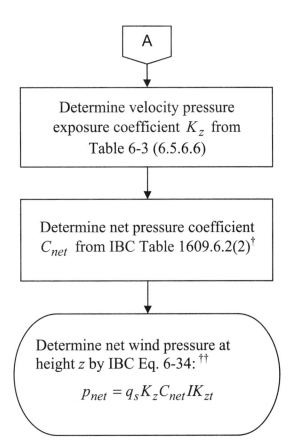

A

Determine velocity pressure exposure coefficient K_z from Table 6-3 (6.5.6.6)

Determine net pressure coefficient C_{net} from IBC Table 1609.6.2(2)[†]

Determine net wind pressure at height z by IBC Eq. 6-34: [††]

$$p_{net} = q_s K_z C_{net} I K_{zt}$$

[†] Where C_{net} has more than one value in IBC Table 1609.6.2(2), the more severe wind load condition shall be used for design.

[††] Wind pressures are to be applied to the building envelope wall and roof surfaces in accordance with IBC 1609.6.4.4.

5.3 EXAMPLES

The following sections contain examples that illustrate the wind design provisions of IBC 1609 and Chapter 6 in ASCE/SEI 7-05.

5.3.1 Example 5.1 – Warehouse Building using Method 1, Simplified Procedure

For the one-story warehouse illustrated in Figure 5.1, determine design wind pressures on (1) the main wind-force-resisting system in both directions, (2) a solid precast wall panel and (3) an open-web joist purlin using Method 1, Simplified Procedure.

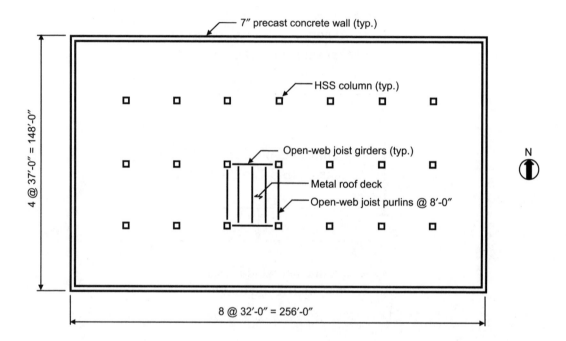

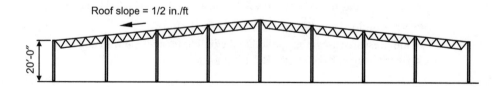

Figure 5.1 Plan and Elevation of Warehouse Building

EXAMPLES

DESIGN DATA

Location: St. Louis, MO

Surface Roughness: C (open terrain with scattered obstructions less than 30 ft in height)

Topography: Not situated on a hill, ridge or escarpment

Occupancy: Less than 300 people congregate in one area and the building is not used to store hazardous or toxic materials

SOLUTION

Part 1: Determine design wind pressures on MWFRS

- **Step 1:** Use Flowchart 5.1 to check if the building meets all of the conditions of 6.4.1.1 so that Method 1 can be used to determine the wind pressures on the MWFRS.

    1. The building is a simple diaphragm building as defined in 6.2, since the windward and leeward wind loads are transmitted through the metal deck roof (diaphragm) to the precast walls (MWFRS), and there are no structural separations in the MWFRS. **O.K.**

    2. Three conditions must be checked to determine if a building is a low-rise building:

        a. Mean roof height = 20 ft < 60 ft **O.K.**[1]

        b. Mean roof height = 20 ft < least horizontal dimension = 148 ft **O.K.**

        c. The enclosure classification of the building, which depends on the number and size of openings in the precast walls.

            Assume that there are two Type A precast panels on each of the east and west walls and two Type B precast panels on each of the north and south walls (see Figure 5.2). The openings in these walls are door openings. All other precast panels do not have door openings, and assume that there are no openings in the roof.

            By definition, an open building is a building having each wall at least 80 percent open. This building is not open including when all of the doors are open at the same time. Therefore, it enclosed or partially enclosed. **O.K.**[2]

        Thus, the building is a low-rise building. **O.K.**

---

[1] For buildings with roof angles less than or equal to 10 degrees, the mean roof height is equal to the roof eave height (see definition of mean roof height in 6.2); the roof angle in this example is approximately 2.4 degrees.

[2] A more specific enclosure classification is determined in item 3 of this part of the solution.

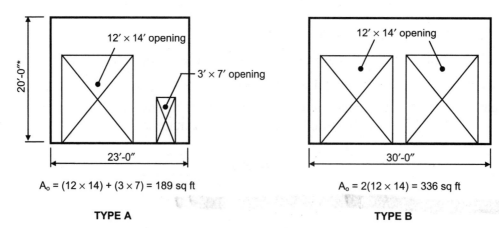

*Elevation of top of walls on north and south faces vary with roof slope.

Figure 5.2 Door Openings in Precast Wall Panels

3. A building is defined as enclosed when it does not comply with the requirements for an open or partially enclosed building. It has been previously established that the building is not open (see item 2 above).

   If all of the doors are closed, the building is enclosed. Also, if all of the doors are open at the same time, the building is enclosed: the first item under the definition of a partially enclosed building is not satisfied, i.e., the total area of openings A_o in a wall that receives positive external pressure is less than 1.1 times the sum of the areas of openings A_{oi} in the balance of the building envelope.[3]

   Assume that one 3 ft by 7 ft door is open and all other doors are closed. Check if the building is partially enclosed per the definition in 6.2:

   a. $A_o = 3 \times 7 = 21$ sq ft $> 1.1 A_{oi} = 0$ where $A_{oi} = 0$, since it is assumed that there are no other doors that are open.

   b. $A_o = 21$ sq ft > 4 sq ft (governs) or $0.01 A_g = 0.01 \times 20 \times 148 = 29.6$ sq ft and $A_{oi} / A_{gi} = 0 < 0.20$.

   Since both of these conditions are satisfied, the building is partially enclosed when one 3 ft by 7 ft door is open and all other doors are closed. However, to illustrate the use of Figure 6-2, assume the building is enclosed.[4]

---

[3] For example, $A_o = 2 \times 336 = 672$ sq ft $< 1.1 A_{oi} = 1.1[672 + (2 \times 378)] = 1,571$ sq ft

[4] Although 6.4.1.1 and Figure 6-2 state that Figure 6-2 is for enclosed buildings, this figure can also be used to determine the wall pressures on a partially enclosed building, since the internal pressures cancel out. However, tabulated pressures must be modified to obtain the correct roof pressures for a partially enclosed building.

EXAMPLES   5-35

The wind-borne debris provisions of 6.5.9.3 are not applicable to buildings located outside of wind-borne debris regions (i.e., specific areas within hurricane prone regions). **O.K.**

4. The building is regularly-shaped, i.e., it does not have any unusual geometric irregularities in spatial form. **O.K.**

5. A flexible building is defined in 6.2 as one in which the fundamental natural frequency of the building n_1 is less than 1 Hz. Although it is evident by inspection that the building is not flexible, the natural frequency will be determined and compared to 1 Hz.

In lieu of obtaining the natural frequency of the building from a dynamic analysis, Eq. C6-16 in the commentary of ASCE/SEI 7 is used to determine an approximate value of n_1 in the N-S direction for concrete shearwall systems:

$$n_1 = 385(C_w)^{0.5} / H$$

where

$$C_w = \frac{100}{A_B} \sum_{i=1}^{n} \left(\frac{H}{h_i}\right)^2 \frac{A_i}{\left[1 + 0.83\left(\frac{h_i}{D_i}\right)^2\right]}$$

A_B = base area of the building = $148 \times 256 = 37{,}888$ sq ft

H, h_i = building height and wall height, respectively = 20 ft

A_i = area of shearwall = $(7/12) \times 148 = 86.3$ sq ft[5]

D_i = length of shearwall = 148 ft

$$C_w = \frac{2 \times 100}{37{,}888} \frac{86.3}{\left[1 + 0.83\left(\frac{20}{148}\right)^2\right]} = 0.45$$

$$n_1 = 385(0.45)^{0.5} / 20 = 12.9 \text{ Hz} \gg 1 \text{ Hz}$$

Similar calculations in the E-W direction yield $n_1 = 17.1$ Hz $\gg$ 1 Hz. Thus, the building is not flexible. **O.K.**

---

[5] Openings in precast wall panels are not considered.

6. The building does not have response characteristics that make it subject to across-wind loading or other similar effects, and it is not sited at a location where channeling effects or buffeting in the wake of upwind obstructions need to be considered. **O.K.**

7. The building has a symmetrical cross-section in each direction and has a relatively flat roof. **O.K.**

8. The building is exempted from torsional load cases as indicated in Note 5 of Figure 6-10 (the building is one-story with a height h less than 30 ft and it has a flexible roof diaphragm). **O.K.**

Since all of the conditions of 6.4.1.1 are satisfied, Method 1 may be used to determine the design wind pressures on the MWFRS.

- Step 2: Use Flowchart 5.3 to determine the net design wind pressures p_s on the MWFRS.

   1. Determine basic wind speed V from Figure 6-1 or Figure 1609.

      From either of these figures, $V = 90$ mph for St. Louis, MO.

   2. Determine importance factor I from Table 6-1 based on occupancy category from IBC Table 1604.5.

      From IBC Table 1604.5, the Occupancy Category is II, based on the occupancy given in the design data. From Table 6-1, $I = 1.0$.

   3. Determine exposure category.

      In the design data, the surface roughness is given as C. It is assumed that Exposures B and D are not applicable, so Exposure C applies (see 6.5.6.3).

   4. Determine adjustment factor for height and exposure λ from Figure 6-2.

      For a mean roof height of 20 ft and Exposure C, $\lambda = 1.29$.

   5. Determine topographic factor K_{zt}.

      As noted in the design data, the building is not situated on a hill, ridge or escarpment. Thus, topographic factor $K_{zt} = 1.0$ (6.5.7.2).

   6. Determine simplified design wind pressures p_{s30} from Figure 6-2 for Zones A through H on the building.

Wind pressures p_{s30} can be read directly from Figure 6-2 for $V = 90$ mph and a roof angle between 0 and 5 degrees. Since the roof is essentially flat, only Load Case 1 is considered (see Note 4 in Figure 6-2). These pressures, which are based on Exposure B, $h = 30$ ft, $K_{zt} = 1.0$, and $I = 1.0$, are given in Table 5.2.

Table 5.2 Wind Pressures p_{s30} on MWFRS

Horizontal pressures (psf)				Vertical pressures (psf)			
A	B	C	D	E	F	G	H
12.8	-6.7	8.5	-4.0	-15.4	-8.8	-10.7	-6.8

7. Determine net design wind pressures $p_s = \lambda K_{zt} I p_{s30}$ by Eq. 6-1 for Zones A through H.

$$p_s = 1.29 \times 1.0 \times 1.0 \times p_{s30} = 1.29 p_{s30}$$

The horizontal pressures in Table 5.3 represent the combination of the windward and leeward net (sum of internal and external) pressures. Similarly, the vertical pressures represent the net (sum of internal and external) pressures.

Table 5.3 Wind Pressures p_s on MWFRS

Horizontal pressures (psf)				Vertical pressures (psf)			
A	B	C	D	E	F	G	H
16.5	-8.6	11.0	-5.2	-19.9	-11.4	-13.8	-8.8

8. The net design pressures p_s in Table 5.3 are to be applied to the surfaces of the building in accordance with Figure 6-2.

According to Note 7 in Figure 6-2, the total horizontal load must not be less than that determined by assuming $p_s = 0$ in Zones B and D. Since the net pressures in Zones B and D in this example act in the direction opposite to those in A and C, they decrease the horizontal load. Thus, the pressures in Zones B and D are set equal to 0 when analyzing the structure for wind in the transverse direction.

According to Note 2 in Figure 6-2, the load patterns for the transverse and longitudinal directions are to be applied to each corner of the building, i.e., each corner of the building must be considered a reference corner. Eight different load cases need to be examined (four in the transverse direction and four in the longitudinal direction). One load pattern in the transverse direction and one in the longitudinal direction are illustrated in Figure 5.3.

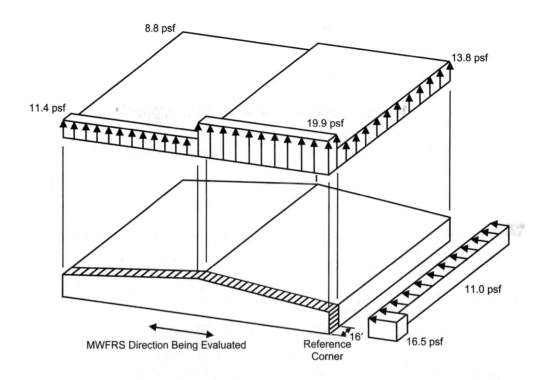
Transverse

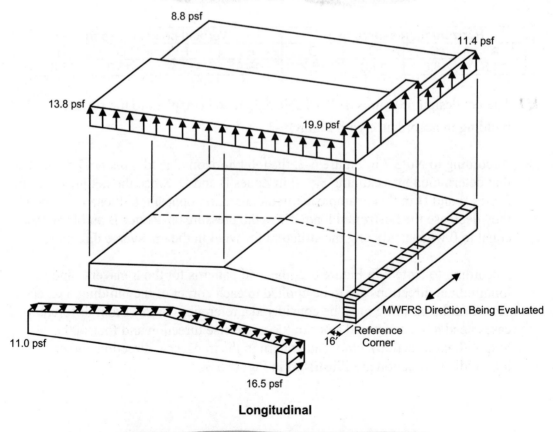
Longitudinal

Figure 5.3 Design Wind Pressures on MWFRS

The width of the end zone $2a$ in this example is equal to 16 ft, where a = least of 0.1(least horizontal dimension) = 0.1 × 148 = 14.8 ft or $0.4h$ = 0.4 × 20 = 8 ft (governs). This value of a is greater than 0.04 (least horizontal dimension) = 0.04 × 148 = 5.9 ft or 3 ft (see Note 10a in Figure 6-2).

The minimum design wind load case of 6.4.2.1.1 must also be considered: the load effects from the design wind pressures calculated above must not be less than the load effects assuming that p_s = + 10 psf in Zones A through D and p_s = 0 psf in Zones E through H (see Figure C6-1 for application of load).

Part 2: Determine design wind pressures on a solid precast wall panel

- Step 1: Use Flowchart 5.2 to check if the building meets all of the conditions of 6.4.1.2 so that Method 1 can be used to determine the wind pressures on the C&C.

  1. Mean roof height h = 20 ft < 60 ft **O.K.**

  2. The building is assumed to be enclosed (see Part 1, Step 1, item 3) and the wind-borne debris provisions of 6.5.9.3 are not applicable **O.K.**

  3. The building is regularly-shaped, i.e., it does not have any unusual geometric irregularities in spatial form. **O.K.**

  4. The building does not have response characteristics that make it subject to across-wind loading or other similar effects, and it is not sited at a location where channeling effects or buffeting in the wake of upwind obstructions need to be considered. **O.K.**

  5. The roof is essentially flat. **O.K.**

  Since all of the conditions of 6.4.1.2 are satisfied, Method 1 may be used to determine the design wind pressures on the C&C.

- Step 2: Use Flowchart 5.4 to determine the net design wind pressures p_{net} on the C&C.

  1. The basic wind speed V, the importance factor I, the exposure category, the adjustment factor for height and exposure λ, and the topographic factor K_{zt} have all been determined previously (Part 1, Step 2, items 1 through 5) and are used in calculating the wind pressures on the precast walls, which are C&C.[6]

  2. Determine net design wind pressures p_{net30} from Figure 6-3 for Zones 4 and 5, which are the interior and end zones of walls, respectively.

---

[6] λ from Figure 6-2 for MWFRS is the same as that in Figure 6-3 for C&C.

Wind pressures p_{net30} can be read directly from Figure 6-3 for $V = 90$ mph and an effective wind area.

The effective wind area is defined as the span length multiplied by an effective width that need not be less than one-third the span length: $20 \times (20/3) = 133.3$ sq ft.[7]

According to Note 4 in Figure 6-3, tabulated pressures may be interpolated for effective wind areas between those given, or the value associated with the lower effective wind area may be used. The latter of these two options is utilized in this example. The pressures p_{net30} in Table 5.4 are obtained from Figure 6-3 for $V = 90$ mph and an effective wind area of 100 sq ft, and are based on Exposure B, $h = 30$ ft, $K_{zt} = 1.0$, and $I = 1.0$.

Table 5.4 Wind Pressures p_{net30} on Precast Walls

Zone	p_{net30} (psf)	
4	12.4	-13.6
5	12.4	-15.1

3. Determine net design wind pressures $p_{net} = \lambda K_{zt} I p_{net30}$ by Eq. 6-2 for Zones 4 and 5.

$$p_{net} = 1.29 \times 1.0 \times 1.0 \times p_{net30} = 1.29 p_{net30}$$

The pressures in Table 5.5 represent the net (sum of internal and external) pressures that are applied normal to the precast walls.

Table 5.5 Wind Pressures p_{net} on Precast Walls

Zone	p_{net} (psf)	
4	16.0	-17.5
5	16.0	-19.5

The width of the end zone (Zone 5) $a = 8$ ft (see Part 1, Step 2, item 8).

In Zones 4 and 5, the computed positive and negative (absolute) pressures are greater than the minimum values prescribed in 6.4.2.2.1 of +10 psf and -10 psf, respectively.

---

[7] The smallest span length corresponding to the east and west walls is used, since this results in larger pressures.

Part 3: Determine design wind pressures on an open-web joist purlin

- **Step 1:** Use Flowchart 5.2 to check if the building meets all of the conditions of 6.4.1.2 so that Method 1 can be used to determine the wind pressures on the C&C.

  It was shown in Part 2, Step 1 that Method 1 may be used to determine the wind forces on the C&C.

- **Step 2:** Use Flowchart 5.4 to determine the net design wind pressures p_{net} on the C&C.

  1. The basic wind speed V, the importance factor I, the exposure category, the adjustment factor for height and exposure λ, and the topographic factor K_{zt} have all been determined previously (Part 1, Step 2, items 1 through 5) and are used in calculating the wind pressures on the open-web joist purlins, which are subject to C&C pressures.

  2. Determine net design wind pressures p_{net30} from Figure 6-3 for Zones 1, 2 and 3, which are the interior, end and corner zones of the roof, respectively.

     Effective wind area is equal to the larger of the purlin tributary area = 37×8 = 296 sq ft or the span length multiplied by an effective width that need not be less than one-third the span length = $37 \times (37/3)$ = 456.3 sq ft (governs).

     The pressures p_{net30} in Table 5.6 are obtained from Figure 6-3 for V = 90 mph, a roof angle between 0 and 7 degrees, and an effective wind area of 100 sq ft.[8] These pressures are based on Exposure B, h = 30 ft, K_{zt} = 1.0, and I = 1.0.

     Table 5.6 Wind Pressures p_{net30} on Open-web Joist Purlins

Zone	p_{net30} (psf)	
1	4.7	-13.3
2	4.7	-15.8
3	4.7	-15.8

  3. Determine net design wind pressures $p_{net} = \lambda K_{zt} I p_{net30}$ by Eq. 6-2 for Zones 1, 2 and 3.

     $$p_{net} = 1.29 \times 1.0 \times 1.0 \times p_{net30} = 1.29 p_{net30}$$

---

[8] Where actual effective wind areas are greater than 100 sq ft, the tabulated pressure values associated with an effective wind area of 100 sq ft are applicable.

The pressures in Table 5.7 represent the net (sum of internal and external) pressures that are applied normal to the open-web joist purlins and that act over the tributary area of each purlin, which is equal to $37 \times 8 = 296$ sq ft.

Table 5.7 Wind Pressures p_{net} on Open-web Joist Purlins

Zone	p_{net} (psf)	
1	6.1	-17.2
2	6.1	-20.4
3	6.1	-20.4

The width of the end and corner zones (Zones 2 and 3) $a = 8$ ft (see Part 1, Step 2, item 8). The positive net design pressures in Zones 1, 2 and 3 must be increased to the minimum value of 10 psf in accordance with 6.4.2.2.1. Figure 5.4 contains the loading diagrams for typical open-web joist purlins located within the various zones of the roof.

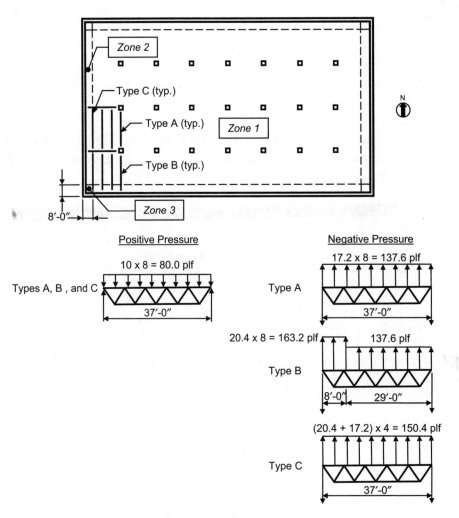

Figure 5.4 Open-web Joist Purlin Loading Diagrams

EXAMPLES 5-43

5.3.2 Example 5.2 – Warehouse Building using Low-rise Building Provisions of Method 2, Analytical Procedure

For the one-story warehouse in Example 5.1, determine design wind pressures on (1) the main wind-force-resisting system in both directions, (2) a solid precast wall panel, and (3) an open-web joist purlin using the low-rise building provisions of Method 2, Analytical Procedure. See Figure 5.1 for building dimensions.

DESIGN DATA

Location:	St. Louis, MO
Surface Roughness:	C (open terrain with scattered obstructions less than 30 ft in height)
Topography:	Not situated on a hill, ridge, or escarpment
Occupancy:	Less than 300 people congregate in one area and the building is not used to store hazardous or toxic materials

SOLUTION

Part 1: Determine design wind pressures on MWFRS

- **Step 1:** Check if the low-rise building provisions of 6.5.12.2.2 can be used to determine the design wind pressures on the MWFRS.

   The provisions of 6.5.12.2.2 may be used to determine design wind pressures provided the building is a regular-shaped low-rise building as defined in 6.2. It was shown in Example 5.1 (Part 1, Step 1, item 2) that this warehouse building is a low-rise building, and the building is regularly-shaped. Also, the building does not have response characteristics that make it subject to across-wind loading or other similar effects, and it is not sited at a location where channeling effects or buffeting in the wake of upwind obstructions need to be considered.

   The low-rise building provisions of Method 2, Analytical Procedure, can be used to determine the design wind pressures on the MWFRS.

- Step 2: Use Flowchart 5.7 to determine the design wind pressures on the MWFRS.

   1. It is assumed in Example 5.1 (Part 1, Step 1, item 3) that the building is enclosed.

   2. Determine velocity pressure q_h using Flowchart 5.5.

      a. Determine basic wind speed V from Figure 6-1 or Figure 1609.

      From either of these figures, V = 90 mph for St. Louis, MO.

b. Determine wind directionality factor K_d from Table 6-4.

   For the MWFRS of a building structure, $K_d = 0.85$.

c. Determine importance factor I from Table 6-1 based on occupancy category from IBC Table 1604.5.

   From IBC Table 1604.5, the Occupancy Category is II. From Table 6-1, $I = 1.0$.

d. Determine exposure category.

   In the design data, the surface roughness is given as C. It is assumed that Exposures B and D are not applicable, so Exposure C applies (see 6.5.6.3).

e. Determine topographic factor K_{zt}.

   As noted in the design data, the building is not situated on a hill, ridge or escarpment. Thus, topographic factor $K_{zt} = 1.0$ (6.5.7.2).

f. Determine velocity pressure exposure coefficient K_h from Table 6-3.

   For Exposure C and a mean roof height of 20 ft, $K_h = 0.90$ from Table 6-3.

g. Determine velocity pressure q_h at the mean roof height by Eq. 6-15.

$$q_h = 0.00256 K_h K_{zt} K_d V^2 I$$
$$= 0.00256 \times 0.90 \times 1.0 \times 0.85 \times 90^2 \times 1.0 = 15.9 \text{ psf}$$

3. Determine external pressure coefficients (GC_{pf}) from Figure 6-10 for building surfaces 1 through 6, 1E, 2E, 3E and 4E.

   External pressure coefficients (GC_{pf}) can be read directly from Table 6-10 using a roof angle between 0 and 5 degrees for wind in the transverse direction. For wind in the longitudinal direction, the pressure coefficients corresponding to a roof angle of 0 degrees are to be used (see Note 7 in Figure 6-10). The pressure coefficients summarized in Table 5.8 are applicable in both the transverse and longitudinal directions in this example.

4. Determine internal pressure coefficients (GC_{pi}) from Figure 6-5.

   For an enclosed building, $(GC_{pi}) = +0.18, -0.18$.

Table 5.8 External Pressure Coefficients (GC_{pf}) for MWFRS

Surface	(GC_{pf})
1	0.40
2	-0.69
3	-0.37
4	-0.29
5	-0.45
6	-0.45
1E	0.61
2E	-1.07
3E	-0.53
4E	-0.43

5. Determine design wind pressure p by Eq. 6-18 on surfaces 1 through 6, 1E, 2E, 3E and 4E.

$$p = q_h[(GC_{pf}) - (GC_{pi})] = 15.9[(GC_{pf}) - (\pm 0.18)]$$

Calculation of design wind pressures is illustrated for surface 1:

For positive internal pressure: $p = 15.9(0.40 - 0.18) = 3.5$ psf

For negative internal pressure: $p = 15.9[0.40 - (-0.18)] = 9.2$ psf

A summary of the design wind pressures is given in Table 5.9. Pressures are applicable to wind in the transverse and longitudinal directions and are provided for both positive and negative internal pressures.

The end zone width $2a$ is equal to 16 ft, where a = least of 0.1 (least horizontal dimension) = $0.1 \times 148 = 14.8$ ft or $0.4h = 0.4 \times 20 = 8$ ft (governs). This value of a is greater than 0.04 (least horizontal dimension) = $0.04 \times 148 = 5.9$ ft or 3 ft (see Note 9 in Figure 6-10).

The design wind pressures summarized in Table 5.9 act normal to the surface.

Table 5.9 Design Wind Pressures p on MWFRS

Surface	(GC_{pf})	Design Pressure, p (psf)	
		$(GC_{pi}) = +0.18$	$(GC_{pi}) = -0.18$
1	0.40	3.5	9.2
2	-0.69	-13.8	-8.1
3	-0.37	-8.8	-3.0
4	-0.29	-7.5	-1.8
5	-0.45	-10.0	-4.3
6	-0.45	-10.0	-4.3
1E	0.61	6.8	12.6
2E	-1.07	-19.9	-14.2
3E	-0.53	-11.3	-5.6
4E	-0.43	-9.7	-4.0

According to Note 8 in Figure 6-10, when the roof pressure coefficients (GC_{pf}) are negative in Zones 2 or 2E, they shall be applied in Zone 2/2E for a distance from the edge of the roof equal to 50 percent of the horizontal dimension of the building that is parallel to the direction of the MWFRS being designed or 2.5 times the eave height h_e at the windward wall, whichever is less. The remainder of Zone 2/2E extending to the ridge line must use the pressure coefficients (GC_{pf}) for Zone 3/3E.

For this building:

Transverse direction: $0.5 \times 256 = 128$ ft

Longitudinal direction: $0.5 \times 148 = 74$ ft

$2.5 h_e = 2.5 \times 20 = 50$ ft (governs in both directions)

Therefore, in the transverse direction, Zone 2/2E applies over a distance of 50 ft from the edge of the windward roof, and Zone 3/3E applies over a distance of $128 - 50 = 78$ ft in what is normally considered to be Zone 2/2E. In the longitudinal direction, Zone 3/3E is applied over a distance of $74 - 50 = 24$ ft.

The design pressures are to be applied on the building in accordance with the eight load cases illustrated in Figure 6-10. As shown in the figure, each corner of the building is considered a reference corner for wind loading in both the transverse and longitudinal directions.

According to Note 4 in Figure 6-10, combinations of external and internal pressures are to be evaluated to obtain the most severe loading. Thus, when both positive and negative pressures are considered, a total of 16 separate loading conditions must be evaluated for this building.[9]

Illustrated in Figures 5.5 and 5.6 are the design wind pressures for one load case in the transverse direction and one load case in the longitudinal direction, respectively, including positive and negative internal pressure.

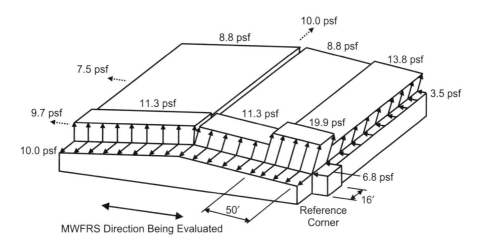

(a) Positive Internal Pressure

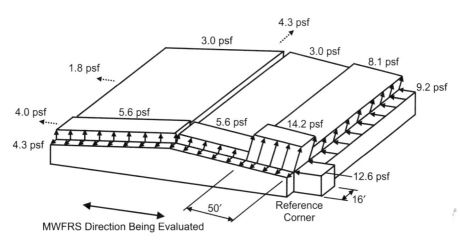

(b) Negative Internal Pressure

Figure 5.5 Design Wind Pressures on MWFRS in Transverse Direction

---

[9] In general, the number of load cases can be reduced for symmetrical buildings.

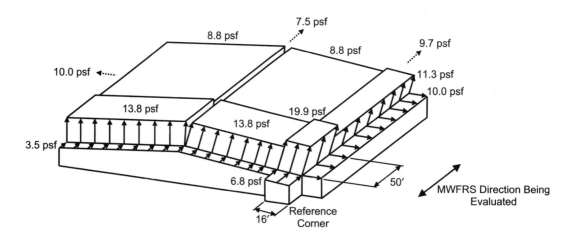

(a) Positive Internal Pressure

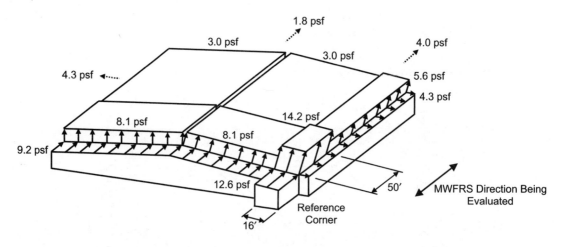

(b) Negative Internal Pressure

Figure 5.6 Design Wind Pressures on MWFRS in Longitudinal Direction

Torsional load cases, which are given in Figure 6-10, must be considered in addition to the basic load cases noted above, unless one or more of the conditions under the exception in Note 5 of the figure are satisfied. Since this building is one story with a mean roof height h less than 30 ft, the first condition is satisfied, and torsional load cases need not be considered. The building also satisfies the third condition, as it is two stories or less in height and has a flexible diaphragm.

The minimum design loading of 6.1.4.1 must also be investigated (see Figure C6-1).

EXAMPLES   5-49

Part 2: Determine design wind pressures on solid precast wall panel

- Step 1: Check if the low-rise provisions of 6.5.12.4.1 can be used to determine the design wind pressures on the C&C.

  The provisions of 6.5.12.4.1 may be used to determine design wind pressures on C&C of low-rise buildings defined in 6.2 and for any regularly-shaped building with a height less than or equal to 60 ft.

  Use 6.5.12.4.1 to determine the design wind pressures on the C&C.

- Step 2: Use Flowchart 5.8 to determine the design wind pressures on the precast walls, which are C&C.

  1. It is assumed in Example 5.1 (Part 1, Step 1, item 3) that the building is enclosed.

  2. Determine velocity pressure q_h using Flowchart 5.5.

     Velocity pressure was determined in Part 1, Step 2, item 2 of this example and is equal to 15.9 psf.

  3. Determine external pressure coefficients (GC_p) from Figure 6-11A for Zones 4 and 5.

     Pressure coefficients for Zones 4 and 5 can be determined from Figure 6-11A based on the effective wind area.

     The effective wind area is defined as the span length multiplied by an effective width that need not be less than one-third the span length: $20 \times (20/3) = 133.3$ sq ft.[10]

     The pressure coefficients from the figure are summarized in Table 5.10.

     Table 5.10 External Pressure Coefficients (GC_p) for Precast Walls

Zone	(GC_p)	
	Positive	Negative
4	0.80	-0.90
5	0.80	-1.00

---

[10] The smallest span length corresponding to the east and west walls is used, since this results in larger pressures.

Note 5 in Figure 6-11A states that values of (GC_p) for walls are to be reduced by 10 percent when the roof angle is less than or equal to 10 degrees. Modified values of (GC_p) based on Note 5 are provided in Table 5.11.

Table 5.11 Modified External Pressure Coefficients (GC_p) for Precast Walls

Zone	(GC_p)	
	Positive	Negative
4	0.72	-0.81
5	0.72	-0.90

The width of the end zone a = least of 0.1 (least horizontal dimension) = 0.1 × 148 = 14.8 ft or $0.4h$ = 0.4 × 20 = 8 ft (governs), which is greater than 0.04 (least horizontal dimension) = 0.04 × 148 = 5.9 ft or 3 ft (see Note 6 in Figure 6-11A).

4. Determine internal pressure coefficients (GC_{pi}) from Figure 6-5.

For an enclosed building, $(GC_{pi}) = +0.18, -0.18$.

5. Determine design wind pressure p by Eq. 6-22 on Zones 4 and 5.

$$p = q_h[(GC_p) - (GC_{pi})] = 15.9[(GC_p) - (\pm 0.18)]$$

Calculation of design wind pressures is illustrated for Zone 4:

For positive (GC_p): $p = 15.9[0.72 - (-0.18)] = 14.3$ psf

For negative (GC_p): $p = 15.9[-0.81 - (+0.18)] = -15.7$ psf

These pressures act perpendicular to the face of the precast walls.

The maximum design wind pressures for positive and negative internal pressures are summarized in Table 5.12.

Table 5.12 Design Wind Pressures p on Precast Walls

Zone	(GC_p)	Design Pressure, p (psf)
4	0.72	14.3
	-0.81	-15.7
5	0.72	14.3
	-0.90	-17.2

EXAMPLES   5-51

In Zones 4 and 5, the computed positive and negative pressures are greater than the minimum values prescribed in 6.1.4.2 of +10 psf and -10 psf, respectively.

Part 3: Determine design wind pressures on an open-web joist purlin

- Step 1: Check if the low-rise provisions of 6.5.12.4.1 can be used to determine the design wind pressures on the C&C.

  As shown in Part 2, Step 1 of this example, 6.5.12.4.1 may be used to determine the design wind pressures on the C&C.

- Step 2: Use Flowchart 5.8 to determine the design wind pressures on the open-web joist purlins, which are C&C.

  1. It is assumed in Example 5.1 (Part 1, Step 1, item 3) that the building is enclosed.

  2. Determine velocity pressure q_h using Flowchart 5.5.

     Velocity pressure was determined in Part 1, Step 2, item 2 of this example and is equal to 15.9 psf.

  3. Determine external pressure coefficients (GC_p) from Figure 6-11B for Zones 1, 2 and 3 for gable roofs with a roof slope less than or equal to 7 degrees.

     Pressure coefficients for Zones 1, 2 and 3 can be determined from Figure 6-11B based on the effective wind area.

     Effective wind area = larger of $37 \times 8 = 296$ sq ft or $37 \times (37/3) = 456.3$ sq ft (governs).

     The pressure coefficients from the figure are summarized in Table 5.13.

     Table 5.13 External Pressure Coefficients (GC_p) for Open-web Joist Purlins

Surface	(GC_p)	
	Positive	Negative
1	0.20	-0.90
2	0.20	-1.10
3	0.20	-1.10

  4. Determine internal pressure coefficients (GC_{pi}) from Figure 6-5.

     For an enclosed building, $(GC_{pi}) = +0.18, -0.18$.

5. Determine design wind pressure p by Eq. 6-22 on Zones 1, 2 and 3.

$$p = q_h[(GC_p) - (GC_{pi})] = 15.9[(GC_p) - (\pm 0.18)]$$

The maximum design wind pressures for positive and negative internal pressures are summarized in Table 5.14.

The pressures in Table 5.14 are applied normal to the open-web joist purlins and act over the tributary area of each purlin, which is equal to $37 \times 8 = 296$ sq ft. If the tributary area were greater than 700 sq ft, the purlins could have been designed using the provisions for MWFRSs (6.5.12.1.3).

The positive pressures on Zones 1, 2 and 3 must be increased to the minimum value of 10 psf per 6.1.4.2.

Table 5.14 Design Wind Pressures p on Open-web Joist Purlins

Zone	(GC_p)	Design Pressure, p (psf)
1	0.20	6.0
	-0.90	-17.2
2 & 3	0.20	6.0
	-1.10	-20.4

The pressures determined by this method are the same as those determined by the simplified method in Example 5.1. Thus, the loading diagrams in Figure 5.4 are applicable in this example.

5.3.3 Example 5.3 – Warehouse Building using Provisions of Method 2, Analytical Procedure

For the one-story warehouse in Example 5.1, determine design wind pressures on (1) the main wind-force-resisting system in both directions, (2) a solid precast wall panel and (3) an open-web joist purlin using the provisions of Method 2, Analytical Procedure. See Figure 5.1 for building dimensions.

DESIGN DATA

Location:	St. Louis, MO
Surface Roughness:	C (open terrain with scattered obstructions less than 30 ft in height)
Topography:	Not situated on a hill, ridge or escarpment
Occupancy:	Less than 300 people congregate in one area and the building is not used to store hazardous or toxic materials

EXAMPLES   5-53

SOLUTION

Part 1: Determine design wind pressures on MWFRS

- Step 1: Check if the provisions of 6.5 can be used to determine the design wind pressures on the MWFRS.

  The provisions of 6.5 may be used to determine design wind pressures provided the conditions of 6.5.1 and 6.5.2 are satisfied. It is clear that these conditions are satisfied for this regular-shaped building that does not have response characteristics that make it subject to across-wind loading or other similar effects, and is not sited at a location where channeling effects or buffeting in the wake of upwind obstructions need to be considered.

  The provisions of Method 2, Analytical Procedure, can be used to determine the design wind pressures on the MWFRS.

- Step 2: Use Flowchart 5.7 to determine the design wind pressures on the MWFRS.

  1. It is assumed in Example 5.1 (Part 1, Step 1, item 3) that the building is enclosed.

  2. Determine whether the building is rigid or flexible and the corresponding gust effect factor from Flowchart 5.6.

     It was determined in Example 5.1, Step 1, item 5 that the building is rigid.

     According to 6.5.8.1, gust effect factor G for rigid buildings may be taken as 0.85 or can be calculated by Eq. 6-4. For simplicity, use $G = 0.85$.

  3. Determine velocity pressure q_z for windward walls along the height of the building and q_h for leeward walls, side walls, and roof using Flowchart 5.5.

     a. Determine basic wind speed V from Figure 6-1 or Figure 1609.

        From either of these figures, $V = 90$ mph for St. Louis, MO.

     b. Determine wind directionality factor K_d from Table 6-4.

        For the MWFRS of a building structure, $K_d = 0.85$.

     c. Determine importance factor I from Table 6-1 based on occupancy category from IBC Table 1604.5.

        From IBC Table 1604.5, the Occupancy Category is II. From Table 6-1, $I = 1.0$.

d. Determine exposure category.

In the design data, the surface roughness is given as C. It is assumed that Exposures B and D are not applicable, so Exposure C applies (see 6.5.6.3).

e. Determine topographic factor K_{zt}.

As noted in the design data, the building is not situated on a hill, ridge or escarpment. Thus, topographic factor $K_{zt} = 1.0$ (6.5.7.2).

f. Determine velocity pressure exposure coefficients K_z and K_h from Table 6-3.

Values of K_z and K_h for Exposure C are summarized in Table 5.15.

Table 5.15 Velocity Pressure Exposure Coefficient K_z

Height above ground level, z (ft)	K_z
20	0.90
15	0.85

g. Determine velocity pressure q_z and q_h by Eq. 6-15.

$$q_z = 0.00256 K_z K_{zt} K_d V^2 I = 0.00256 \times K_z \times 1.0 \times 0.85 \times 90^2 \times 1.0 = 17.63 K_z \text{ psf}$$

A summary of the velocity pressures is given in Table 5.16.

Table 5.16 Velocity Pressure q_z

Height above ground level, z (ft)	K_z	q_z (psf)
20	0.90	15.9
15	0.85	15.0

4. Determine pressure coefficients C_p for the walls and roof from Figure 6-6.

For wind in the E-W (transverse) direction:

Windward wall: $C_p = 0.8$ for use with q_z

Leeward wall ($L/B = 256/148 = 1.73$): $C_p = -0.35$ (from linear interpolation) for use with q_h

Side wall: $C_p = -0.7$ for use with q_h

Roof (normal to ridge with $\theta < 10$ degrees and $h/L = 20/256 = 0.08 < 0.5$)[11]:

$C_p = -0.9, -0.18$ from windward edge to $h = 20$ ft for use with q_h

$C_p = -0.5, -0.18$ from 20 ft to $2h = 40$ ft for use with q_h

$C_p = -0.3, -0.18$ from 40 ft to 256 ft for use with q_h

For wind in the N-S (longitudinal) direction:

Windward wall: $C_p = 0.8$ for use with q_z

Leeward wall ($L/B = 148/256 = 0.58$): $C_p = -0.5$ for use with q_h

Side wall: $C_p = -0.7$ for use with q_h

Roof (parallel to ridge with $h/L = 20/148 = 0.14 < 0.5$):

$C_p = -0.9, -0.18$ from windward edge to $h = 20$ ft for use with q_h

$C_p = -0.5, -0.18$ from 20 ft to $2h = 40$ ft for use with q_h

$C_p = -0.3, -0.18$ from 40 ft to 148 ft for use with q_h

5. Determine q_i for the walls and roof using Flowchart 5.5.

In accordance with 6.5.12.2.1, $q_i = q_h = 15.9$ psf for windward walls, side walls, leeward walls and roofs of enclosed buildings.

6. Determine internal pressure coefficients (GC_{pi}) from Figure 6-5.

For an enclosed building, $(GC_{pi}) = +0.18, -0.18$.

7. Determine design wind pressures p_z and p_h by Eq. 6-17.

Windward walls:

$p_z = q_z GC_p - q_h(GC_{pi})$
$= (0.85 \times 0.8 \times q_z) - 15.9(\pm 0.18)$
$= (0.68 q_z \mp 2.9)$ psf (external $\pm$ internal pressure)

Leeward wall, side walls and roof:

---

[11] The smaller uplift pressures on the roof due to $C_p = -0.18$ may govern the design when combined with roof live load or snow loads. This pressure is not shown in this example, but in general must be considered.

$$p_h = q_h GC_p - q_h(GC_{pi})$$
$$= (15.9 \times 0.85 \times C_p) - 15.9(\pm 0.18)$$
$$= (13.5C_p \mp 2.9) \text{ psf} \quad (\text{external} \pm \text{internal pressure})$$

A summary of the maximum design wind pressures in the E-W and N-S directions is given in Tables 5.17 and 5.18, respectively.

Table 5.17 Design Wind Pressures p in E-W (Transverse) Direction

Location	Height above ground level, z (ft)	q (psf)	External pressure qGC_p (psf)	Internal pressure $q_h(GC_{pi})$ (psf)	Net pressure p (psf)	
					$+(GC_{pi})$	$-(GC_{pi})$
Windward wall	20	15.9	10.8	±2.9	7.9	13.7
	15	15.0	10.2	±2.9	7.3	13.1
Leeward wall	All	15.9	-4.7	±2.9	-7.6	-1.8
Side walls	All	15.9	-9.5	±2.9	-12.4	-6.6
Roof	20	15.9	-12.2*	±2.9	-15.1	-9.3
	20	15.9	-6.8†	±2.9	-9.7	-3.9
	20	15.9	-4.1‡	±2.9	-7.0	-1.2

* from windward edge to 20 ft
† from 20 ft to 40 ft
‡ from 40 ft to 256 ft

Table 5.18 Design Wind Pressures p in N-S (Longitudinal) Direction

Location	Height above ground level, z (ft)	q (psf)	External pressure qGC_p (psf)	Internal pressure $q_h(GC_{pi})$ (psf)	Net pressure p (psf)	
					$+(GC_{pi})$	$-(GC_{pi})$
Windward wall	20	15.9	10.8	±2.9	7.9	13.7
	15	15.0	10.2	±2.9	7.3	13.1
Leeward wall	All	15.9	-6.8	±2.9	-9.7	-3.9
Side walls	All	15.9	-9.5	±2.9	-12.4	-6.6
Roof	20	15.9	-12.2*	±2.9	-15.1	-9.3
	20	15.9	-6.8†	±2.9	-9.7	-3.9
	20	15.9	-4.1‡	±2.9	-7.0	-1.2

* from windward edge to 20 ft
† from 20 ft to 40 ft
‡ from 40 ft to 148 ft

Illustrated in Figures 5.7 and 5.8 are the net design wind pressures in the E-W (transverse) and N-S (longitudinal) directions, respectively, for positive and negative internal pressure.

The MWFRS of buildings whose wind loads have been determined by 6.5.12.2.1 must be designed for the wind load cases defined in Figure 6-9 (6.5.12.3).

In Case 1, the full design wind pressures act on the projected area perpendicular to each principal axis of the structure. These pressures are assumed to act separately along each principal axis. The wind pressures on the windward and leeward walls depicted in Figures 5.7 and 5.8 fall under Case 1.

According to the exception in 6.5.12.3, one-story buildings with $h \leq 30$ ft need only be designed for Load Case 1 and Load Case 3.

In Case 3, 75 percent of the wind pressures on the windward and leeward walls of Case 1, which are shown in Figures 5.7 and 5.8, act simultaneously on the building (see Figure 6-9). This load case, which needs to be considered in addition to the load cases in Figures 5.7 and 5.8, accounts for the effects due to wind along the diagonal of the building.

Finally, the minimum design wind loading prescribed in 6.1.4.1 must be considered as a load case in addition to those load cases described above (see Figure C6-1).

Part 2: Determine design wind pressures on solid precast wall panel

The design wind pressures on the precast walls panels are the same as those determined in Part 2 of Example 5.2.

Part 3: Determine design wind pressures on an open-web joist purlin

The design wind pressures on the open-web joist purlins are the same as those determined in Part 3 of Example 5.2.

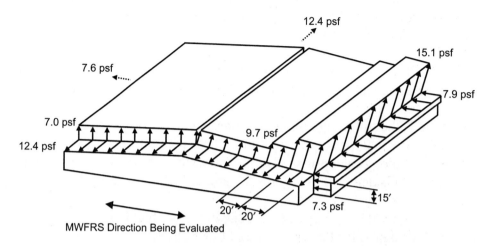

(a) Positive Internal Pressure

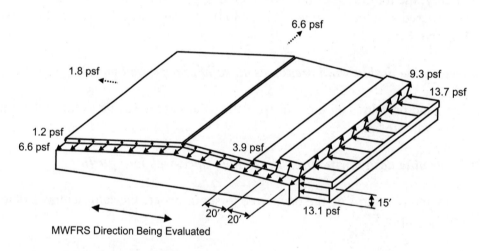

(b) Negative Internal Pressure

Figure 5.7 Design Wind Pressures on MWFRS in E-W (Transverse) Direction

EXAMPLES

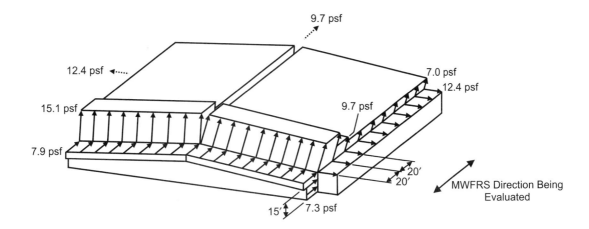

(a) Positive Internal Pressure

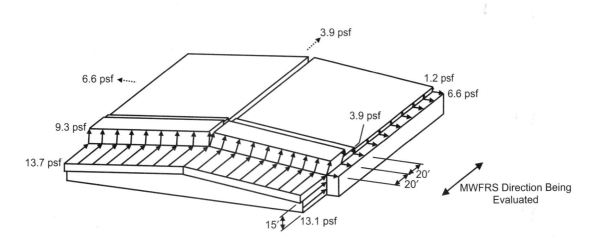

(b) Negative Internal Pressure

Figure 5.8 Design Wind Pressures on MWFRS in N-S (Longitudinal) Direction

5.3.4 Example 5.4 – Warehouse Building using Alternate All-heights Method

For the one-story warehouse in Example 5.1, determine design wind pressures on (1) the main wind-force-resisting system in both directions, (2) a solid precast wall panel and (3) an open-web joist purlin using the Alternate All-heights Method of IBC 1609.6. See Figure 5.1 for building dimensions.

DESIGN DATA

Location:	St. Louis, MO
Surface Roughness:	C (open terrain with scattered obstructions less than 30 ft in height)
Topography:	Not situated on a hill, ridge or escarpment
Occupancy:	Less than 300 people congregate in one area and the building is not used to store hazardous or toxic materials

---

SOLUTION

Part 1: Determine design wind pressures on MWFRS

- **Step 1:** Check if the provisions of IBC 1609.6 can be used to determine the design wind pressures on this building.

  The provisions of IBC 1609.6 may be used to determine design wind pressures on this regularly-shaped building provided the conditions of IBC 1609.6.1 are satisfied:

  1. The height of the building is 20 ft, which is less than 75 ft, and the height-to-least-width ratio = 20/148 = 0.14 < 4. Also, it was shown in Example 5.1 (Part 1, Step 1, item 5) that the fundamental frequency n_1 > 1 Hz in both directions. **O.K.**

  2. As was discussed in Example 5.1 (Part 1, Step 1, item 6), this building is not sensitive to dynamic effects. **O.K.**

  3. This building is not located on a site where channeling effects or buffeting in the wake of upwind obstructions need to be considered. **O.K.**

  4. As was shown in Example 5.1 (Part 1, Step 1, item 1), the building meets the requirements of a simple diaphragm building as defined in 6.2. **O.K.**

  5. The fifth condition is not applicable in this example.

  The provisions of the Alternate All-heights Method of IBC 1609.6 can be used to determine the design wind pressures on the MWFRS.[12]

---

[12] This method can also be used to determine design wind pressures on the C&C (IBC 1609.6).

EXAMPLES   5-61

- **Step 2: Use Flowchart 5.10 to determine the design wind pressures on the MWFRS.**

    1. Determine basic wind speed V from Figure 6-1 or Figure 1609.

        From either of these figures, $V = 90$ mph for St. Louis, MO.

    2. Determine the wind stagnation pressure q_s from IBC Table 1609.6.2(1).

        For $V = 90$ mph, $q_s = 20.7$ psf.

    3. Determine importance factor I from Table 6-1 based on occupancy category from IBC Table 1604.5.

        From IBC Table 1604.5, the Occupancy Category is II. From Table 6-1, $I = 1.0$.

    4. Determine exposure category.

        In the design data, the surface roughness is given as C. It is assumed that Exposures B and D are not applicable, so Exposure C applies (see 6.5.6.3).

    5. Determine topographic factor K_{zt}.

        As noted in the design data, the building is not situated on a hill, ridge or escarpment. Thus, topographic factor $K_{zt} = 1.0$ (6.5.7.2).

    6. Determine velocity pressure exposure coefficients K_z from Table 6-3.

        Values of K_z for Exposure C are summarized in Table 5.19.

        Table 5.19 Velocity Pressure Exposure Coefficient K_z

Height above ground level, z (ft)	K_z
20	0.90
15	0.85

    7. Determine net pressure coefficients C_{net} for the walls and roof from IBC Table 1609.6.2(2) assuming the building is enclosed.

        For wind in the E-W (transverse) direction:

        Windward wall: $C_{net} = 0.43$ for positive internal pressure
        $C_{net} = 0.73$ for negative internal pressure

Leeward wall: $C_{net} = -0.51$ for positive internal pressure
$C_{net} = -0.21$ for negative internal pressure
Side walls: $C_{net} = -0.66$ for positive internal pressure
$C_{net} = -0.35$ for negative internal pressure
Leeward roof (wind perpendicular to ridge):
$C_{net} = -0.66$ for positive internal pressure
$C_{net} = -0.35$ for negative internal pressure
Windward roof (wind perpendicular to ridge with roof slope $< 2:12$):
$C_{net} = -1.09, -0.28$ for positive internal pressure
$C_{net} = -0.79, 0.02$ for negative internal pressure

For wind in the N-S (longitudinal) direction:

Windward wall: $C_{net} = 0.43$ for positive internal pressure
$C_{net} = 0.73$ for negative internal pressure
Leeward wall: $C_{net} = -0.51$ for positive internal pressure
$C_{net} = -0.21$ for negative internal pressure
Side walls: $C_{net} = -0.66$ for positive internal pressure
$C_{net} = -0.35$ for negative internal pressure
Roof (wind parallel to ridge):
$C_{net} = -1.09$ for positive internal pressure
$C_{net} = -0.79$ for negative internal pressure

8. Determine net design wind pressures p_{net} by IBC Eq. 16-34.

$$p_{net} = q_s K_z C_{net} I K_{zt} = 20.7 K_z C_{net}$$

A summary of the net design wind pressures in the E-W and N-S directions is given in Tables 5.20 and 5.21, respectively. Illustrated in Figures 5.9 and 5.10 are the net design wind pressures in the E-W (transverse) and N-S (longitudinal) directions, respectively, for positive and negative internal pressure.[13]

The MWFRS of buildings whose wind loads have been determined by IBC 1609.6 must be designed for the wind load cases defined in ASCE/SEI Figure 6-9 (IBC 1609.6.4.1). In Case 1, the full design wind pressures act on the projected area perpendicular to each principal axis of the structure. These pressures are assumed to act separately along each principal axis. The wind

---

[13] For wind in the E-W direction, only the Condition 1 wind pressures on the roof are illustrated in Figures 5.9 and 5.10. Although these pressures are not shown in the figures, they must be considered in the overall design.

pressures on the windward and leeward walls depicted in Figures 5.9 and 5.10 fall under Case 1.

Table 5.20 Net Design Wind Pressures p_{net} in E-W (Transverse) Direction

Location		Height above ground level, z (ft)	K_z	C_{net}		Net design pressure p_{net} (psf)	
				+ Internal pressure	− Internal pressure	+ Internal pressure	− Internal pressure
Windward wall		20	0.90	0.43	0.73	8.0	13.6
		15	0.85	0.43	0.73	7.6	12.8
Leeward wall		All	0.90	−0.51	−0.21	−9.5	−3.9
Side walls		All	0.90	−0.66	−0.35	−12.3	−6.5
Roof	Windward	20	0.90	−1.09	−0.79	−20.3	−14.7
		20	0.90	−0.28	0.02	−5.2	0.4
	Leeward	20	0.90	−0.66	−0.35	−12.3	−6.5

Table 5.21 Net Design Wind Pressures p_{net} in N-S (Longitudinal) Direction

Location	Height above ground level, z (ft)	K_z	C_{net}		Net design pressure p_{net} (psf)	
			+ Internal pressure	− Internal pressure	+ Internal pressure	− Internal pressure
Windward wall	20	0.90	0.43	0.73	8.0	13.6
	15	0.85	0.43	0.73	7.6	12.8
Leeward wall	All	0.90	−0.51	−0.21	−9.5	−3.9
Side walls	All	0.90	−0.66	−0.35	−12.3	−6.5
Roof	20	0.90	−1.09	−0.79	−20.3	−14.7

According to IBC 1609.6.4.2, windward wall pressures are based on height z, and leeward walls, side walls and roof pressures are based on mean roof height h.

IBC 1609.6.4.1 requires consideration of torsional effects as indicated in Figure 6-9. According to the exception in 6.5.12.3, one-story buildings with $h \leq 30$ ft need only be designed for Load Case 1 and Load Case 3.

In Case 3, 75 percent of the wind pressures on the windward and leeward walls of Case 1, which are shown in Figures 5.9 and 5.10, act simultaneously on the building (see ASCE/SEI Figure 6-9). This load case, which needs to be considered in addition to the load cases in Figures 5.9 and 5.10, accounts for the effects due to wind along the diagonal of the building.

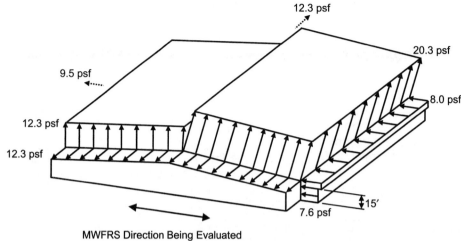

(a) Positive Internal Pressure

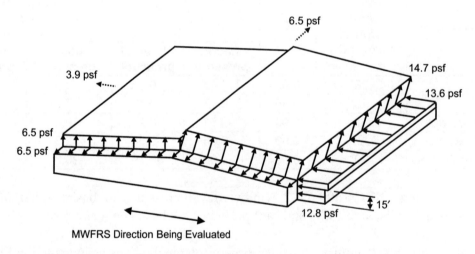

(b) Negative Internal Pressure

Figure 5.9 Design Wind Pressures on MWFRS in E-W (Transverse) Direction

EXAMPLES

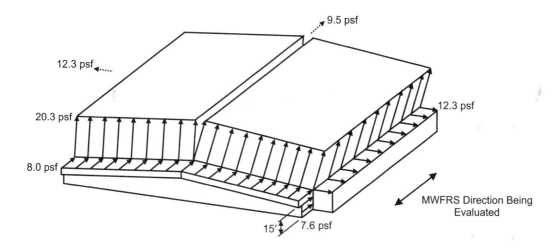

(a) Positive Internal Pressure

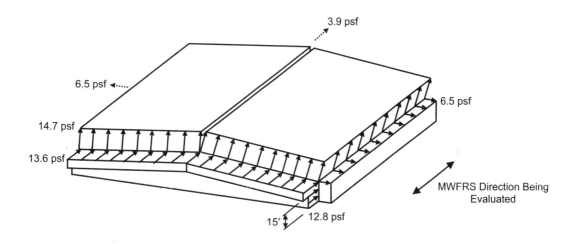

(b) Negative Internal Pressure

Figure 5.10 Design Wind Pressures on MWFRS in N-S (Longitudinal) Direction

Finally, the minimum design wind loading prescribed in IBC 1609.6.3 must be considered as a load case in addition to those load cases described above. The minimum 10 psf wind pressure acts on the area of the building projected on a plane normal to the direction of wind (see ASCE/SEI 6.1.4 and ASCE/SEI Figure C6-1).

Part 2: Determine design wind pressures on solid precast wall panel

- **Step 1**: Check if the provisions of IBC 1609.6 can be used to determine the design wind pressures on the C&C.

  It was shown in Part 1 of this example that the provisions of IBC 1609.6 can be used to determine the design wind pressures on the C&C.

- **Step 2**: Use Flowchart 5.10 to determine the design wind pressures on the C&C.

  1. through 6. These items are the same as those shown in Part 1 of this example.

  7. Determine net pressure coefficients C_{net} for Zones 4 and 5 in ASCE/SEI Figure 6-11A from IBC Table 1609.6.2(2).

  The effective wind area is defined as the span length multiplied by an effective width that need not be less than one-third the span length: $20 \times (20/3) = 133.3$ sq ft.[14]

  The net pressure coefficients from IBC Table 1609.6.2(2) for C&C (walls) not in areas of discontinuity (item 4, $h \leq 60$ ft, Zone 4) and in areas of discontinuity (item 5, $h \leq 60$ ft, Zone 5) are summarized in Table 5.22.[15]

  Table 5.22 Net Pressure Coefficients C_{net} for Precast Walls

Zone	C_{net} Positive	C_{net} Negative
4	0.94	-1.03
5	0.94	-1.21

  The width of the end zone a = least of 0.1 (least horizontal dimension) = $0.1 \times 148 = 14.8$ ft or $0.4h = 0.4 \times 20 = 8$ ft (governs), which is greater than 0.04 (least horizontal dimension) = $0.04 \times 148 = 5.9$ ft or 3 ft (see Note 6 in Figure 6-11A).

---

[14] The smallest span length corresponding to the east and west walls is used, since this results in larger pressures.

[15] Linear interpolation was used to determine the values of C_{net} in Table 5.22 [see note a in IBC Table 1609.6.2(2)].

8. Determine net design wind pressures p_{net} by IBC Eq. 16-34.

$$p_{net} = q_s K_z C_{net} I K_{zt} = (20.7 \times 0.90) C_{net} = 18.6 C_{net}$$

where $K_z = K_h = 0.90$ from Table 5.19.

A summary of the design wind pressures on the precast walls is given in Table 5.23. These pressures act perpendicular to the face of the precast walls.

Table 5.23 Design Wind Pressures on Precast Walls

Zone	C_{net}	Design Pressure, p_{net} (psf)
4	0.94	17.5
4	-1.03	-19.2
5	0.94	17.5
5	-1.21	-22.5

In Zones 4 and 5, the computed positive and negative pressures are greater than the minimum values prescribed in IBC 1609.6.3 of +10 psf and −10 psf, respectively.

Part 3: Determine design wind pressures on an open-web joist purlin

- Step 1: Check if the provisions of IBC 1609.6 can be used to determine the design wind pressures on the C&C.

It was shown in Part 1 of this example that the provisions of IBC 1609.6 can be used to determine the design wind pressures on the C&C.

- Step 2: Use Flowchart 5.10 to determine the design wind pressures on the C&C.

1. through 6. These items are the same as those shown in Part 1 of this example.

7. Determine net pressure coefficients C_{net} for Zones 1, 2 and 3 in ASCE/SEI Figure 6-11C from IBC Table 1609.6.2(2).

Effective wind area = larger of $37 \times 8 = 296$ sq ft or $37 \times (37/3) = 456.3$ sq ft (governs).

The net pressure coefficients from IBC Table 1609.6.2(2) for C&C (roofs) not in areas of discontinuity (item 2, gable roof with flat < slope < 6:12, Zone 1) and in areas of discontinuity (item 3, gable roof with flat < slope < 6:12, Zones 2 and 3) are summarized in Table 5.24.

Table 5.24 Net Pressure Coefficients C_{net} for Open-web Joist Purlins

Zone	C_{net}	
	Positive	Negative
1	0.41	-0.92
2	0.41	-1.17
3	0.41	-1.85

8. Determine net design wind pressures p_{net} by IBC Eq. 16-34.

$$p_{net} = q_s K_z C_{net} I K_{zt} = (20.7 \times 0.90) C_{net} = 18.6 C_{net}$$

where $K_z = K_h = 0.90$ from Table 5.19.

A summary of the design wind pressures on the open-web joist purlins is given in Table 5.25.

Table 5.25 Design Wind Pressures on Open-web Joist Purlins

Zone	C_{net}	Design Pressure, p_{net} (psf)
1	0.41	7.6
	-0.92	-17.1
2	0.41	7.6
	-1.17	-21.8
3	0.41	7.6
	-1.85	-34.4

The pressures in Table 5.25 are applied normal to the open-web joist purlins and act over the tributary area of each purlin, which is equal to $37 \times 8 = 296$ sq ft. If the tributary area were greater than 700 sq ft, the purlins could have been designed using the provisions for MWFRSs (6.5.12.1.3).

The positive pressures on Zones 1, 2 and 3 must be increased to the minimum value of 10 psf in accordance with IBC 1609.6.3.

5.3.5 Example 5.5 – Residential Building using Method 2, Analytical Procedure

For the three-story residential building illustrated in Figure 5.11, determine design wind pressures on (1) the main wind-force-resisting system in both directions, (2) a typical wall stud in the third story, (3) a typical roof truss and (4) a typical roof sheathing panel using Method 2, Analytical Procedure. Note that door and window openings are not shown in the figure.

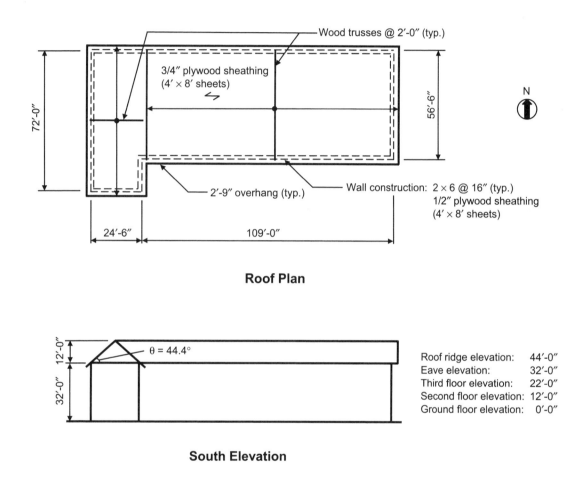

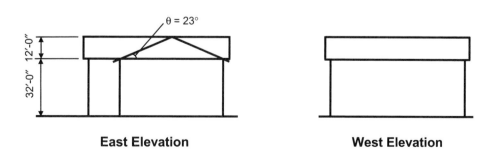

Figure 5.11 Roof Plan and Elevations of Three-story Residential Building

DESIGN DATA

Location: Sacramento, CA

Surface Roughness: B (suburban area with numerous closely spaced obstructions having the size of single-family dwellings and larger)

Topography: Not situated on a hill, ridge or escarpment

Occupancy: Residential building where less than 300 people congregate in one area

SOLUTION

Part 1: Determine design wind pressures on MWFRS

- Step 1: Check if the provisions of 6.5 can be used to determine the design wind pressures on the MWFRS.

  The provisions of 6.5 may be used to determine design wind pressures provided the conditions of 6.5.1 and 6.5.2 are satisfied. It is clear that these conditions are satisfied for this residential building that does not have response characteristics that make it subject to across-wind loading or other similar effects, and that is not sited at a location where channeling effects or buffeting in the wake of upwind obstructions need to be considered.

  The provisions of Method 2, Analytical Procedure, can be used to determine the design wind pressures on the MWFRS.[16]

- Step 2: Use Flowchart 5.7 to determine the design wind pressures on the MWFRS.

  1. It is assumed in this example that the building is enclosed.

  2. Determine whether the building is rigid or flexible and the corresponding gust effect factor from Flowchart 5.6.

     Assume that the fundamental frequency of the building n_1 has been determined to be greater than 1 Hz. Thus, the building is rigid.

     According to 6.5.8.1, gust effect factor G for rigid buildings may be taken as 0.85 or can be calculated by Eq. 6-4. For simplicity, use $G = 0.85$.

  3. Determine velocity pressure q_z for windward walls along the height of the building and q_h for leeward walls, side walls and roof using Flowchart 5.5.

---

[16] Even though the building is less than 60 ft in height, it is not recommended to use the low-rise provisions of 6.5.12.2, since L-, T-, and U-shaped buildings are considered to be outside of the scope of that method.

a. Determine basic wind speed V from Figure 6-1 or Figure 1609.

   From either of these figures, $V = 85$ mph for Sacramento, CA.

b. Determine wind directionality factor K_d from Table 6-4.

   For the MWFRS of a building structure, $K_d = 0.85$.

c. Determine importance factor I from Table 6-1 based on occupancy category from IBC Table 1604.5.

   From IBC Table 1604.5, the Occupancy Category is II for a residential building. From Table 6-1, $I = 1.0$.

d. Determine exposure category.

   In the design data, the surface roughness is given as B. Assume that Exposure B is applicable in all directions (6.5.6.3).

e. Determine topographic factor K_{zt}.

   As noted in the design data, the building is not situated on a hill, ridge or escarpment. Thus, topographic factor $K_{zt} = 1.0$ (6.5.7.2).

f. Determine velocity pressure exposure coefficients K_z and K_h from Table 6-3.

   According to Note 1 in Table 6-3, values of K_z and K_h under Case 2 for Exposure B must be used for MWFRSs in buildings that are not designed using Figure 6-10 for low-rise buildings. Values of K_z and K_h are summarized in Table 5.26.

   Mean roof height $= \dfrac{44 + 32}{2} = 38$ ft

Table 5.26 Velocity Pressure Exposure Coefficient K_z

Height above ground level, z (ft)	K_z
38	0.75
30	0.70
25	0.66
20	0.62
15	0.57

g. Determine velocity pressure q_z and q_h by Eq. 6-15.

$$q_z = 0.00256 K_z K_{zt} K_d V^2 I = 0.00256 \times K_z \times 1.0 \times 0.85 \times 85^2 \times 1.0 = 15.72 K_z \text{ psf}$$

A summary of the velocity pressures is given in Table 5.27.

Table 5.27 Velocity Pressure q_z

Height above ground level, z (ft)	K_z	q_z (psf)
38	0.75	11.8
30	0.70	11.0
25	0.66	10.4
20	0.62	9.8
15	0.57	9.0

4. Determine pressure coefficients C_p for the walls and roof from Figure 6-6.

Since the building is not symmetric, all four wind directions normal to the walls must be considered.

Figure 5.12 provides identification marks for each surface of the building.

Tables 5.28 through 5.31 contain the external pressure coefficients for wind in all four directions.

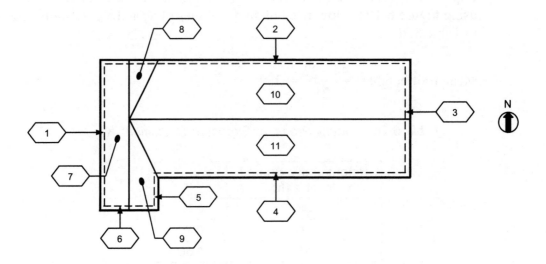

Figure 5.12 Identification Marks for Building Surfaces

Table 5.28 External Pressure Coefficients C_p for Wind from West to East

Surface(s)	Type		C_p	Use with
1	Windward wall		0.80	q_z
1	Overhang		0.80*	q_h
2, 4, 6	Side wall		-0.70	q_h
3, 5	Leeward wall		-0.32**	q_h
7	Windward roof		0.40†	q_h
8, 9	Leeward roof		-0.60†	q_h
10, 11	Roof parallel to ridge‡	Edge to 38'	-0.90, -0.18	q_h
		38' to 76'	-0.50, -0.18	
		76' to end	-0.30, -0.18	

* See 6.5.11.4.1
** Obtained by linear interpolation using $L/B = 133.5/72 = 1.9$
† Normal to ridge with θ = 44.4 degrees and $h/L = 38/133.5 = 0.3$
‡ The smaller uplift pressures on the roof due to $C_p = -0.18$ may govern the design when combined with roof live load or snow loads. This pressure is not shown in this example, but in general must be considered.

Table 5.29 External Pressure Coefficients C_p for Wind from East to West

Surface(s)	Type		C_p	Use with
1	Leeward wall		-0.32*	q_h
2, 4, 6	Side wall		-0.70	q_h
3, 5	Windward wall		0.80	q_z
3, 5	Overhang		0.80**	q_h
7	Leeward roof		-0.60†	q_h
8, 9	Windward roof		0.40†	q_h
10, 11	Roof parallel to ridge‡	Edge to 38'	-0.90, -0.18	q_h
		38' to 76'	-0.50, -0.18	
		76' to end	-0.30, -0.18	

* Obtained by linear interpolation using $L/B = 133.5/72 = 1.9$
** See 6.5.11.4.1
† Normal to ridge with θ = 44.4 degrees and $h/L = 38/133.5 = 0.3$
‡ The smaller uplift pressures on the roof due to $C_p = -0.18$ may govern the design when combined with roof live load or snow loads. This pressure is not shown in this example, but in general must be considered.

Table 5.30 External Pressure Coefficients C_p for Wind from North to South

Surface(s)	Type		C_p	Use with
1, 3, 5	Side wall		-0.70	q_h
2	Windward wall		0.80	q_z
2	Overhang		0.80*	q_h
4, 6	Leeward wall		-0.50**	q_h
7, 8	Roof parallel to ridge†	Edge to 38'	-0.90, -0.18	q_h
		38' to 76'	-0.50, -0.18	
		76' to end	-0.30, -0.18	
9, 11	Leeward roof		-0.60‡	q_h
10	Windward roof		0.11, -0.35‡	q_h

* See 6.5.11.4.1
** $L/B = 72/133.5 = 0.54$
† For surface 8, $C_p = -0.90$ for the entire length. The smaller uplift pressures on the roof due to $C_p = -0.18$ may govern the design when combined with roof live load or snow loads. This pressure is not shown in this example, but in general must be considered.
‡ Normal to ridge with θ = 23 degrees and $h/L = 38/72 = 0.53$. On windward roof, values were obtained by linear interpolation.

Table 5.31 External Pressure Coefficients C_p for Wind from South to North

Surface(s)	Type		C_p	Use with
1, 3, 5	Side wall		-0.70	q_h
2	Leeward wall		-0.50*	q_h
4, 6	Windward wall		0.80	q_z
4, 6	Overhang		0.80**	q_h
7, 9	Roof parallel to ridge†	Edge to 38'	-0.90, -0.18	q_h
		38' to 76'	-0.50, -0.18	
		76' to end	-0.30, -0.18	
8, 10	Leeward roof		-0.60‡	q_h
11	Windward roof		0.11, -0.35‡	q_h

* $L/B = 72/133.5 = 0.54$
** See 6.5.11.4.1
† For surface 9, $C_p = -0.90$ for the entire length. The smaller uplift pressures on the roof due to $C_p = -0.18$ may govern the design when combined with roof live load or snow loads. This pressure is not shown in this example, but in general must be considered.
‡ Normal to ridge with θ = 23 degrees and $h/L = 38/72 = 0.53$. Values were obtained by linear interpolation.

5. Determine q_i for the walls and roof using Flowchart 5.5.

   In accordance with 6.5.12.2.1, $q_i = q_h = 11.8$ psf for windward walls, side walls, leeward walls and roofs of enclosed buildings.

6. Determine internal pressure coefficients (GC_{pi}) from Figure 6-5.

   For an enclosed building, $(GC_{pi}) = +0.18, -0.18$.

7. Determine design wind pressures p_z and p_h by Eq. 6-17.

   Windward walls:

   $$p_z = q_z GC_p - q_h(GC_{pi})$$
   $$= (0.85 \times 0.8 \times q_z) - 11.8(\pm 0.18)$$
   $$= (0.68 q_z \mp 2.1) \text{ psf } (\text{external} \pm \text{internal pressure})$$

   Leeward wall, side walls and roof:

   $$p_h = q_h GC_p - q_h(GC_{pi})$$
   $$= (11.8 \times 0.85 \times C_p) - 11.8(\pm 0.18)$$
   $$= (10.0 C_p \mp 2.1) \text{ psf } (\text{external} \pm \text{internal pressure})$$

   Overhangs:

   $$p_h = q_h GC_p$$
   $$= 11.8 \times 0.85 \times 0.8 = 8.0 \text{ psf}$$

   A summary of the maximum design wind pressures in all four wind directions is given in Tables 5.32 through 5.35.

   The MWFRS of buildings whose wind loads have been determined by 6.5.12.2.1 must be designed for the wind load cases defined in Figure 6-9 (6.5.12.3). Since the building in this example is not symmetrical, all four wind directions must be considered when combining wind loads according to Figure 6-9.

   In Case 1, the full design wind pressures act on the projected area perpendicular to each principal axis of the structure. These pressures are assumed to act separately along each principal axis. The wind pressures on the windward and leeward walls given in Tables 5.32 through 5.35 fall under Case 1.

Table 5.32 Design Wind Pressures p for Wind from West to East

Surface(s)	Height above ground level, z (ft)	q (psf)	External pressure qGC_p (psf)	Internal pressure $q_h(GC_{pi})$ (psf)	Net pressure p (psf) $+(GC_{pi})$	Net pressure p (psf) $-(GC_{pi})$
1	38	11.8	8.0	±2.1	5.9	10.1
1	30	11.0	7.5	±2.1	5.4	9.6
1	25	10.4	7.1	±2.1	5.0	9.2
1	20	9.8	6.7	±2.1	4.6	8.8
1	15	9.0	6.1	±2.1	4.0	8.2
2, 4, 6	All	11.8	-7.0	±2.1	-9.1	-4.9
3, 5	All	11.8	-3.2	±2.1	-5.3	-1.1
7	38	11.8	4.0	±2.1	1.9	6.1
8, 9	38	11.8	-6.0	±2.1	-8.1	-3.9
10, 11	38	11.8	-9.0*	±2.1	-11.1	-6.9
10, 11	38	11.8	-5.0†	±2.1	-7.1	-2.9
10, 11	38	11.8	-3.0‡	±2.1	-5.1	-0.9

* from windward edge to 38 ft
† from 38 ft to 76 ft
‡ from 76 ft to end

Table 5.33 Design Wind Pressures p for Wind from East to West

Surface(s)	Height above ground level, z (ft)	q (psf)	External pressure qGC_p (psf)	Internal pressure $q_h(GC_{pi})$ (psf)	Net pressure p (psf) $+(GC_{pi})$	Net pressure p (psf) $-(GC_{pi})$
1	All	11.8	-3.2	±2.1	-5.3	-1.1
2, 4, 6	All	11.8	-7.0	±2.1	-9.1	-4.9
3, 5	38	11.8	8.0	±2.1	5.9	10.1
3, 5	30	11.0	7.5	±2.1	5.4	9.6
3, 5	25	10.4	7.1	±2.1	5.0	9.2
3, 5	20	9.8	6.7	±2.1	4.6	8.8
3, 5	15	9.0	6.1	±2.1	4.0	8.2
7	38	11.8	-6.0	±2.1	-8.1	-3.9
8, 9	38	11.8	4.0	±2.1	1.9	6.1
10, 11	38	11.8	-9.0*	±2.1	-11.1	-6.9
10, 11	38	11.8	-5.0†	±2.1	-7.1	-2.9
10, 11	38	11.8	-3.0‡	±2.1	-5.1	-0.9

* from windward edge to 38 ft
† from 38 ft to 76 ft
‡ from 76 ft to end

Table 5.34 Design Wind Pressures p for Wind from North to South

Surface(s)	Height above ground level, z (ft)	q (psf)	External pressure qGC_p (psf)	Internal pressure $q_h(GC_{pi})$ (psf)	Net pressure p (psf) $+(GC_{pi})$	Net pressure p (psf) $-(GC_{pi})$
1, 3, 5	All	11.8	-7.0	±2.1	-9.1	-4.9
2	38	11.8	8.0	±2.1	5.9	10.1
2	30	11.0	7.5	±2.1	5.4	9.6
2	25	10.4	7.1	±2.1	5.0	9.2
2	20	9.8	6.7	±2.1	4.6	8.8
2	15	9.0	6.1	±2.1	4.0	8.2
4, 6	All	11.8	-5.0	±2.1	-7.1	-2.9
7, 8	38	11.8	-9.0*	±2.1	-11.1	-6.9
7, 8	38	11.8	-5.0†	±2.1	-7.1	-2.9
7, 8	38	11.8	-3.0‡	±2.1	-5.1	-0.9
9, 11	38	11.8	-6.0	±2.1	-8.1	-3.9
10	38	11.8	1.1	±2.1	-1.0	3.2
10	38	11.8	-3.5	±2.1	-5.6	-1.4

* from windward edge to 38 ft
† from 38 ft to 76 ft
‡ from 76 ft to end

Table 5.35 Design Wind Pressures p for Wind from South to North

Surface(s)	Height above ground level, z (ft)	q (psf)	External pressure qGC_p (psf)	Internal pressure $q_h(GC_{pi})$ (psf)	Net pressure p (psf) $+(GC_{pi})$	Net pressure p (psf) $-(GC_{pi})$
1, 3, 5	All	11.8	-7.0	±2.1	-9.1	-4.9
2	All	11.8	-5.0	±2.1	-7.1	-2.9
4, 6	38	11.8	8.0	±2.1	5.9	10.1
4, 6	30	11.0	7.5	±2.1	5.4	9.6
4, 6	25	10.4	7.1	±2.1	5.0	9.2
4, 6	20	9.8	6.7	±2.1	4.6	8.8
4, 6	15	9.0	6.1	±2.1	4.0	8.2
7, 9	38	11.8	-9.0*	±2.1	-11.1	-6.9
7, 9	38	11.8	-5.0†	±2.1	-7.1	-2.9
7, 9	38	11.8	-3.0‡	±2.1	-5.1	-0.9
8, 10	38	11.8	-6.0	±2.1	-8.1	-3.9
11	38	11.8	1.1	±2.1	-1.0	3.2
11	38	11.8	-3.5	±2.1	-5.6	-1.4

* from windward edge to 38 ft
† from 38 ft to 76 ft
‡ from 76 ft to end

In Case 2, 75 percent of the design wind pressures on the windward and leeward walls are applied on the projected area perpendicular to each principal axis of the building along with a torsional moment. The wind pressures and torsional moment are applied separately for each principal axis.

The wind pressures and torsional moment at the mean roof height for Case 2 are as follows:[17]

For west-to-east wind: $0.75 P_{WX} = 0.75 \times 10.1 = 7.6$ psf (surface 1)
$0.75 P_{LX} = 0.75 \times 1.1 = 0.8$ psf (surfaces 3 and 5)
$M_T = 0.75(P_{WX} + P_{LX}) B_X e_X$
$= 0.75(10.1 + 1.1) \times 72 \times (\pm 0.15 \times 72)$
$= \pm 6,532$ ft-lb/ft

For north-to-south wind: $0.75 P_{WY} = 0.75 \times 10.1 = 7.6$ psf (surface 2)
$0.75 P_{LY} = 0.75 \times 2.9 = 2.2$ psf (surfaces 4 and 6)
$M_T = 0.75(P_{WY} + P_{LY}) B_Y e_Y$
$= 0.75(10.1 + 2.9) \times 133.5 \times (\pm 0.15 \times 133.5)$
$= \pm 26,065$ ft-lb/ft

Similar calculations can be made for east-to-west wind and for south-to-north wind. All four of these load combinations must be considered for Case 2, since the building is not symmetrical. Similar calculations can also be made at other elevations below the roof.

In Case 3, 75 percent of the wind pressures of Case 1 are applied to the building simultaneously. This accounts for wind along the diagonal of the building. Like in Case 2, four load combinations must be considered for Case 3.

In Case 4, 75 percent of the wind pressures and torsional moments defined in Case 2 act simultaneously on the building. As with all of the other cases, four load combinations must be considered for this load case as well.

Finally, the minimum design wind loading prescribed in 6.1.4.1 must be considered as a load case in addition to those load cases described above.

Part 2: Determine design wind pressures on wall stud

Use Flowchart 5.8 to determine the design wind pressures on the walls studs in the third story, which are C&C.

---

[17] Since the internal pressures cancel out in the horizontal direction, it makes no difference whether the net pressures based on positive or negative internal pressure are used in the load cases for the design of the MWFRS. In this example, the net pressures based on negative internal pressure are used.

EXAMPLES 5-79

1. It is assumed that the building is enclosed.

2. Determine velocity pressure q_h using Flowchart 5.5.

   Velocity pressure was determined in Part 1, Step 2, item 3 of this example and is equal to 11.8 psf.[18]

3. Determine external pressure coefficients (GC_p) from Figure 6-11A for Zones 4 and 5.

   Pressure coefficients for Zones 4 and 5 can be determined from Figure 6-11A based on the effective wind area.

   The effective wind area is the larger of the tributary area of a wall stud and the span length multiplied by an effective width that need not be less than one-third the span length.

   Effective wind area = larger of $10 \times (16/12) = 13.3$ sq ft or $10 \times (10/3) = 33.3$ sq ft (governs).

   The pressure coefficients from the figure are summarized in Table 5.36.

   Table 5.36 External Pressure Coefficients (GC_p) for Wall Studs

Zone	(GC_p)	
	Positive	Negative
4	0.9	-1.0
5	0.9	-1.2

   The width of the end zone a = least of 0.1 (least horizontal dimension) = $0.1 \times 72 = 7.2$ ft (governs) or $0.4h = 0.4 \times 38 = 15.2$ ft. This value of a is greater than 0.04 (least horizontal dimension) = $0.04 \times 72 = 2.9$ ft or 3 ft (see Note 6 in Figure 6-11A).

4. Determine internal pressure coefficients (GC_{pi}) from Figure 6-5.

   For an enclosed building, $(GC_{pi}) = +0.18, -0.18$.

---

[18] In accordance with Note 1 of Table 6-3, values of K_h under Case 1 are to be used for C&C. At a mean roof height of 38 ft, K_h under Cases 1 and 2 are the same. Thus, q_h used in the design of the MWFRS can be used in the design of the C&C in this example.

5. Determine design wind pressure p by Eq. 6-22 in Zones 4 and 5.

$$p = q_h[(GC_p) - (GC_{pi})] = 11.8[(GC_p) - (\pm 0.18)]$$

Calculation of design wind pressures is illustrated for Zone 4:

For positive (GC_p): $p = 11.8[0.9 - (-0.18)] = 12.7$ psf

For negative (GC_p): $p = 11.8[-1.0 - (+0.18)] = -13.9$ psf

The maximum design wind pressures for positive and negative internal pressures are summarized in Table 5.37.

Table 5.37 Design Wind Pressures p on Wall Studs

Zone	(GC_p)	Design Pressure, p (psf)
4	0.9	12.7
	-1.0	-13.9
5	0.9	12.7
	-1.2	-16.3

The pressures in Table 5.37 are applied normal to the wall studs and act over the tributary area of each stud.

In Zones 4 and 5, the computed positive and negative pressures are greater than the minimum values prescribed in 6.1.4.2 of +10 psf and -10 psf, respectively.

Part 3: Determine design wind pressures on roof trusses

Use Flowchart 5.8 to determine the design wind pressures on the roof trusses spanning in the N-S and E-W directions, which are C&C.

1. It is assumed in that the building is enclosed.

2. Determine velocity pressure q_h using Flowchart 5.5.

   Velocity pressure was determined in Part 1, Step 2, item 3 of this example and is equal to 11.8 psf.[19]

---

[19] In accordance with Note 1 of Table 6-3, values of K_h under Case 1 are to be used for C&C. At a mean roof height of 38 ft, K_h under Cases 1 and 2 are the same. Thus, q_h used in the design of the MWFRS can be used in the design of the C&C in this example.

EXAMPLES 5-81

3. Determine external pressure coefficients (GC_p) from appropriate figures for roof trusses spanning in the N-S and E-W directions.

   a. Trusses spanning in the N-S direction

   Pressure coefficients for Zones 1, 2 and 3 can be determined from Figure 6-11C ($7° < \theta = 23° < 27°$) based on the effective wind area assuming a 4 ft by 2 ft panel size. Included are the pressure coefficients for the overhanging portions of the trusses.

   Effective wind area = larger of $56.5 \times 2 = 113$ sq ft or $56.5 \times (56.5/3) = 1,064.1$ sq ft (governs).

   The pressure coefficients from Figure 6-11C are summarized in Table 5.38. According to Note 5 in the table, values of (GC_p) for roof overhangs include pressure contributions from both the upper and lower surfaces of the overhang.

Table 5.38 External Pressure Coefficients (GC_p) for Roof Trusses Spanning in the N-S Direction

Zone	(GC_p) Positive	(GC_p) Negative
1	0.3	-0.8
2	0.3	-1.2
2 (overhang)	---	-2.2
3	0.3	-2.0
3 (overhang)	---	-2.5

   b. Trusses spanning in the E-W direction

   Pressure coefficients for Zones 1, 2 and 3 can be determined from Figure 6-11D ($27° < \theta = 44.4° < 45°$) based on the effective wind area. Included are the pressure coefficients for the overhanging portions of the trusses.

   Effective wind area = larger of $24.5 \times 2 = 49$ sq ft or $24.5 \times (24.5/3) = 200$ sq ft (governs).

   The pressure coefficients from Figure 6-11D are summarized in Table 5.39. According to Note 5 in the table, values of (GC_p) for roof overhangs include pressure contributions from both the upper and lower surfaces of the overhang.

Table 5.39 External Pressure Coefficients (GC_p) for Roof Trusses Spanning in the E-W Direction

Zone	(GC_p)	
	Positive	Negative
1	0.8	-0.8
2	0.8	-1.0
2 (overhang)	---	-1.8
3	0.8	-1.0
3 (overhang)	---	-1.8

4. Determine internal pressure coefficients (GC_{pi}) from Figure 6-5.

   For an enclosed building, $(GC_{pi}) = +0.18, -0.18$.

5. Determine design wind pressure p by Eq. 6-22 in Zones 1, 2 and 3.

   $$p = q_h[(GC_p) - (GC_{pi})] = 11.8[(GC_p) - (\pm 0.18)]$$

   a. Trusses spanning in the N-S direction

   The maximum design wind pressures for positive and negative internal pressures are summarized in Table 5.40.

   b. Trusses spanning in the E-W direction

   The maximum design wind pressures for positive and negative internal pressures are summarized in Table 5.41.

   The pressures in Tables 5.40 and 5.41 are applied normal to the roof trusses and act over the tributary area of each truss. If the tributary areas were greater than 700 sq ft, the trusses could have been designed using the provisions for MWFRSs (6.5.12.1.3).

   The positive pressures in Zones 1, 2 and 3 for roof trusses spanning in the N-S direction must be increased to the minimum value of 10 psf in accordance with 6.1.4.2.

Table 5.40 Design Wind Pressures p on Roof Trusses Spanning in the N-S Direction

Zone	(GC_p)	Design Pressure, p (psf)
1	0.3	5.7
1	-0.8	-11.6
2	0.3	5.7
2	-1.2	-16.3
2 (overhang)	-2.2	-26.0*
3	0.3	5.7
3	-2.0	-25.7
3 (overhang)	-2.5	-29.5*

* Net overhang pressure = $q_h(GC_p)$

Table 5.41 Design Wind Pressures p on Roof Trusses Spanning in the E-W Direction

Zone	(GC_p)	Design Pressure, p (psf)
1	0.8	11.6
1	-0.8	-11.6
2 & 3	0.8	11.6
2 & 3	-1.0	-13.9
2 & 3 (overhang)	-1.8	-21.2*

* Net overhang pressure = $q_h(GC_p)$

Figure 5.13 contains the loading diagrams for typical trusses located within various zones of the roof.

Part 4: Determine design wind pressures on roof sheathing panel

Use Flowchart 5.8 to determine the design wind pressures on the roof sheathing panels, which are C&C.

1. It is assumed in that the building is enclosed.

2. Determine velocity pressure q_h using Flowchart 5.5.

Velocity pressure was determined in Part 1, Step 2, item 3 of this example and is equal to 11.8 psf.[20]

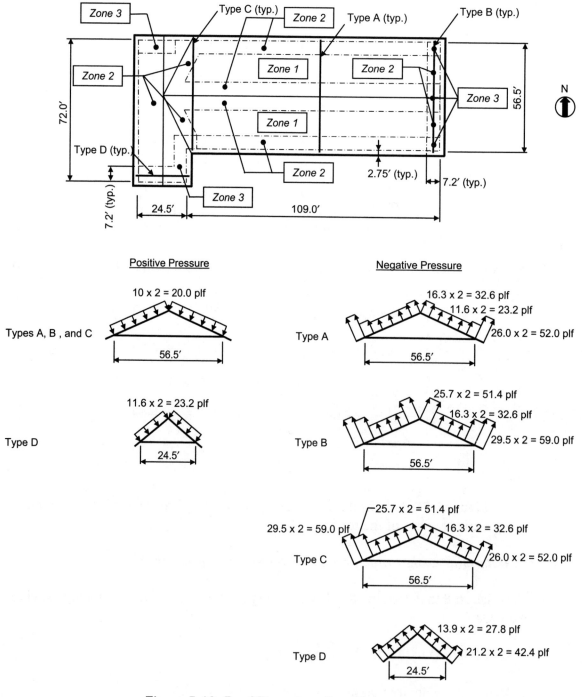

Figure 5.13 Roof Truss Loading Diagrams

---

[20] In accordance with Note 1 of Table 6-3, values of K_h under Case 1 are to be used for C&C. At a mean roof height of 38 ft, K_h under Cases 1 and 2 are the same. Thus, q_h used in the design of the MWFRS can be used in the design of the C&C in this example.

3. Determine external pressure coefficients (GC_p) from appropriate figures for roof panels on the east and west wings.

   a. Roof panels on the east wing

   Pressure coefficients for Zones 1, 2, and 3 can be determined from Figure 6-11C ($7° < \theta = 23° < 27°$) based on the effective wind area assuming a 4 ft by 2 ft panel size. Included are the pressure coefficients for the overhanging portions of panels.

   Effective wind area = larger of $2 \times 4 = 8$ sq ft (governs) or $2 \times (2/3) = 1.33$ sq ft.

   The pressure coefficients from Figure 6-11C are summarized in Table 5.42. According to Note 5 in the table, values of (GC_p) for roof overhangs include pressure contributions from both the upper and lower surfaces of the overhang.

Table 5.42 External Pressure Coefficients (GC_p) for Roof Panels on the East Wing

Zone	(GC_p)	
	Positive	Negative
1	0.5	-0.9
2	0.5	-1.7
2 (overhang)	---	-2.2
3	0.5	-2.6
3 (overhang)	---	-3.7

   b. Roof panels on the west wing

   Pressure coefficients for Zones 1, 2, and 3 can be determined from Figure 6-11D ($27° < \theta = 44.4° < 45°$) based on the effective wind area. Included are the pressure coefficients for the overhanging portions of the trusses.

   Effective wind area = larger of $2 \times 4 = 8$ sq ft (governs) or $2 \times (2/3) = 1.33$ sq ft.

   The pressure coefficients from Figure 6-11D are summarized in Table 5.43. According to Note 5 in the table, values of (GC_p) for roof overhangs include pressure contributions from both the upper and lower surfaces of the overhang.

Table 5.43 External Pressure Coefficients (GC_p) for Roof Panels on the West Wing

Zone	(GC_p) Positive	(GC_p) Negative
1	0.9	-1.0
2	0.9	-1.2
2 (overhang)	---	-2.0
3	0.9	-1.2
3 (overhang)	---	-2.0

4. Determine internal pressure coefficients (GC_{pi}) from Figure 6-5.

   For an enclosed building, $(GC_{pi}) = +0.18, -0.18$.

5. Determine design wind pressure p by Eq. 6-22 in Zones 1, 2 and 3.

   $$p = q_h[(GC_p) - (GC_{pi})] = 11.8[(GC_p) - (\pm 0.18)]$$

   a. Roof panels on the east wing

   The maximum design wind pressures for positive and negative internal pressures are summarized in Table 5.44.

Table 5.44 Design Wind Pressures p on Roof Panels on the East Wing

Zone	(GC_p)	Design Pressure, p (psf)
1	0.5	8.0
1	-0.9	-12.7
2	0.5	8.0
2	-1.7	-22.2
2 (overhang)	-2.2	-26.0*
3	0.5	8.0
3	-2.6	-32.8
3 (overhang)	-3.7	-43.7*

* Net overhang pressure = $q_h(GC_p)$

b. Roof panels on the west wing

The maximum design wind pressures for positive and negative internal pressures are summarized in Table 5.45.

Table 5.45 Design Wind Pressures p on Roof Panels on the West Wing

Zone	(GC_p)	Design Pressure, p (psf)
1	0.9	12.7
1	-1.0	-13.9
2 & 3	0.9	12.7
2 & 3	-1.2	-16.3
2 & 3 (overhang)	-2.0	-23.6*

* Net overhang pressure = $q_h(GC_p)$

The pressures in Tables 5.44 and 5.45 are applied normal to the roof panels and act over the tributary area of each panel.

The positive pressures in Zones 1, 2 and 3 for roof panels on the east wing must be increased to the minimum value of 10 psf in accordance with 6.1.4.2.

5.3.6 Example 5.6 – Six-Story Hotel using Method 2, Analytical Procedure

For the six-story hotel illustrated in Figure 5.14, determine (1) design wind pressures on the main wind-force-resisting system in both directions and (2) design wind forces on the rooftop equipment using Method 2, Analytical Procedure. Note that door and window openings are not shown in the figure.

DESIGN DATA

Location:	Miami, FL
Surface Roughness:	C (adjacent to water in hurricane prone region)
Topography:	Not situated on a hill, ridge or escarpment
Occupancy:	Residential building where less than 300 people congregate in one area

SOLUTION

Part 1: Determine design wind pressures on MWFRS

- Step 1: Check if the provisions of 6.5 can be used to determine the design wind pressures on the MWFRS.

The provisions of 6.5 may be used to determine design wind pressures provided the conditions of 6.5.1 and 6.5.2 are satisfied. It is clear that these conditions are satisfied for this residential building that does not have response characteristics that make it subject to across-wind loading or other similar effects, and that is not sited at a location where channeling effects or buffeting in the wake of upwind obstructions need to be considered.

The provisions of Method 2, Analytical Procedure, can be used to determine the design wind pressures on the MWFRS.[21]

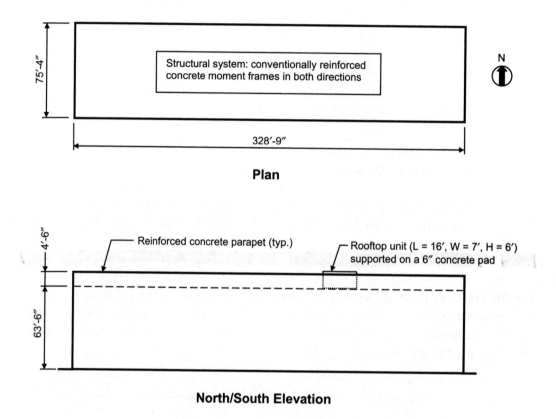

Figure 5.14 Plan and Elevation of Six-story Hotel

- Step 2: Use Flowchart 5.7 to determine the design wind pressures on the MWFRS.

  1. From Figure 6-1 or Figure 1609, the basic wind speed V is equal to 145 mph. Thus, the building is located within a wind-borne debris region, since the basic wind speed is greater than 120 mph.[22]

---

[21] The mean roof height is greater than 60 ft, so the low-rise provisions of 6.5.12.2 cannot be used.
[22] In this example, the building is located within one mile of the coastal mean high water line where $V >$ 110 mph. This satisfies another condition of a wind-borne debris region.

EXAMPLES   5-89

According to 6.5.9.3, glazing in buildings located in wind-borne debris regions must be protected with an approved impact-resistant covering or it must be impact-resistant glazing.[23] It is assumed that impact-resistant glazing is used over the entire height of the building, so the building is classified as enclosed.

2. Determine the design wind pressure effects of the parapet on the MWFRS.

   a. Determine the velocity pressure q_p at the top of the parapet from Flowchart 5.5.

      i. Determine basic wind speed V from Figure 6-1 or Figure 1609.

         As was determined in item 1, $V = 145$ mph for Miami, FL.

      ii. Determine wind directionality factor K_d from Table 6-4.

         For building structures, $K_d = 0.85$

      iii. Determine importance factor I from Table 6-1 based on occupancy category from IBC Table 1604.5.

         From IBC Table 1604.5, the Occupancy Category is II for a residential building. From Table 6-1, $I = 1.0$ for buildings in hurricane prone regions with $V > 100$ mph.

      iv. Determine exposure category.

         In the design data, the surface roughness is given as C. Assume that Exposure C is applicable in all directions (6.5.6.3).

      v. Determine topographic factor K_{zt}.

         As noted in the design data, the building is not situated on a hill, ridge or escarpment. Thus, topographic factor $K_{zt} = 1.0$ (6.5.7.2).

      vi. Determine velocity pressure exposure coefficient K_h from Table 6-3.

         For Exposure C at a height of 68 ft at the top of the parapet, $K_h = 1.16$ by linear interpolation.

---

[23] Impact-resistant glazing must conform to ASTM E 1886 and ASTM E 1996 or other approved test methods and performance criteria. Glazing in Occupancy Category II, III, or IV buildings located over 60 ft above the ground and over 30 ft above aggregate surface roofs located within 1,500 ft of the building shall be permitted to be unprotected (6.5.9.3).

vii. Determine velocity pressure q_p evaluated at the top of the parapet by Eq. 6-15.

$$q_p = 0.00256 K_h K_{zt} K_d V^2 I$$
$$= 0.00256 \times 1.16 \times 1.0 \times 0.85 \times 145^2 \times 1.0 = 53.1 \text{ psf}$$

b. Determine combined net pressure coefficient GC_{pn} for the parapets.

In accordance with 6.5.12.2.4, $GC_{pn} = 1.5$ for windward parapet and $GC_{pn} = -1.0$ for leeward parapet.

c. Determine combined net design pressure p_p on the parapet by Eq. 6-20.

$$p_p = q_p GC_{pn}$$
$$= 53.1 \times 1.5 = 79.7 \text{ psf on windward parapet}$$
$$= 53.1 \times (-1.0) = -53.1 \text{ psf on leeward parapet}$$

The forces on the windward and leeward parapets can be obtained by multiplying the pressures by the height of the parapet:

Windward parapet: $F = 79.7 \times 4.5 = 358.7$ plf
Leeward parapet: $F = 53.1 \times 4.5 = 239.0$ plf

3. Determine whether the building is rigid or flexible and the corresponding gust effect factor by Flowchart 5.6.

In lieu of determining the natural frequency n_1 of the building from a dynamic analysis, Eq. C6-15 in the commentary of ASCE/SEI 7 is used to compute n_1 for concrete moment-resisting frames:

$$n_1 = \frac{43.5}{H^{0.9}} = \frac{43.5}{63.5^{0.9}} = 1.04 \text{ Hz}$$

Since $n_1 > 1.0$ Hz, the building is defined as a rigid building. For simplicity, use $G = 0.85$ (6.5.8.1).[24]

4. Determine velocity pressure q_z for windward walls along the height of the building and q_h for leeward walls, side walls, and roof using Flowchart 5.5. Most of the quantities needed to compute q_z and q_h were determined in item 2 above.

---

[24] According to C6.2, most rigid buildings have a height to minimum width ratio less than 4. In this example, $63.5/75.33 = 0.84 < 4$.

The velocity exposure coefficients K_z and K_h are summarized in Table 5.46.

Table 5.46 Velocity Pressure Exposure Coefficient K_z

Height above ground level, z (ft)	K_z
63.5	1.14
60	1.13
50	1.09
40	1.04
30	0.98
25	0.94
20	0.90
15	0.85

Velocity pressures q_z and q_h are determined by Eq. 6-15:

$$q_z = 0.00256 K_z K_{zt} K_d V^2 I$$
$$= 0.00256 \times K_z \times 1.0 \times 0.85 \times 145^2 \times 1.0 = 45.75 K_z \text{ psf}$$

A summary of the velocity pressures is given in Table 5.47.

Table 5.47 Velocity Pressure q_z

Height above ground level, z (ft)	K_z	q_z (psf)
63.5	1.14	52.2
60	1.13	51.7
50	1.09	49.9
40	1.04	47.6
30	0.98	44.8
25	0.94	43.0
20	0.90	41.2
15	0.85	38.9

5. Determine pressure coefficients C_p for the walls and roof from Figure 6-6.

For wind in the E-W direction:
   Windward wall: $C_p = 0.8$ for use with q_z
   Leeward wall ($L/B = 328.75/75.33 = 4.4$): $C_p = -0.2$ for use with q_h

Side wall: $C_p = -0.7$ for use with q_h

Roof (normal to ridge with $\theta < 10$ degrees and parallel to ridge for all θ with $h/L = 63.5/328.75 = 0.19 < 0.5)^{25}$:

$C_p = -0.9, -0.18$ from windward edge to $h = 63.5$ ft for use with q_h

$C_p = -0.5, -0.18$ from 63.5 ft to $2h = 127.0$ ft for use with q_h

$C_p = -0.3, -0.18$ from 127.0 ft to 328.75 ft for use with q_h

For wind in the N-S direction:

Windward wall: $C_p = 0.8$ for use with q_z

Leeward wall ($L/B = 75.33/328.75 = 0.23$): $C_p = -0.5$ for use with q_h

Side wall: $C_p = -0.7$ for use with q_h

Roof (normal to ridge with $\theta < 10$ degrees and parallel to ridge for all θ with $h/L = 63.5/75.33 = 0.84$):

$C_p = -1.0, -0.18$ from windward edge to $h/2 = 31.75$ ft for use with q_h [26]

$C_p = -0.76, -0.18$ from 31.75 ft to $h = 63.5$ ft for use with q_h

$C_p = -0.64, -0.18$ from 63.5 ft to 75.33 ft for use with q_h

6. Determine q_i for the walls and roof using Flowchart 5.5.

   In accordance with 6.5.12.2.1, $q_i = q_h = 52.2$ psf for windward walls, side walls, leeward walls, and roofs of enclosed buildings.

7. Determine internal pressure coefficients (GC_{pi}) from Figure 6-5.

   For an enclosed building, $(GC_{pi}) = +0.18, -0.18$.

8. Determine design wind pressures p_z and p_h by Eq. 6-17.

   Windward walls:

---

[25] The smaller uplift pressures on the roof due to $C_p = -0.18$ may govern the design when combined with roof live load or snow loads. This pressure is not shown in this example, but in general must be considered.

[26] $C_p = 1.3$ may be reduced based on area over which it is applicable = $(63.5/2) \times 328.75 = 10{,}438$ sq ft > 1,000 sq ft. Reduction factor = 0.8 from Figure 6-6. Thus, $C_p = 0.8 \times (-1.3) = -1.04$ was used in the linear interpolation to determine C_p for $h/L = 0.84$.

$$p_z = q_z GC_p - q_h(GC_{pi})$$
$$= (0.85 \times 0.8 \times q_z) - 52.2(\pm 0.18)$$
$$= (0.68 q_z \mp 9.4) \text{ psf} \quad (\text{external} \pm \text{internal pressure})$$

Leeward wall, side walls and roof:

$$p_h = q_h GC_p - q_h(GC_{pi})$$
$$= (52.2 \times 0.85 \times C_p) - 52.2(\pm 0.18)$$
$$= (44.4 C_p \mp 9.4) \text{ psf} \quad (\text{external} \pm \text{internal pressure})$$

A summary of the maximum design wind pressures in the E-W and N-S directions is given in Tables 5.48 and 5.49, respectively.

Table 5.48 Design Wind Pressures p in the E-W Direction

Location	Height above ground level, z (ft)	q (psf)	External pressure qGC_p (psf)	Internal pressure $q_h(GC_{pi})$ (psf)	Net pressure p (psf) $+(GC_{pi})$	Net pressure p (psf) $-(GC_{pi})$
Windward	63.5	52.2	35.5	±9.4	26.1	44.9
	60	51.7	35.2	±9.4	25.8	44.6
	50	49.9	33.9	±9.4	24.5	43.3
	40	47.6	32.4	±9.4	23.0	41.8
	30	44.8	30.5	±9.4	21.1	39.9
	25	43.0	29.2	±9.4	19.8	38.6
	20	41.2	28.0	±9.4	18.6	37.4
	15	38.9	26.5	±9.4	17.1	35.9
Leeward	All	52.2	-8.9	±9.4	-18.3	-0.5
Side	All	52.2	-31.1	±9.4	-40.5	-21.7
Roof	63.5	52.2	-40.0*	±9.4	-49.4	-30.6
	63.5	52.2	-22.2†	±9.4	-31.6	-12.8
	63.5	52.2	-13.3‡	±9.4	-22.7	-3.9

* from windward edge to 63.5 ft
† from 63.5 ft to 127.0 ft
‡ from 127.0 ft to 328.75 ft

Illustrated in Figures 5.15 and 5.16 are the external design wind pressures in the E-W and N-S directions, respectively. Included in the figures are the forces on the windward and leeward parapets, which add to the overall wind forces in the direction of analysis. When considering horizontal wind forces on the MWFRS, it is clear that the effects from the internal pressure cancel out. On the roof, the effects from internal pressure add directly to those from the external pressure.

Table 5.49 Design Wind Pressures p in the N-S Direction

Location	Height above ground level, z (ft)	q (psf)	External pressure qGC_p (psf)	Internal pressure $q_h(GC_{pi})$ (psf)	Net pressure p (psf) $+(GC_{pi})$	Net pressure p (psf) $-(GC_{pi})$
Windward	63.5	52.2	35.5	±9.4	26.1	44.9
Windward	60	51.7	35.2	±9.4	25.8	44.6
Windward	50	49.9	33.9	±9.4	24.5	43.3
Windward	40	47.6	32.4	±9.4	23.0	41.8
Windward	30	44.8	30.5	±9.4	21.1	39.9
Windward	25	43.0	29.2	±9.4	19.8	38.6
Windward	20	41.2	28.0	±9.4	18.6	37.4
Windward	15	38.9	26.5	±9.4	17.1	35.9
Leeward	All	52.2	-22.2	±9.4	-31.6	-12.8
Side	All	52.2	-31.1	±9.4	-40.5	-21.7
Roof	63.5	52.2	-44.4*	±9.4	-53.8	-35.0
Roof	63.5	52.2	-33.7†	±9.4	-43.1	-24.3
Roof	63.5	52.2	-28.4‡	±9.4	-37.8	-19.0

* from windward edge to 31.75 ft
† from 31.75 ft to 63.5 ft
‡ from 63.5 ft to 75.33 ft

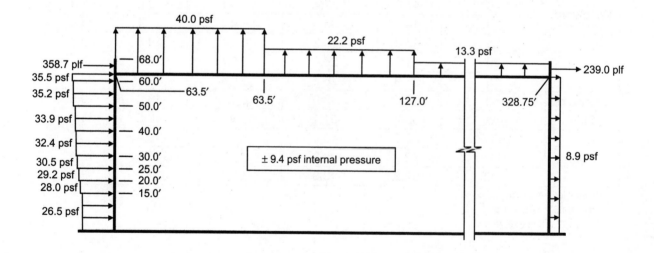

Figure 5.15 Design Wind Pressures in the E-W Direction

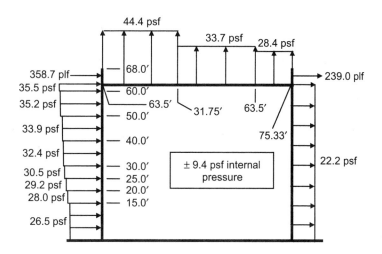

Figure 5.16 Design Wind Pressures in the N-S Direction

The MWFRS of buildings whose wind loads have been determined by 6.5.12.2.1 must be designed for the wind load cases defined in Figure 6-9 (6.5.12.3). In Case 1, the full design wind pressures act on the projected area perpendicular to each principal axis of the structure. These pressures are assumed to act separately along each principal axis. The wind pressures on the windward and leeward walls depicted in Figures 5.15 and 5.16 fall under Case 1.

In Case 2, 75 percent of the design wind pressures on the windward and leeward walls are applied on the projected area perpendicular to each principal axis of the building along with a torsional moment. The wind pressures and torsional moments, both of which vary over the height of the building, are applied separately for each principal axis.

As an example of the calculations that need to be performed over the height of the building, the wind pressures and torsional moment at the mean roof height for Case 2 are as follows:

For E-W wind: $0.75 P_{WX} = 0.75 \times 35.5 = 26.6$ psf (windward wall)

$0.75 P_{LX} = 0.75 \times 8.9 = 6.7$ psf (leeward wall)

$M_T = 0.75(P_{WX} + P_{LX}) B_X e_X$

$= 0.75(35.5 + 8.9) \times 75.33 \times (\pm 0.15 \times 75.33)$

$= \pm 28,345$ ft-lb/ft

For N-S wind: $0.75 P_{WY} = 0.75 \times 35.5 = 26.6$ psf (windward wall)

$0.75 P_{LY} = 0.75 \times 22.2 = 16.7$ psf (leeward wall)

$M_T = 0.75(P_{WY} + P_{LY}) B_Y e_Y$

$= 0.75(35.5 + 22.2) \times 328.75 \times (\pm 0.15 \times 328.75)$

$= \pm 701,552$ ft-lb/ft

In Case 3, 75 percent of the wind pressures of Case 1 are applied to the building simultaneously. This accounts for wind along the diagonal of the building.

In Case 4, 75 percent of the wind pressures and torsional moments defined in Case 2 act simultaneously on the building.

Figure 5.17 illustrates Load Cases 1 through 4 for MWFRS wind pressures acting on the projected area at the mean roof height. Note that internal pressures are always equal and opposite to each other and therefore not included. Similar loading diagrams can be obtained at other locations below the mean roof height.

Finally, the minimum design wind loading prescribed in 6.1.4.1 must be considered as a load case in addition to those load cases described above.

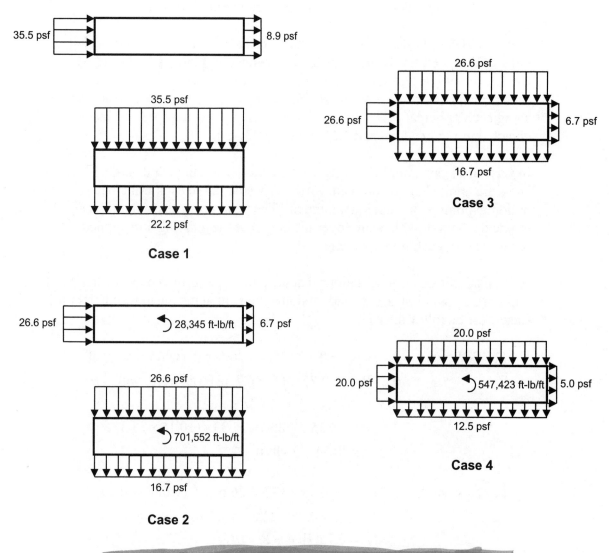

Figure 5.17 Load Cases 1 through 4 at the Mean Roof Height

Part 2: Determine design wind forces on rooftop equipment

Use Flowchart 5.9 to determine the wind force on the rooftop equipment.

1. Determine velocity pressure q_z evaluated at height z of the centroid of area A_f of the rooftop unit.

   The distance from the ground level to the centroid of the rooftop unit = 63.5 + 0.5 + (6/2) = 67 ft

   From Table 6-3, $K_z = 1.16$ (by linear interpolation) for Exposure C at a height of 67 ft above the ground level.

   Velocity pressures q_z is determined by Eq. 6-15:

   $$q_z = 0.00256 K_z K_{zt} K_d V^2 I$$
   $$= 0.00256 \times 1.16 \times 1.0 \times 0.90 \times 145^2 \times 1.0 = 56.2 \text{ psf}$$

   where $K_d = 0.90$ for square-shaped rooftop equipment (see Table 6-4).

2. Determine gust effect factor G from Flowchart 5.6.

   The gust effect factor G is equal to 0.85 (see Part 1, Step 2, item 3 of this example).

3. Determine force coefficient C_f from Figure 6-21 for rooftop equipment.

   $h = 67$ ft

   Least horizontal dimension D of rooftop unit = 7 ft

   $h/D = 67/7 = 9.6$

   From Figure 6-21, $C_f = 1.5$ by linear interpolation for square cross-section.

4. Check if A_f is less than $0.1Bh$.

   For the smaller face, $A_f = 6 \times 7 = 42$ sq ft
   For the larger face, $A_f = 6 \times 16 = 96$ sq ft

   $0.1Bh = 0.1 \times 75.33 \times 63.5 = 478.4$ sq ft $> A_f$

5. Determine design wind force F by Eq. 6-28.

   Since $A_f < 0.1Bh$, F is equal to 1.9 times the value obtained by Eq. 6-28:[27]

   $$F = 1.9 q_z G C_f A_f$$

   $$= 1.9 \times 56.2 \times 0.85 \times 1.5 \times 42 / 1{,}000 = 5.7 \text{ kips on the smaller face}$$

   $$= 1.9 \times 56.2 \times 0.85 \times 1.5 \times 96 / 1{,}000 = 13.1 \text{ kips on the larger face}$$

   These forces act perpendicular to the respective faces of the equipment.

5.3.7 Example 5.7 – Six-Story Hotel Located on an Escarpment using Method 2, Analytical Procedure

For the six-story hotel in Example 5.6, determine (1) design wind pressures on the main wind-force-resisting system in both directions and (2) design wind forces on the rooftop equipment using Method 2, Analytical Procedure assuming the structure is located on an escarpment.

DESIGN DATA

Location:	Miami, FL
Surface Roughness:	C (adjacent to water in hurricane prone region)
Topography:	2-D Escarpment (see Figure 5.18)
Occupancy:	Residential building where less than 300 people congregate in one area

SOLUTION

Part 1: Determine design wind pressures on MWFRS

- Step 1: Check if the provisions of 6.5 can be used to determine the design wind pressures on the MWFRS.

  It was shown in Part 1, Step 1 of Example 5.6 that Method 2 can be used to determine the design wind pressures on the MWFRS.

---

[27] The mean roof height of the building h is equal to 63.5 ft, which is slightly greater than the 60-ft limit prescribed in 6.5.15.1. The force is increased by 1.9 for conservatism.

EXAMPLES 5-99

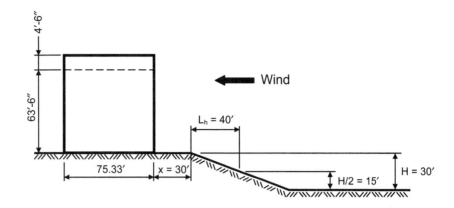

Figure 5.18 Six-story Hotel on Escarpment

- Step 2: Use Flowchart 5.7 to determine the design wind pressures on the MWFRS.

1. It was determined in Part 1, Step 2, item 1 of Example 5.6 that the basic wind speed V is equal to 145 mph and that the building is enclosed.

2. Determine the design wind pressure effects of the parapet on the MWFRS.

   a. Determine the velocity pressure q_p at the top of the parapet from Flowchart 5.5.

      i. Determine basic wind speed V from Figure 6-1 or Figure 1609.

         As noted in item 1, $V = 145$ mph for Miami, FL.

      ii. Determine wind directionality factor K_d from Table 6-4.

         For building structures, $K_d = 0.85$.

      iii. Determine importance factor I from Table 6-1 based on occupancy category from IBC Table 1604.5.

         From Example 5.5, $I = 1.0$.

      iv. Determine exposure category.

         In the design data, the surface roughness is given as C. Assume that Exposure C is applicable in all directions (6.5.6.3).

      v. Determine topographic factor K_{zt}.

Check if all five conditions of 6.5.7.1 are satisfied:

- Assume that the topography is such that conditions 1 and 2 are satisfied.

- Condition 3 is satisfied since the building is located near the crest of the escarpment.

- $H/L_h = 30/40 = 0.75 > 0.2$, so condition 4 is satisfied.

- $H = 30$ ft > 15 ft for Exposure C, so condition 5 is satisfied.

Since all five conditions of 6.5.7.1 are satisfied, wind speed-up effects at the escarpment must be considered in the design, and K_{zt} must be determined by Eq. 6-3:

$$K_{zt} = (1 + K_1 K_2 K_3)^2$$

where the multipliers K_1, K_2 and K_3 are given in Figure 6-4 for Exposure C. Also given in the figure are parameters and equations that can be used to determine the multipliers for any exposure category.

It was determined above that $H/L_h = 0.75$. According to Note 2 in Figure 6-4, where $H/L_h > 0.5$, use $H/L_h = 0.5$ when evaluating K_1 and substitute $2H$ for L_h when evaluating K_2 and K_3.

From Figure 6-4, $K_1 = 0.43$ for a 2-D escarpment with $H/L_h = 0.5$. This multiplier is related to the shape of the topographic feature and the maximum wind speed-up near the crest.

The multiplier K_2 accounts for the reduction in speed-up with distance upwind or downwind of the crest. Since $x/2H = 30/60 = 0.5$, $K_2 = 0.88$ for a 2-D escarpment from Figure 6-4.

The multiplier K_3 accounts for the reduction in speed-up with height z above the local ground surface. Even though the velocity pressure q_p is evaluated at the top of the parapet, the multiplier K_3 is conservatively determined at the height z corresponding to the centroid of the parapet, which is equal to $63.5 + (4.5/2) = 65.75$ ft. Thus, $z/2H = 65.75/60 = 1.1$ and $K_3 = 0.07$ from Figure 6-4 (by linear interpolation) for a 2-D escarpment.

Therefore, $K_{zt} = [1 + (0.43 \times 0.88 \times 0.07)]^2 = 1.05$.

EXAMPLES 5-101

vi. Determine velocity pressure exposure coefficient K_h from Table 6-3.

For Exposure C at a height of 68 ft at the top of the parapet, $K_h = 1.16$ by linear interpolation.

vii. Determine velocity pressure q_p evaluated at the top of the parapet by Eq. 6-15.

$$q_p = 0.00256 K_h K_{zt} K_d V^2 I$$
$$= 0.00256 \times 1.16 \times 1.05 \times 0.85 \times 145^2 \times 1.0 = 55.7 \text{ psf}$$

This velocity pressure is 5 percent greater than that determined in Example 5.6 where the building is not on an escarpment.

b. Determine combined net pressure coefficient GC_{pn}.

In accordance with 6.5.12.2.4, $GC_{pn} = 1.5$ for windward parapet and $GC_{pn} = -1.0$ for leeward parapet.

c. Determine combined net design pressure p_p on the parapet by Eq. 6-20.

$$p_p = q_p GC_{pn}$$
$$= 55.7 \times 1.5 = 83.6 \text{ psf on windward parapet}$$
$$= 55.7 \times (-1.0) = -55.7 \text{ psf on leeward parapet}$$

The forces on the windward and leeward parapets can be obtained by multiplying the pressures by the height of the parapet:

Windward parapet: $F = 83.6 \times 4.5 = 376.2$ plf
Leeward parapet: $F = 55.7 \times 4.5 = 250.7$ plf

These forces are 5 percent greater than those determined in Example 5.6.

3. Determine whether the building is rigid or flexible and the corresponding gust effect factor by Flowchart 5.6.

It was determined in Part 1, Step 2, item 3 of Example 5.6 that the building is rigid and $G = 0.85$.

4. Determine velocity pressure q_z for windward walls along the height of the building and q_h for leeward walls, side walls and roof using Flowchart 5.5.

Velocity pressures q_z and q_h are determined by Eq. 6-15:

$$q_z = 0.00256 K_z K_{zt} K_d V^2 I$$
$$= 0.00256 \times K_z \times K_{zt} \times 0.85 \times 145^2 \times 1.0 = 45.75 K_z K_{zt} \text{ psf}$$

The velocity exposure coefficient K_z and the topographic factor K_{zt} vary with height above the local ground surface. Values of K_z were determined in Example 5.6 (see Table 5.46) and are repeated in Table 5.50 for convenience. From item 2 above, $K_{zt} = [1 + (0.43 \times 0.88 \times K_3)]^2 = [1 + 0.38 K_3]^2$.

Values of K_{zt} are given in Table 5.50 as a function of $z/2H = z/60$ where z is taken midway between the height range.[28] Also given in the table is a summary of the velocity pressure q_z over the height of the building.

Table 5.50 Velocity Pressure q_z

Height above ground level, z (ft)	K_z	z/2H *	K_3	K_{zt}	q_z (psf)
63.5	1.14	1.03	0.08	1.06	55.3
60	1.13	0.92	0.10	1.08	55.8
50	1.09	0.75	0.15	1.12	55.9
40	1.04	0.58	0.24	1.19	56.6
30	0.98	0.46	0.32	1.26	56.5
25	0.94	0.38	0.39	1.32	56.8
20	0.90	0.29	0.48	1.40	57.7
15	0.85	0.13	0.72	1.62	63.0

*z is taken midway between the height range

As an example, determine K_{zt} at a height $z = 60$ ft above the local ground level. The multiplier K_3 is determined based on a height z taken midway between the range of 60 ft and 50 ft, i.e., $z = 55$ ft. Thus, $z/2H = 55/60 = 0.92$, and by linear interpolation from Figure 6-4, $K_3 = 0.10$ for a 2-D escarpment.[29] Then,

$$K_{zt} = [1 + (0.38 \times 0.10)]^2 = 1.08$$

5. Determine pressure coefficients C_p for the walls and roof from Figure 6-6.

---

[28] It is unconservative to use the top height of the range when determining K_3.

[29] K_3 may also be computed from the equation given in Fig. 6-4: $K_3 = e^{-\gamma z / L_h} = e^{-(2.5 \times 0.92)} = 0.10$.

The pressure coefficients are the same as those determined in Part 1, Step 2, item 5 of Example 5.6.

6. Determine q_i for the walls and roof using Flowchart 5.5.

   In accordance with 6.5.12.2.1, $q_i = q_h = 55.3$ psf for windward walls, side walls, leeward walls and roofs of enclosed buildings.

7. Determine internal pressure coefficients (GC_{pi}) from Figure 6-5.

   For an enclosed building, $(GC_{pi}) = +0.18, -0.18$.

8. Determine design wind pressures p_z and p_h by Eq. 6-17.

   Windward walls:

   $$p_z = q_z GC_p - q_h(GC_{pi})$$
   $$= (0.85 \times 0.8 \times q_z) - 55.3(\pm 0.18)$$
   $$= (0.68 q_z \mp 10.0) \text{ psf} \quad (\text{external} \pm \text{internal pressure})$$

   Leeward wall, side walls and roof:

   $$p_h = q_h GC_p - q_h(GC_{pi})$$
   $$= (55.3 \times 0.85 \times C_p) - 55.3(\pm 0.18)$$
   $$= (47.0 C_p \mp 10.0) \text{ psf} \quad (\text{external} \pm \text{internal pressure})$$

   A summary of the maximum design wind pressures in the E-W and N-S directions is given in Tables 5.51 and 5.52, respectively.

   The percent increase in the external pressure on the windward wall of the building due to the escarpment is summarized in Table 5.53.

   The external pressures on the leeward wall, side wall and roof as well as the internal pressure increase by 6 percent, since these pressures depend on the velocity pressure at the roof height q_h.

   Load Cases 1 through 4, depicted in Figure 6-9, must be investigated for the windward and leeward pressures, similar to that shown in Example 5.6.

Table 5.51 Design Wind Pressures p in the E-W Direction

Location	Height above ground level, z (ft)	q (psf)	External pressure qGC_p (psf)	Internal pressure $q_h(GC_{pi})$ (psf)	Net pressure p (psf) $+(GC_{pi})$	Net pressure p (psf) $-(GC_{pi})$
Windward	63.5	55.3	37.6	±10.0	27.6	47.6
	60	55.8	37.9	±10.0	27.9	47.9
	50	55.9	38.0	±10.0	28.0	48.0
	40	56.6	38.5	±10.0	28.5	48.5
	30	56.5	38.4	±10.0	28.4	48.4
	25	56.8	38.6	±10.0	28.6	48.6
	20	57.7	39.2	±10.0	29.2	49.2
	15	63.0	42.8	±10.0	32.8	52.8
Leeward	All	55.3	-9.4	±10.0	-19.4	-0.6
Side	All	55.3	-32.9	±10.0	-42.9	-22.9
Roof	63.5	55.3	-42.3*	±10.0	-52.3	-32.3
	63.5	55.3	-23.5†	±10.0	-33.5	-13.5
	63.5	55.3	-14.1‡	±10.0	-24.1	-4.1

* from windward edge to 63.5 ft
† from 63.5 ft to 127.0 ft
‡ from 127.0 ft to 328.75 ft

Table 5.52 Design Wind Pressures p in the N-S Direction

Location	Height above ground level, z (ft)	q (psf)	External pressure qGC_p (psf)	Internal pressure $q_h(GC_{pi})$ (psf)	Net pressure p (psf) $+(GC_{pi})$	Net pressure p (psf) $-(GC_{pi})$
Windward	63.5	55.3	37.6	±10.0	27.6	47.6
	60	55.8	37.9	±10.0	27.9	47.9
	50	55.9	38.0	±10.0	28.0	48.0
	40	56.6	38.5	±10.0	28.5	48.5
	30	56.5	38.4	±10.0	28.4	48.4
	25	56.8	38.6	±10.0	28.6	48.6
	20	57.7	39.2	±10.0	29.2	49.2
	15	63.0	42.8	±10.0	32.8	52.8
Leeward	All	55.3	-23.5	±10.0	-33.5	-13.5
Side	All	55.3	-32.9	±10.0	-42.9	-22.9
Roof	63.5	55.3	-48.9*	±10.0	-58.9	-38.9
	63.5	55.3	-32.9†	±10.0	-42.9	-22.9

* from windward edge to 31.75 ft
† from 31.75 ft to 75.33 ft

Table 5.53 Comparison of External Design Wind Pressures on the Windward Wall with and without an Escarpment

Height above ground level, z (ft)	External pressure qGC_p (psf)		Percent Increase
	Without Escarpment	With Escarpment	
63.5	35.5	37.6	6
60	35.2	37.9	8
50	33.9	38.0	12
40	32.4	38.5	19
30	30.5	38.4	26
25	29.2	38.6	32
20	28.0	39.2	40
15	26.5	42.8	62

Part 2: Determine design wind forces on rooftop equipment

The calculations in Part 2 of Example 5.6 can be modified to account for the speed-up at the escarpment. In particular, the topographic factor K_{zt} must be determined at the centroid of the rooftop unit, which is 67 ft above the local ground level.

In this case, $z/2H = 67/60 = 1.12$ and by linear interpolation from Figure 6-4, $K_3 = 0.07$.

Thus, $K_{zt} = [1+(0.38 \times 0.07)]^2 = 1.05$.

The velocity pressure q_z is equal to 1.05 times that determined in Example 5.6, i.e., $q_z = 1.05 \times 56.2 = 59.0$ psf.

Consequently, the design wind forces on the rooftop units supported on the building located on the escarpment are 5 percent greater than those determined in Example 5.6:

$F = 1.9 q_z G C_f A_f$
   $= 1.05 \times 5.7 = 6.0$ kips on the smaller face
   $= 1.05 \times 13.1 = 13.8$ kips on the larger face

5.3.8 Example 5.8 – Six-Story Hotel using Alternate All-heights Method

For the six-story hotel in Example 5.6, determine (1) design wind pressures on the main wind-force-resisting system in both directions and (2) design wind forces on the rooftop equipment using the Alternate All-heights Method of IBC 1609.6.

SOLUTION

Part 1: Determine design wind pressures on MWFRS

- Step 1: Check if the provisions of IBC 1609.6 can be used to determine the design wind pressures on this building.

   The provisions of IBC 1609.6 may be used to determine design wind pressures on this regularly-shaped building provided the conditions of IBC 1609.6.1 are satisfied:

   1. The height of the building is 63 ft-6 in., which is less than 75 ft, and the height-to-least-width ratio = 63.5/75.33 = 0.84 < 4. Also, it was shown in Example 5.6 (Part 1, Step 2, item 3) that the fundamental frequency n_1 = 1.04 Hz > 1 Hz. **O.K.**

   2. As was discussed in Example 5.6 (Part 1, Step 1), this building is not sensitive to dynamic effects. **O.K.**

   3. This building is not located on a site where channeling effects or buffeting in the wake of upwind obstructions need to be considered. **O.K.**

   4. This building meets the requirements of a simple diaphragm building as defined in 6.2, since the windward and leeward wind loads are transmitted through the reinforced concrete floor slabs (diaphragms) to the reinforced concrete moment frames (MWFRS), and there are no structural separations in the MWFRS. **O.K.**

   5. The fifth condition is applicable to the rooftop equipment only.

   The provisions of the Alternate All-heights Method of IBC 1609.6 can be used to determine the design wind pressures on the MWFRS.[30]

- Step 2: Use Flowchart 5.10 to determine the design wind pressures on the MWFRS.

   1. Determine basic wind speed V from Figure 6-1 or Figure 1609.

      From either of these figures, V = 145 mph for Miami, FL. Thus, the building is located within a wind-borne debris region, since the basic wind speed is greater than 120 mph.[31] According to ASCE/SEI 6.5.9.3 and IBC 1609.1.2, glazing in buildings located in wind-borne debris regions must be protected with an approved impact-resistant covering or it must be impact-resistant glazing.[32] It is

---

[30] This method can also be used to determine design wind pressures on the C&C (IBC 1609.6).
[31] In this example, the building is located within one mile of the coastal mean high water line where V > 110 mph. This satisfies another condition of a wind-borne debris region.
[32] Impact-resistant glazing must conform to ASTM E 1886 and ASTM E 1996 or other approved test methods and performance criteria. Glazing in Occupancy Category II, III, or IV buildings located over 60 ft above the ground and over 30 ft above aggregate surface roofs located within 1,500 ft of the building shall be permitted to be unprotected (6.5.9.3). Also see IBC 1609.1.2 for other exceptions.

assumed that impact-resistant glazing is used over the entire height of the building, so the building is classified as enclosed.

2. Determine the wind stagnation pressure q_s from IBC Table 1609.6.2(1), footnote a.

   For $V = 145$ mph, $q_s = 0.00256 \times 145^2 = 53.8$ psf.

3. Determine importance factor I from Table 6-1 based on occupancy category from IBC Table 1604.5.

   From IBC Table 1604.5, the Occupancy Category is II for a residential building. From Table 6-1, $I = 1.0$ for buildings in hurricane prone regions with $V > 100$ mph.

4. Determine exposure category.

   In the design data, the surface roughness is given as C. Assume that Exposure C is applicable in all directions (see 6.5.6.3).

5. Determine topographic factor K_{zt}.

   As noted in the design data, the building is not situated on a hill, ridge or escarpment. Thus, topographic factor $K_{zt} = 1.0$ (6.5.7.2).

6. Determine velocity pressure exposure coefficients K_z from Table 6-3.

   Values of K_z for Exposure C are summarized in Table 5.54.

   Table 5.54 Velocity Pressure Exposure Coefficient K_z

Height above ground level, z (ft)	K_z
68	1.16
63.5	1.14
60	1.13
50	1.09
40	1.04
30	0.98
25	0.94
20	0.90
15	0.85

7. Determine net pressure coefficients C_{net} for the walls, roof from IBC Table 1609.6.2(2) assuming the building is enclosed.

Windward wall: $C_{net} = 0.43$ for positive internal pressure
$C_{net} = 0.73$ for negative internal pressure
Leeward wall: $C_{net} = -0.51$ for positive internal pressure
$C_{net} = -0.21$ for negative internal pressure
Side walls: $C_{net} = -0.66$ for positive internal pressure
$C_{net} = -0.35$ for negative internal pressure
Roof: $C_{net} = -1.09$ for positive internal pressure
$C_{net} = -0.79$ for negative internal pressure
Parapet: $C_{net} = 1.28$ for windward
$C_{net} = -0.85$ for leeward

These net pressure coefficients are applicable for wind in both the N-S and E-W directions.

8. Determine net design wind pressures p_{net} by IBC Eq. 16-34.

$$p_{net} = q_s K_z C_{net} I K_{zt}$$

Windward walls: $p_{net} = 53.8 K_z C_{net}$

Leeward walls, side walls and roof: $p_{net} = 53.8 \times 1.14 \times C_{net} = 61.3 C_{net}$

Parapet: $p_{net} = 53.8 \times 1.16 \times C_{net} = 62.4 C_{net}$

A summary of the maximum design wind pressures is given in Tables 5.55.

Illustrated in Figures 5.19 are the net design wind pressures in the N-S direction with positive internal pressure.[33] Included in the figure are the following forces on the windward and leeward parapets, which add to the overall wind forces in the direction of analysis:

On the windward parapet: $p_{net} = 62.4 \times 1.28 = 79.9 \text{ psf}$
$F = 79.9 \times 4.5 = 359.6 \text{ plf}$

On the leeward parapet: $p_{net} = 62.4 \times (-0.85) = -53.0 \text{ psf}$
$F = -53.0 \times 4.5 = -238.5 \text{ plf}$

---

[33] Wind pressures in the E-W direction are the same as in the N-S direction.

EXAMPLES

Table 5.55 Net Design Wind Pressures p_{net}

Location	Height above ground level, z (ft)	K_z	C_{net}		Net design pressure p_{net} (psf)	
			+ Internal Pressure	- Internal Pressure	+ Internal Pressure	- Internal Pressure
Windward	63.5	1.14	0.43	0.73	26.4	44.8
	60	1.13	0.43	0.73	26.1	44.4
	50	1.09	0.43	0.73	25.2	42.8
	40	1.04	0.43	0.73	24.1	40.9
	30	0.98	0.43	0.73	22.7	38.5
	25	0.94	0.43	0.73	21.8	36.9
	20	0.90	0.43	0.73	20.8	35.4
	15	0.85	0.43	0.73	19.7	33.4
Leeward	All	1.14	-0.51	-0.21	-31.3	-12.9
Side	All	1.14	-0.66	-0.35	-40.5	-21.5
Roof	60	1.14	-1.09	-0.79	-66.8	-48.4

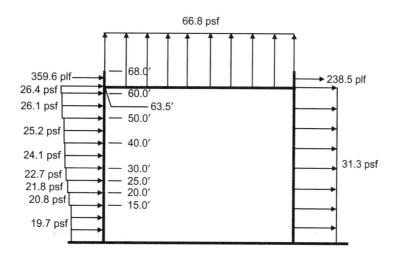

Figure 5.19 Net Design Wind Pressures in the N-S Direction

The MWFRS of buildings whose wind loads have been determined by IBC 1609.6 must be designed for the wind load cases defined in ASCE/SEI Figure 6-9 (IBC 1609.6.4.1). In Case 1, the full design wind pressures act on the projected area perpendicular to each principal axis of the structure. These pressures are assumed to act separately along each principal axis. The wind pressures on the windward and leeward walls depicted in Figures 5.19 fall under Case 1.

The calculations for the additional load cases that need to be considered in this example are similar to those shown in Example 5.6.

Part 2: Determine design wind forces on rooftop equipment

According to item 5 under IBC 1609.6.1, the applicable provisions in ASCE/SEI Chapter 6 are to be used to determine the design wind forces on rooftop equipment.

Calculations for the rooftop equipment in this example are provided in Part 2 of Example 5.6.

5.3.9 Example 5.9 – Fifteen-Story Office Building using Method 2, Analytical Procedure

For the 15-story office building depicted in Figure 5.20, determine (1) design wind pressures on the main wind-force-resisting system in both directions and (2) design wind forces on the cladding using Method 2, Analytical Procedure.

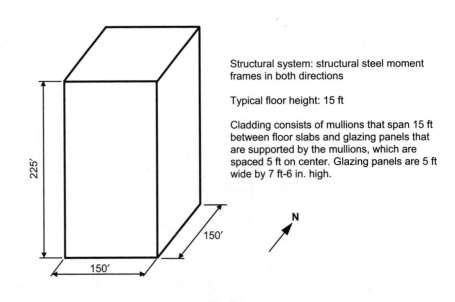

Structural system: structural steel moment frames in both directions

Typical floor height: 15 ft

Cladding consists of mullions that span 15 ft between floor slabs and glazing panels that are supported by the mullions, which are spaced 5 ft on center. Glazing panels are 5 ft wide by 7 ft-6 in. high.

Figure 5.20 Fifteen-story Office Building

DESIGN DATA

Location:	Chicago, IL
Surface Roughness:	B (suburban area with numerous closely spaced obstructions having the size of single-family dwellings and larger)
Topography:	Not situated on a hill, ridge or escarpment
Occupancy:	Business occupancy where less than 300 people congregate in one area

SOLUTION

Part 1: Determine design wind pressures on MWFRS

- Step 1: Check if the provisions of 6.5 can be used to determine the design wind pressures on the MWFRS.

The provisions of 6.5 may be used to determine design wind pressures provided the conditions of 6.5.1 and 6.5.2 are satisfied. It is clear that these conditions are satisfied for this office building that does not have response characteristics that make it subject to across-wind loading or other similar effects, and that is not sited at a location where channeling effects or buffeting in the wake of upwind obstructions need to be considered.

The provisions of Method 2, Analytical Procedure, can be used to determine the design wind pressures on the MWFRS.

- Step 2: Use Flowchart 5.7 to determine the design wind pressures on the MWFRS.

    1. For illustrative purposes, it is assumed in this example that the building is partially enclosed.[34]

    2. Determine whether the building is rigid or flexible and the corresponding gust effect factor from Flowchart 5.6.

    In lieu of determining the natural frequency n_1 of the building from a dynamic analysis, Eq. C6-14 in the commentary of ASCE/SEI 7 is used to compute n_1 for steel moment-resisting frames:

    $$n_1 = \frac{22.2}{H^{0.8}} = \frac{22.2}{225^{0.8}} = 0.3 \text{ Hz} < 1.0 \text{ Hz}$$

    Since $n_1 < 1.0$ Hz, the building is defined as a flexible building. Thus, the gust effect factor G_f for flexible buildings must be determined by Eq. 6-8 in 6.5.8.2.

    a. Determine g_Q and g_v

    $$g_Q = g_v = 3.4$$

---

[34] For office buildings of this type, it is common to assume that the building is enclosed. Where buildings have operable windows or where it is anticipated that debris may compromise some of the windows during a windstorm, it may be more appropriate to assume that the building is partially enclosed.

b. Determine g_R

$$g_R = \sqrt{2\ln(3{,}600 n_1)} + \frac{0.577}{\sqrt{2\ln(3{,}600 n_1)}}$$

$$= \sqrt{2\ln(3{,}600 \times 0.3)} + \frac{0.577}{\sqrt{2\ln(3{,}600 \times 0.3)}} = 3.9$$

Eq. 6-9

c. Determine $I_{\bar{z}}$

$$\bar{z} = 0.6h = 0.6 \times 225 = 135 \text{ ft} > z_{\min} = 30 \text{ ft} \qquad \text{Table 6-2 for Exposure B}$$

$$I_{\bar{z}} = c\left(\frac{33}{\bar{z}}\right)^{1/6}$$

$$= 0.30\left(\frac{33}{135}\right)^{1/6} = 0.24$$

Eq. 6-5 and Table 6-2 for Exposure B

d. Determine Q

$$L_{\bar{z}} = \ell\left(\frac{\bar{z}}{33}\right)^{\bar{\epsilon}}$$

$$= 320\left(\frac{135}{33}\right)^{1/3} = 511.8 \text{ ft}$$

Eq. 6-7 and Table 6-2 for Exposure B

$$Q = \sqrt{\frac{1}{1 + 0.63\left(\frac{B+h}{L_{\bar{z}}}\right)^{0.63}}}$$

$$= \sqrt{\frac{1}{1 + 0.63\left(\frac{150 + 225}{511.8}\right)^{0.63}}} = 0.81$$

Eq. 6-6

e. Determine R

From Figure 6-1 or Figure 1609, the basic wind speed V is equal to 90 mph for Chicago, IL.

EXAMPLES

$$\bar{V}_{\bar{z}} = \bar{b}\left(\frac{\bar{z}}{33}\right)^{\bar{\alpha}} V\left(\frac{88}{60}\right)$$

$$= 0.45\left(\frac{135}{33}\right)^{1/4} \times 90\left(\frac{88}{60}\right)$$

Eq. 6-14 and Table 6-2 for Exposure B

$$= 84.5 \text{ ft/sec}$$

$$N_1 = \frac{n_1 L_{\bar{z}}}{\bar{V}_{\bar{z}}}$$

$$= \frac{0.3 \times 511.8}{84.5} = 1.8$$

Eq. 6-12

$$R_n = \frac{7.47 N_1}{(1 + 10.3 N_1)^{5/3}}$$

$$= \frac{7.47 \times 1.8}{[1 + (10.3 \times 1.8)]^{5/3}} = 0.10$$

Eq. 6-11

$$\eta_h = \frac{4.6 n_1 h}{\bar{V}_{\bar{z}}}$$

$$= \frac{4.6 \times 0.3 \times 225}{84.5} = 3.7$$

$$R_h = \frac{1}{\eta_h} - \frac{1}{2\eta_h^2}\left(1 - e^{-2\eta_h}\right)$$

$$= \frac{1}{3.7} - \frac{1}{2 \times 3.7^2}(1 - e^{-2 \times 3.7}) = 0.23$$

Eq. 6-13(a)

$$\eta_B = \frac{4.6 n_1 B}{\bar{V}_{\bar{z}}}$$

$$= \frac{4.6 \times 0.3 \times 150}{84.5} = 2.5$$

$$R_B = \frac{1}{\eta_B} - \frac{1}{2\eta_B^2}\left(1 - e^{-2\eta_B}\right)$$

$$= \frac{1}{2.5} - \frac{1}{2 \times 2.5^2}(1 - e^{-2 \times 2.5}) = 0.32$$

Eq. 6-13(a)

$$\eta_L = \frac{15.4 n_1 L}{\overline{V}_{\overline{z}}}$$

$$= \frac{15.4 \times 0.3 \times 150}{84.5} = 8.2$$

$$R_L = \frac{1}{\eta_L} - \frac{1}{2\eta_L^2}\left(1 - e^{-2\eta_L}\right)$$

$$= \frac{1}{8.2} - \frac{1}{2 \times 8.2^2}(1 - e^{-2 \times 8.2}) = 0.12$$

Eq. 6-13(a)

Assume damping ratio $\beta = 0.01$ (see C6.5.8 for suggested value for steel buildings).

$$R = \sqrt{\frac{1}{\beta} R_n R_h R_B (0.53 + 0.47 R_L)}$$

$$= \sqrt{\frac{1}{0.01} \times 0.10 \times 0.23 \times 0.32[0.53 + (0.47 \times 0.12)]}$$

$$= 0.66$$

Eq. 6-10

f. Determine G_f

$$G_f = 0.925 \left(\frac{1 + 1.7 I_{\overline{z}} \sqrt{g_Q^2 Q^2 + g_R^2 R^2}}{1 + 1.7 g_v I_{\overline{z}}} \right)$$

$$= 0.925 \left(\frac{1 + (1.7 \times 0.24)\sqrt{(3.4^2 \times 0.81^2) + (3.9^2 \times 0.66^2)}}{1 + (1.7 \times 3.4 \times 0.24)} \right)$$

$$= 0.98$$

Eq. 6-8

3. Determine velocity pressure q_z for windward walls along the height of the building and q_h for leeward walls, side walls and roof using Flowchart 5.5.

   a. Determine basic wind speed V from Figure 6-1 or Figure 1609.

   As was determined above, $V = 90$ mph for Chicago, IL.

   b. Determine wind directionality factor K_d from Table 6-4.

   For the MWFRS of a building structure, $K_d = 0.85$.

EXAMPLES 5-115

c. Determine importance factor I from Table 6-1 based on occupancy category from IBC Table 1604.5.

From IBC Table 1604.5, the Occupancy Category is II for an office building. From Table 6-1, $I = 1.0$.

d. Determine exposure category.

In the design data, the surface roughness is given as B. Assume that Exposure B is applicable in all directions (6.5.6.3).

e. Determine topographic factor K_{zt}.

As noted in the design data, the building is not situated on a hill, ridge or escarpment. Thus, topographic factor $K_{zt} = 1.0$ (6.5.7.2).

f. Determine velocity pressure exposure coefficients K_z and K_h from Table 6-3.

According to Note 1 in Table 6-3, values of K_z and K_h under Case 2 for Exposure B must be used for MWFRSs in buildings that are not designed using Figure 6-10 for low-rise buildings. Values of K_z and K_h are summarized in Table 5.56.

Table 5.56 Velocity Pressure Exposure Coefficient K_z for MWFRS

Height above ground level, z (ft)	K_z
225	1.24
200	1.20
180	1.17
160	1.13
140	1.09
120	1.04
100	0.99
90	0.96
80	0.93
70	0.89
60	0.85
50	0.81
40	0.76
30	0.70
25	0.66
20	0.62
15	0.57

g. Determine velocity pressure q_z and q_h by Eq. 6-15.

$$q_z = 0.00256 K_z K_{zt} K_d V^2 I = 0.00256 \times K_z \times 1.0 \times 0.85 \times 90^2 \times 1.0 = 17.63 K_z \text{ psf}$$

A summary of the velocity pressures is given in Table 5.57.

4. Determine pressure coefficients C_p for the walls and roof from Figure 6-6.

The pressure coefficients will be the same in both the N-S and E-W directions, since the building is square in plan.

Windward wall: $C_p = 0.8$ for use with q_z

Table 5.57 Velocity Pressure q_z for MWFRS

Height above ground level, z (ft)	K_z	q_z (psf)
225	1.24	21.9
200	1.20	21.2
180	1.17	20.6
160	1.13	19.9
140	1.09	19.2
120	1.04	18.3
100	0.99	17.5
90	0.96	16.9
80	0.93	16.4
70	0.89	15.7
60	0.85	15.0
50	0.81	14.3
40	0.76	13.4
30	0.70	12.3
25	0.66	11.6
20	0.62	10.9
15	0.57	10.1

Leeward wall ($L/B = 150/150 = 1.0$): $C_p = -0.5$ for use with q_h

Side wall: $C_p = -0.7$ for use with q_h

Roof (normal to ridge with $\theta < 10$ degrees and parallel to ridge for all θ with $h/L = 225/150 = 1.5 > 1.0$)[35]:

$C_p = -1.04, -0.18$ from windward edge to $h/2 = 112.5$ ft for use with q_h[36]

$C_p = -0.7, -0.18$ from 112.5 ft to 150 ft for use with q_h

5. Determine q_i for the walls and roof using Flowchart 5.5.

   According to 6.5.12.2.1, it is permitted to take $q_i = q_h = 21.9$ psf for windward walls, side walls, leeward walls and roofs of partially enclosed buildings.

6. Determine internal pressure coefficients (GC_{pi}) from Figure 6-5.

   For a partially enclosed building, $(GC_{pi}) = +0.55, -0.55$.

7. Determine design wind pressures p_z and p_h by Eq. 6-19.

   Windward walls:

   $$p_z = q_z G_f C_p - q_h (GC_{pi})$$
   $$= (0.98 \times 0.8 \times q_z) - 21.9(\pm 0.55)$$
   $$= (0.78 q_z \mp 12.1) \text{ psf (external} \pm \text{internal pressure)}$$

   Leeward wall, side walls and roof:

   $$p_h = q_h G_f C_p - q_h (GC_{pi})$$
   $$= (21.9 \times 0.98 \times C_p) - 21.9(\pm 0.55)$$
   $$= (21.5 C_p \mp 12.1) \text{ psf (external} \pm \text{internal pressure)}$$

   A summary of the maximum design wind pressures that are valid in both the N-S and E-W directions is given in Table 5.58.

   Illustrated in Figure 5.21 are the external design wind pressures in the N-S or E-W directions. When considering horizontal wind forces on the MWFRS, it is clear that the effects from the internal pressure cancel out. On the roof, the effects from internal pressure add directly to those from the external pressure.

---

[35] The smaller uplift pressures on the roof due to $C_p = -0.18$ may govern the design when combined with roof live load or snow loads. This pressure is not shown in this example, but in general must be considered.

[36] $C_p = 1.3$ may be reduced based on area over which it is applicable = $(225/2) \times 150 = 16,875$ sq ft > 1,000 sq ft. Reduction factor = 0.8 from Figure 6-6. Thus, $C_p = 0.8 \times (-1.3) = -1.04$.

The MWFRS of buildings whose wind loads have been determined by 6.5.12.2.1 must be designed for the four wind load cases defined in Figure 6-9 (6.5.12.3). In Case 1, the full design wind pressures act on the projected area perpendicular to each principal axis of the structure. These pressures are assumed to act separately along each principal axis. The wind pressures on the windward and leeward walls depicted in Figure 5.18 fall under Case 1.

In Case 2, 75 percent of the design wind pressures on the windward and leeward walls are applied on the projected area perpendicular to each principal axis of the building along with a torsional moment. The wind pressures and torsional moments, both of which vary over the height of the building, are applied separately for each principal axis.

Table 5.58 Design Wind Pressures p for MWFRS

Location	Height above ground level, z (ft)	q (psf)	External pressure qGC_p (psf)	Internal pressure $q_h(GC_{pi})$ (psf)
Windward	225	21.9	17.1	±12.1
	200	21.2	16.5	±12.1
	180	20.6	16.1	±12.1
	160	19.9	15.5	±12.1
	140	19.2	15.0	±12.1
	120	18.3	14.3	±12.1
	100	17.5	13.7	±12.1
	90	16.9	13.2	±12.1
	80	16.4	12.8	±12.1
	70	15.7	12.3	±12.1
	60	15.0	11.7	±12.1
	50	14.3	11.2	±12.1
	40	13.4	10.5	±12.1
	30	12.3	9.6	±12.1
	25	11.6	9.1	±12.1
	20	10.9	8.5	±12.1
	15	10.1	7.9	±12.1
Leeward	All	21.9	−10.8	±12.1
Side	All	21.9	−15.1	±12.1
Roof	225	21.9	−22.4*	±12.1
	225	21.9	−15.1†	±12.1

* from windward edge to 112.5 ft
† from 112.5 ft to 150.0 ft

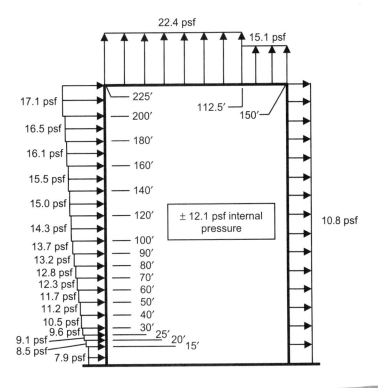

Figure 5.21 Design Wind Pressures in the N-S or E-W Directions

As an example of the calculations that need to be performed over the height of the building, the wind pressures and torsional moment at the mean roof height for Case 2 are as follows:

$$0.75 P_{WX} = 0.75 \times 17.1 = 12.8 \text{ psf (windward wall)}$$

$$0.75 P_{LX} = 0.75 \times 10.8 = 8.1 \text{ psf (leeward wall)}$$

For flexible buildings, the eccentricity e that is used to determine the torsional moment M_T is given by Eq. 6-21. Assuming that the elastic shear center and the center of mass coincide (i.e., $e_R = 0$),

$$e = \frac{e_Q + 1.7 I_{\bar{z}} \sqrt{(g_Q Q e_Q)^2 + (g_R R e_R)^2}}{1 + 1.7 I_{\bar{z}} \sqrt{(g_Q Q)^2 + (g_R R)^2}}$$

$$= \frac{(0.15 \times 150) + (1.7 \times 0.24)\sqrt{[3.4 \times 0.81 \times (0.15 \times 150)]^2 + 0}}{1 + (1.7 \times 0.24)\sqrt{(3.4 \times 0.81)^2 + (3.9 \times 0.66)^2}} = 18.8 \text{ ft}$$

The eccentricity determined by Eq. 6-21 is less than that for a rigid building, which is equal to $0.15 \times 150 = 22.5$ ft (see Figure 6-9). For conservatism, an eccentricity of 22.5 ft is used in this example.

$$M_T = 0.75(P_{WX} + P_{LX})B_X e_X$$
$$= 0.75(17.1 + 10.8) \times 150 \times (\pm 0.15 \times 150)$$
$$= \pm 70{,}622 \text{ ft-lb/ft}$$

In Case 3, 75 percent of the wind pressures of Case 1 are applied to the building simultaneously. This accounts for wind along the diagonal of the building. In Case 4, 75 percent of the wind pressures and torsional moments defined in Case 2 act simultaneously on the building.

Figure 5.22 illustrates Load Cases 1 through 4 for MWFRS wind pressures acting on the projected area at the mean roof height. Similar loading diagrams can be obtained at other locations below the mean roof height. The minimum design wind loading prescribed in 6.1.4.1 must be considered as a load case in addition to those load cases described above.

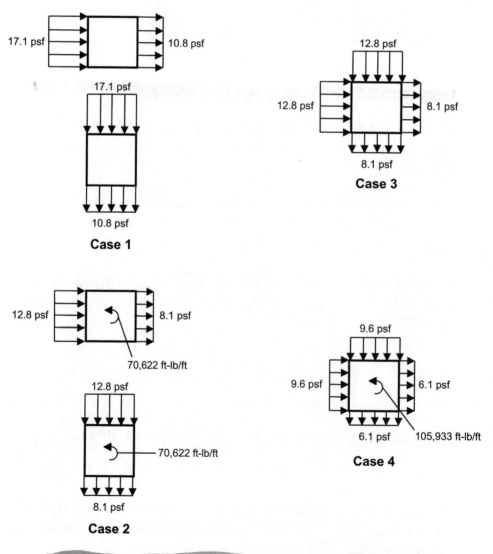

Figure 5.22 Load Cases 1 through 4 at the Mean Roof Height

The above wind pressure calculations assume a uniform design pressure over the incremental heights above ground level, which are given in Table 6-3. Alternatively, wind pressures can be computed at each floor level, and a uniform pressure is assumed between midstory heights above and below the floor level under consideration.

Shown in Figure 5.23 are the wind pressures computed at each floor level for this example building. It can be shown that the base shears in Figures 5.21 and 5.23 are virtually the same.

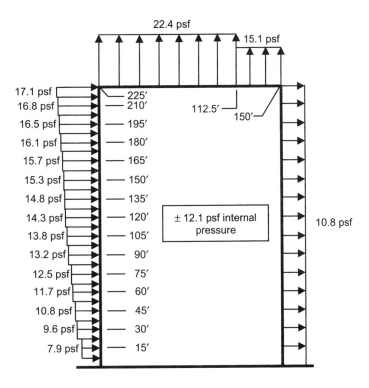

Figure 5.23 Design Wind Pressures Computed at the Floor Levels in the N-S or E-W Directions

Part 2: Determine design wind pressures on cladding

Use Flowchart 5.8 to determine the design wind pressures on the cladding.

1. It is assumed that the building is partially enclosed.

2. Determine velocity pressures q_z and q_h using Flowchart 5.5.

   For buildings with a mean roof height greater than 60 ft, the velocity pressures on the C&C on the windward walls vary with height above the base of the building. Most of the quantities needed to determine q_z and q_h were determined in Part 1, Step 2, item 3 of this example.

The velocity exposure coefficients K_z and K_h are summarized in Table 5.59.

Table 5.59 Velocity Pressure Exposure Coefficient K_z for C&C

Height above ground level, z (ft)	K_z
225	1.24
200	1.20
180	1.17
160	1.13
140	1.09
120	1.04
100	0.99
90	0.96
80	0.93
70	0.89
60	0.85
50	0.81
40	0.76
30	0.70
25	0.70
20	0.70
15	0.70

According to Note 1 in Table 6-3, values of K_z under Case 1 must be used for C&C in Exposure B. That is why the values of K_z in Table 5.50 (Case 1) differ from those in Table 5.47 (Case 2) for the MWFRS up to a height of 25 ft.

Velocity pressures q_z and q_h are determined by Eq. 6-15:

$$q_z = 0.00256 K_z K_{zt} K_d V^2 I = 0.00256 \times K_z \times 1.0 \times 0.85 \times 90^2 \times 1.0 = 17.63 K_z \text{ psf}$$

A summary of the velocity pressures is given in Table 5.60.

3. Determine external pressure coefficients (GC_p) from Figure 6-17 for Zones 4 and 5.

Pressure coefficients for Zones 4 and 5 can be determined from Figure 6-17 based on the effective wind area.

Table 5.60 Velocity Pressure q_z for C&C

Height above ground level, z (ft)	K_z	q_z (psf)
225	1.24	21.9
200	1.20	21.2
180	1.17	20.6
160	1.13	19.9
140	1.09	19.2
120	1.04	18.3
100	0.99	17.5
90	0.96	16.9
80	0.93	16.4
70	0.89	15.7
60	0.85	15.0
50	0.81	14.3
40	0.76	13.4
30	0.70	12.3
25	0.70	12.3
20	0.70	12.3
15	0.70	12.3

In general, the effective wind area is the larger of the tributary area and the span length multiplied by an effective width that need not be less than one-third the span length.

Effective wind area for the mullions = larger of $15 \times 5 = 75$ sq ft or $15 \times (15/3) = 75$ sq ft.

Effective wind area for the glazing panels = larger of $7.5 \times 5 = 37.5$ sq ft (governs) or $5 \times (5/3) = 8.3$ sq ft.

The pressure coefficients from Figure 6-17 for the cladding are summarized in Table 5.61.

Table 5.61 External Pressure Coefficients (GC_p) for C&C

Zone	(GC_p)			
	Mullions		Glazing Panels	
	Positive	Negative	Positive	Negative
4	0.78	-0.82	0.84	-0.86
5	0.78	-1.47	0.84	-1.64

The width of the end zone (Zone 5) $a = 0.1$(least horizontal dimension) $= 0.1 \times 150 = 15$ ft, which is greater than 3 ft (see Note 8 in Figure 6-17).

4. Determine internal pressure coefficients (GC_{pi}) from Figure 6-5.

   For a partially enclosed building, $(GC_{pi}) = +0.55, -0.55$.

5. Determine design wind pressure p by Eq. 6-23 in Zones 4 and 5.

$$p = q(GC_p) - q_h(GC_{pi})$$
$$= q(GC_p) - 21.9(\pm 0.55)$$
$$= [q(GC_p) \mp 12.1] \text{ psf } (\text{external} \pm \text{internal pressure})$$

where $q = q_z$ for positive external pressure and $q = q_h$ for negative external pressure (6.5.12.4.2). Note that $q_i = q_h$ for negative internal pressure in partially enclosed buildings. Also, $q_i = q_h$ may be conservatively used for positive internal pressure.

The maximum design wind pressures for positive and negative internal pressures are summarized in Table 5.62. The maximum positive pressure, which varies with height, is obtained with negative internal pressure. Similarly, the maximum negative pressure, which is a constant over the height of the building, is obtained with positive internal pressure. The pressures in Table 5.62 are applied normal to the cladding and act over the respective tributary area. The computed positive and negative pressures are greater than the minimum values prescribed in 6.1.4.2 of +10 psf and -10 psf, respectively.

Table 5.62 Design Wind Pressures p on C&C

Height above ground level, z (ft)	Design Pressure, p (psf)							
	Mullions				Glazing Panels			
	Zone 4		Zone 5		Zone 4		Zone 5	
	Positive	Negative	Positive	Negative	Positive	Negative	Positive	Negative
225	29.2	-30.1	29.2	-44.3	30.5	-30.9	30.5	-48.0
200	28.6	-30.1	28.6	-44.3	29.9	-30.9	29.9	-48.0
180	28.2	-30.1	28.2	-44.3	29.4	-30.9	29.4	-48.0
160	27.6	-30.1	27.6	-44.3	28.8	-30.9	28.8	-48.0
140	27.1	-30.1	27.1	-44.3	28.2	-30.9	28.2	-48.0
120	26.4	-30.1	26.4	-44.3	27.5	-30.9	27.5	-48.0
100	25.8	-30.1	25.8	-44.3	26.8	-30.9	26.8	-48.0
90	25.3	-30.1	25.3	-44.3	26.3	-30.9	26.3	-48.0
80	24.9	-30.1	24.9	-44.3	25.9	-30.9	25.9	-48.0

(continued)

EXAMPLES

Table 5.62 Design Wind Pressures p on C&C (continued)

Height above ground level, z (ft)	Design Pressure, p (psf)							
	Mullions				Glazing Panels			
	Zone 4		Zone 5		Zone 4		Zone 5	
	Positive	Negative	Positive	Negative	Positive	Negative	Positive	Negative
70	24.3	-30.1	24.3	-44.3	25.3	-30.9	25.3	-48.0
60	23.8	-30.1	23.8	-44.3	24.7	-30.9	24.7	-48.0
50	23.3	-30.1	23.3	-44.3	24.1	-30.9	24.1	-48.0
40	22.6	-30.1	22.6	-44.3	23.4	-30.9	23.4	-48.0
30	21.7	-30.1	21.7	-44.3	22.4	-30.9	22.4	-48.0
25	21.7	-30.1	21.7	-44.3	22.4	-30.9	22.4	-48.0
20	21.7	-30.1	21.7	-44.3	22.4	-30.9	22.4	-48.0
15	21.7	-30.1	21.7	-44.3	22.4	-30.9	22.4	-48.0

5.3.10 Example 5.10 – Agricultural Building using Method 2, Analytical Procedure

For the agricultural building depicted in Figure 5.24, determine (1) design wind pressures on the main wind-force-resisting system in both directions and (2) design wind pressures on the roof trusses using Method 2, Analytical Procedure.

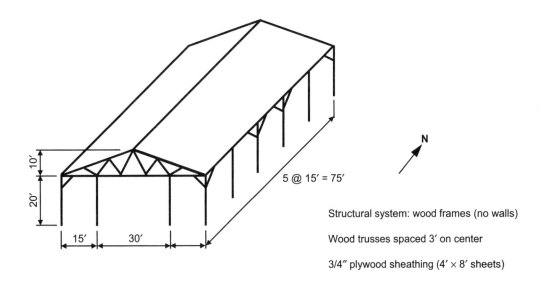

Figure 5.24 Agricultural Building

DESIGN DATA

Location: Ames, IA

Surface Roughness: C (open terrain with scattered obstructions having heights less than 30 ft)

Topography: Not situated on a hill, ridge or escarpment

Occupancy: Utility and miscellaneous occupancy

SOLUTION

Part 1: Determine design wind pressures on MWFRS

- Step 1: Check if the provisions of 6.5 can be used to determine the design wind pressures on the MWFRS.

  The provisions of 6.5 may be used to determine design wind pressures provided the conditions of 6.5.1 and 6.5.2 are satisfied. It is clear that these conditions are satisfied for this agricultural building that does not have response characteristics that make it subject to across-wind loading or other similar effects, and that is not sited at a location where channeling effects or buffeting in the wake of upwind obstructions need to be considered.

  The provisions of Method 2, Analytical Procedure, can be used to determine the design wind pressures on the MWFRS.[37]

- Step 2: Use Flowchart 5.7 to determine the design wind pressures on the MWFRS.

  1. Since the building does not have any walls, it is classified as open.

  2. Determine velocity pressure q_h using Flowchart 5.5.

     a. Determine basic wind speed V from Figure 6-1 or Figure 1609.

        From Figure 6.1 or Figure 1609, $V = 90$ mph for Ames, IA.

     b. Determine wind directionality factor K_d from Table 6-4.

        For the MWFRS of a building structure, $K_d = 0.85$.

---

[37] Even though the building is less than 60 ft in height, Method 1, Simplified Procedure cannot be used to determine the wind pressures, since the building is not enclosed.

EXAMPLES 5-127

c. Determine importance factor I from Table 6-1 based on occupancy category from IBC Table 1604.5.

From IBC Table 1604.5, the Occupancy Category is I for an agricultural facility. From Table 6-1, $I = 0.87$.

d. Determine exposure category.

In the design data, the surface roughness is given as C. Assume that Exposure C is applicable in all directions (6.5.6.3).

e. Determine topographic factor K_{zt}.

As noted in the design data, the building is not situated on a hill, ridge or escarpment. Thus, topographic factor $K_{zt} = 1.0$ (6.5.7.2).

f. Determine velocity pressure exposure coefficient K_h from Table 6-3.

$$\text{Mean roof height} = \frac{20+30}{2} = 25 \text{ ft}$$

From Table 6-3, $K_h = 0.94$ for Exposure C.

g. Determine velocity pressure q_h by Eq. 6-15.

$$q_h = 0.00256 K_h K_{zt} K_d V^2 I = 0.00256 \times 0.94 \times 1.0 \times 0.85 \times 90^2 \times 0.87 = 14.4 \text{ psf}$$

3. Determine gust effect factor G from Flowchart 5.6.

Assuming the building is rigid, G may be computed by Eq. 6-4 or may be taken as 0.85. For simplicity, use $G = 0.85$.

4. Determine net pressure coefficients C_N for the pitched roof.

a. Wind in the E-W direction ($\gamma = 0°, 180°$)

Figure 6-18B is used to determine the net pressure coefficients C_{NW} and C_{NL} on the windward and leeward portions of the roof surface for wind in the E-W direction. These net pressure coefficients include contributions from both the top and bottom surfaces of the roof (see Note 1 in Figure 6-18B).

The wind pressures on the roof depend on the level of wind flow restriction below the roof. Clear wind flow implies that little (less than or equal to 50 percent) or no portion of the cross-section below the roof is blocked by goods or materials (see Note 2 in Figure 6-18B). Obstructed wind flow means that a significant portion (more than 50 percent) of the cross-section is blocked. Since the usage of the space below the roof is not known, wind pressures will be determined for both situations.

A summary of the net pressure coefficients is given in Table 5.63. The roof angle in this example is equal to approximately 18.4 degrees, and the values of C_{NW} and C_{NL} were obtained by linear interpolation (see Note 3 in Figure 6-18B).

Table 5.63 Windward and Leeward Net Pressure Coefficients C_{NW} and C_{NL} for Wind in the E-W Direction

Load Case	Clear Wind Flow		Obstructed Wind Flow	
	C_{NW}	C_{NL}	C_{NW}	C_{NL}
A	1.10	-0.17	-1.20	-1.09
B	0.01	-0.95	-0.69	-1.65

b. Wind in the N-S direction ($\gamma = 90°$)

Figure 6-18D is used to determine the net pressure coefficients C_N at various distances from the windward edge of the roof.

Net pressure coefficients are given in Table 5.64.

Table 5.64 Net Pressure Coefficients C_N for Wind in the N-S Direction

Horizontal Distance from Windward Edge	Load Case	Clear Wind Flow	Obstructed Wind Flow
		C_N	C_N
$\leq h = 25'$	A	-0.8	-1.2
	B	0.8	0.5
$> h = 25', \leq 2h = 50'$	A	-0.6	-0.9
	B	0.5	0.5
$\geq 2h = 50'$	A	-0.3	-0.6
	B	0.3	0.3

5. Determine net design pressure p by Eq. 6-25.

$$p = q_h G C_N = 14.4 \times 0.85 \times C_N = 12.2 C_N \text{ psf}$$

A summary of the net design wind pressures is given in Table 5.65 for wind in the E-W direction and in Table 5.66 for wind in the N-S direction. These pressures act perpendicular to the roof surface.

Table 5.65 Net Design Wind Pressures (psf) for Wind in the E-W Direction

Load Case	Clear Wind Flow		Obstructed Wind Flow	
	Windward	Leeward	Windward	Leeward
A	13.4	-2.1	-14.6	-13.3
B	0.1	-11.6	-8.4	-20.1

Table 5.66 Net Design Wind Pressures (psf) for Wind in the N-S Direction

Horizontal Distance from Windward Edge	Load Case	Clear Wind Flow C_N	Obstructed Wind Flow C_N
$\leq h = 25'$	A	-9.8	-14.6
	B	9.8	6.1
$> h = 25', \leq 2h = 50'$	A	-7.3	-11.0
	B	6.1	6.1
$\geq 2h = 50'$	A	-3.7	-7.3
	B	3.7	3.7

The minimum design wind loading prescribed in 6.1.4.1 must be considered as a load case in addition to Load Cases A and B described above.

Part 2: Determine design wind pressures on roof trusses

Use Flowchart 5.8 to determine the design wind pressures on the roof trusses, which are C&C.

1. Determine velocity pressure q_h using Flowchart 5.5.

   Velocity pressure was determined in Part 1, Step 2, item 2 of this example and is equal to 14.4 psf.

2. Determine gust effect factor G from Flowchart 5.6.

   The gust effect factor was determined in Part 1, Step 2, item 3 of this example and is equal to 0.85.

3. Determine net pressure coefficients C_N for the pitched roof.

   Figure 6-19B is used to determine the net pressure coefficients C_N for Zones 1, 2 and 3 on the roof. In this example, $h/L = 25/60 = 0.42$, which is between the limits of 0.25 and 1.0.

   The magnitude of the net pressure coefficient depends on the effective wind area, which is the larger of the tributary area and the span length multiplied by an effective width that need not be less than one-third the span length.

   Effective wind area = larger of $60 \times 3 = 180$ sq ft or $60 \times (60/3) = 1,200$ sq ft (governs).

   The width of the zones a = least of 0.1 (least horizontal dimension) = $0.1 \times 60 = 6.0$ ft (governs) or $0.4h = 0.4 \times 25 = 10.0$ ft. This value of a is greater than 0.04 (least horizontal dimension) = $0.04 \times 60 = 2.4$ ft or 3 ft (see Note 6 in Figure 6-19B).

   The effective wind area is greater than $4.0a^2 = 4.0 \times 6.0^2 = 144$ sq ft; this information is also needed to determine the net pressure coefficients.

   Like in the case of the MWFRS, the magnitude of C_N depends on whether the wind flow is clear or obstructed. Both situations are examined in this example.

   For clear wind flow, $C_N = 1.15, -1.06$ in Zones 1, 2 and 3. Linear interpolation was used to determine these values for a roof angle of 18.4 degrees and an effective wind area $> 4.0a^2$ (see Note 3 in Figure 6-19B).

   For obstructed wind flow, $C_N = 0.50, -1.51$ in Zones 1, 2, and 3 by linear interpolation.

4. Determine net design wind pressure p by Eq. 6-26.

   $p = q_h G C_N = 14.4 \times 0.85 \times C_N = 12.2 C_N$ psf

   For clear wind flow: $p = 14.0$ psf, -12.9 psf in Zones 1, 2 and 3

   For obstructed wind flow: $p = 6.1$ psf, -18.4 psf in Zones 1, 2 and 3

   In the case of obstructed wind flow, the net positive pressure must be increased to 10 psf to satisfy the minimum requirements of 6.1.4.2.

   Illustrated in Figure 5.25 are the loading diagrams on a typical interior roof truss for clear wind flow. Similar loading diagrams can be obtained for obstructed wind flow.

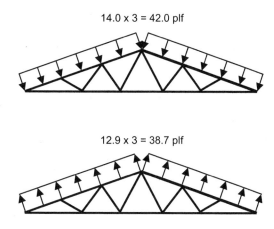

Figure 5.25 Roof Truss Loading Diagrams for Clear Wind Flow

5.3.11 Example 5.11 – Freestanding Masonry Wall using Method 2, Analytical Procedure

Determine the design wind forces on the architectural freestanding masonry screen wall depicted in Figure 5.26 using Method 2, Analytical Procedure (6.5.14).

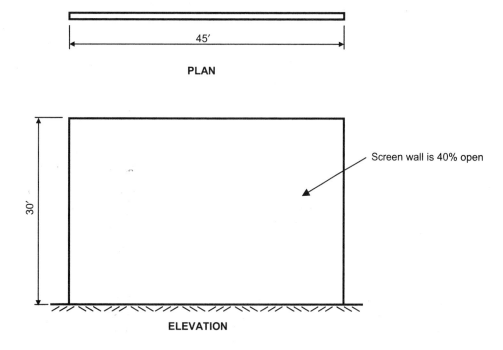

Figure 5.26 Freestanding Masonry Wall

DESIGN DATA

Basic wind speed: 90 mph
Exposure category: B
Topography: Not situated on a hill, ridge or escarpment
Occupancy Category: I (low hazard to human life)

SOLUTION

Use Flowchart 5.9 to determine the design wind force on the freestanding wall.

1. Determine velocity pressure q_h using Flowchart 5.5.

    a. Determine basic wind speed V from Figure 6-1 or Figure 1609.
       The basic wind speed was given in the design data as $V = 90$ mph.

    b. Determine wind directionality factor K_d from Table 6-4.

       For solid signs or solid freestanding walls, $K_d = 0.85$.

    c. Determine importance factor I from Table 6-1 based on occupancy category from IBC Table 1604.5.

       The Occupancy Category is given as I in the design data. From Table 6-1, $I = 0.87$.

    d. Determine exposure category.

       Exposure B is given in the design data.

    e. Determine topographic factor K_{zt}.

       As noted in the design data, the building is not situated on a hill, ridge or escarpment. Thus, topographic factor $K_{zt} = 1.0$ (6.5.7.2).

    f. Determine velocity pressure exposure coefficient K_h from Table 6-3.

       From Table 6-3, $K_h = 0.70$ in Case 2 under Exposure B at a height of 30 ft.

g. Determine velocity pressure q_h by Eq. 6-15.

$$q_h = 0.00256 K_h K_{zt} K_d V^2 I = 0.00256 \times 0.70 \times 1.0 \times 0.85 \times 90^2 \times 0.87 = 12.3 \text{ psf}$$

2. Determine gust effect factor G from Flowchart 5.6.

   Assuming the wall is rigid, G may be computed by Eq. 6-4 or may be taken as 0.85. For simplicity, use $G = 0.85$.

3. Determine net force coefficient C_f from Figure 6-20.

   In general, Cases A, B and C must be considered for freestanding walls. A different loading condition is considered in each case.

   In this example, the aspect ratio $B/s = 45/30 = 1.5 < 2$. Therefore, according to Note 3 in Figure 6-20, only Cases A and B need to be considered.

   From Figure 6-20, $C_f = 1.43$ (by linear interpolation) for Cases A and B with $s/h = 1$ and $B/s = 1.5$ (see Note 5 in Figure 6-20).

   According to Note 2 in Figure 6-20, force coefficients for solid freestanding walls with openings may be multiplied by the reduction factor $[1 - (1-\varepsilon)^{1.5}]$ where ε is equal to the ratio of the solid area to gross area of the wall. In this example, $\varepsilon = 0.6$, and the reduction factor = 0.75. Therefore, $C_f = 0.75 \times 1.43 = 1.07$.

4. Determine design wind force F from Eq. 6-27 (6.5.14).

   $$F = q_h G C_f A_s = 12.3 \times 0.85 \times 1.07 \times (30 \times 45) = 15,102 \text{ lbs}$$

   In Case A, this force acts at a distance equal to 5 percent of the height of the wall above the geometric center of the wall, i.e., the resultant force is located at (30/2) + (0.05 × 30) = 16.5 ft above the ground level (see Figure 6-20).

   In Case B, the resultant force is located 16.5 ft above the ground level and (45/2) − (0.2 × 45) = 13.5 from the windward edge of the wall (see Figure 6-20).

The minimum design wind loading prescribed in 6.1.4.1 must be considered as a load case in addition to Load Cases A and B described above.

CHAPTER 6   EARTHQUAKE LOADS

6.1 INTRODUCTION

According to IBC 1613.1, the effects of earthquake motion on structures and their components are to be determined in accordance with ASCE/SEI 7-05, excluding Chapter 14 (Material Specific Seismic Design and Detailing Requirements) and Appendix 11A (Quality Assurance Provisions).[1] These chapters from ASCE/SEI 7 have been excluded because the IBC includes quality assurance provision in Chapter 17 and structural material provisions in Chapters 19 through 23.

A summary of the chapters of ASCE/SEI 7 that contain earthquake load provisions that are referenced by the 2009 IBC is provided in Table 6.1. The primary focus of the discussion in this chapter of the publication is on Chapters 11, 12, 13, 15, 20, 21 and 22 of ASCE/SEI 7.

Table 6.1 Summary of Chapters in ASCE/SEI 7-05 that Are Referenced by the 2009 IBC for Earthquake Load Provisions

Chapter	Title
11	Seismic Design Criteria
12	Seismic Design Requirements for Building Structures
13	Seismic Design Requirements for Nonstructural Components
15	Seismic Design Requirements for Nonbuilding Structures
16	Seismic Response History Procedures
17	Seismic Design Requirements for Seismically Isolated Structures
18	Seismic Design Requirements for Structures with Damping Systems
19	Soil Structure Interaction for Seismic Design
20	Site Classification Procedure for Seismic Design
21	Site-Specific Ground Motion Procedures for Seismic Design
22	Seismic Ground Motion and Long Period Transition Maps
23	Seismic Design Reference Documents

---

[1] IBC 1613.6 contains modifications to the ASCE/SEI 7-05 earthquake load provisions. Note that the modifications are optional rather than mandatory.

IBC 1613.1 lists four exemptions to seismic design requirements presented in this section:

1. Detached one- and two-family dwellings that are assigned to Seismic Design Category (SDC) A, B or C (i.e, $S_{DS} < 0.5$ and $S_{D1} < 0.2$), or located where $S_S < 0.4$.[2]
2. Conventional light-frame wood construction that conforms to IBC 2308.[3]
3. Agricultural storage structures where human occupancy is incidental.
4. Structures that are covered under other regulations, such as vehicular bridges, electrical transmission towers, hydraulic structures, buried utility lines and nuclear reactors.

ASCE/SEI 11.1.2 contains essentially the same exemptions as the IBC; the only difference occurs in the second exemption, which is stated in ASCE/SEI 11.1.2 as follows: detached one- and two-family wood-frame dwellings not included in Exception 1 that are less than or equal to two stories, satisfying the limitations and constructed in accordance with the *International Residential Code®* (IRC®).

The seismic requirements of the IBC need not be applied to structures that meet at least one of these four criteria.

Seismic requirements for existing buildings are contained in IBC 1613.3. In general, additions, alterations, repairs or change of occupancy of structures are governed by the provisions of IBC Chapter 34, Existing Structures. The seismic resistance requirements of IBC 3403.4.1 and 3404.4.1 must be satisfied where additions and alterations are made to an existing building, respectively.

IBC 3408 provides guidance with respect to the impact a change of occupancy has on an existing building.

6.2 SEISMIC DESIGN CRITERIA

6.2.1 Seismic Ground Motion Values

Mapped Acceleration Parameters

IBC Figures 1613.5(1), 1613.5(2) and ASCE/SEI Figures 22-1, 22-2 contain contour maps of the conterminous United States giving S_S and S_1, which are the mapped maximum considered earthquake (MCE) spectral response accelerations at periods of 0.2 sec and 1.0 sec, respectively, for a Site Class B soil profile and 5-percent damping.[4]

---

[2] Definitions of Seismic Design Category and spectral response accelerations S_S, S_{DS} and S_{D1} are given in subsequent sections of this publication.
[3] Limitations for conventional light-frame wood construction are given in IBC 2308.2.
[4] The MCE spectral response accelerations, which are directly related to base shear, reflect the maximum level of earthquake ground shaking that is considered reasonable for the design of new structures.

IBC Figures 1613.5(3) through 1613.5(14) and ASCE/SEI Figures 22-3 through 22-14 contain similar contour maps for specific regions of the conterminous United States, Alaska, Hawaii, Puerto Rico and U.S. commonwealths and territories.

In lieu of the maps, MCE spectral response accelerations can be obtained from the Ground Motion Parameter Calculator that can be accessed on the United States Geological Survey (USGS) website.[5] Accelerations are output for a specific latitude-longitude or zip code, which is input by the user. More accurate spectral accelerations for a given site are typically obtained by inputting a latitude-longitude, especially in areas where the mapped ground motions are highly variable or where a zip code encompasses a large area.

Where $S_S \leq 0.15$ and $S_1 \leq 0.04$, the structure is permitted to be assigned to SDC A (IBC 1613.5.1). These areas are considered to have very low seismic risk based solely on the mapped ground motions.

Site Class

Six site classes are defined in IBC Table 1613.5.2 and ASCE/SEI Table 20.3-1. A site is to be classified as one of these six site classes based on one of three soil properties (soil shear wave velocity, standard penetration resistance or blow count and soil undrained shear strength) measured over the top 100 ft of the site. Steps for classifying a site are given in IBC 1613.5.5.1 and ASCE/SEI 20.3. Methods of determining the site class where the soil is not homogeneous over the top 100 ft are provided in IBC 1613.5.5 and ASCE/SEI 20.4.

Site Class A is hard rock, which is typically found in the eastern United States, while Site Class B is a softer rock that is commonly found in western parts of the country. Site Classes C, D and E indicate progressively softer soils, while Site Class F indicates soil so poor that a site-specific geotechnical investigation and dynamic site response analysis is required to determine site coefficients. Site-specific ground motion procedures for seismic design are given in ASCE/SEI Chapter 21.

At locations or in cases where soil property measurements to a depth of 100 ft are not feasible, the registered design professional that is responsible for preparing the geotechnical report may estimate soil properties from geological conditions. When soil properties are not known in sufficient detail to determine the site class in accordance with code provisions, Site Class D must be used, unless the building official requires that Site E or F must be used at the site.

Site Coefficients and Adjusted MCE Spectral Response Acceleration Parameters

Once the mapped spectral accelerations and site class have been established, the MCE spectral response acceleration for short periods S_{MS} and at 1-second period S_{M1} adjusted

---

[5] http://earthquake.usgs.gov/research/hazmaps/design/

for site class effects are determined by IBC Eqs. 16-36 and 16-37, respectively, or ASCE/SEI Eqs. 11.4-1 and 11.4-2, respectively:

$$S_{MS} = F_a S_S$$

$$S_{M1} = F_v S_1$$

where F_a = short-period site coefficient determined from IBC Table 1613.5.3(1) or ASCE/SEI Table 11.4-1 and F_v = long-period site coefficient determined from IBC Table 1613.5.3(2) or ASCE/SEI Table 11.4-2. For site classes other than B, an adjustment to the mapped spectral response accelerations is necessary.

Typically, ground motion is amplified in softer soils (Site Classes C through E) and attenuated in stiffer soils (Site Class A). This can be observed in the tables where the magnitudes of F_a and F_v increase going from Site Class A to F for a given mapped ground motion acceleration. The only exception to this occurs for short periods where $S_S \geq 1.0$ and the Site Class changes from D to E. Very soft soils are not capable of amplifying the short-period components of subsurface rock motion; in fact, deamplification occurs in such cases.

Design Spectral Response Acceleration Parameters

Five-percent damped design spectral response accelerations at short periods S_{DS} and at 1-sec period S_{D1} are determined by IBC Eqs. 16-38 and 16-39, respectively, or ASCE/SEI Eqs. 11.4-3 and 11.4-4, respectively:

$$S_{DS} = \frac{2}{3} S_{MS}$$

$$S_{D1} = \frac{2}{3} S_{M1}$$

The design ground motion is 2/3 = 1/1.5 times the soil-modified MCE ground motion; the basis of this factor is that it is highly unlikely that a structure designed by the code provisions will collapse when subjected to ground motion that is 1.5 times as strong as the design ground motion. More information on the two-thirds adjustment factor can be found in Chapter 3 of the NEHRP Commentary (FEMA 450).

6.2.2 Occupancy Category and Importance Factor

Occupancy categories are defined in IBC Table 1604.5 and ASCE/SEI Table 1-1. An importance factor I is assigned to a building or structure in accordance with ASCE/SEI Table 11.5-1 based on its occupancy category. Larger values of I are assigned to high occupancy and essential facilities to increase the likelihood that such structures would suffer less damage and continue to function during and following a design earthquake.

6.2.3 Seismic Design Category

All buildings and structures must be assigned to a Seismic Design Category (SDC) in accordance with IBC 1613.5.6 or ASCE/SEI 11.6. In general, a SDC is a function of occupancy or use and the design spectral accelerations at the site.

The SDC is determined twice: first as a function of S_{DS} by IBC Table 1613.5.6(1) or ASCE/SEI Table 11.6-1 and second as a function of S_{D1} by IBC Table 1613.5.6(2) or ASCE/SEI Table 11.6-2. The more severe seismic design category of the two governs. Where S_1 is less than 0.75, the SDC may be determined by IBC Table 1613.5.6(1) or ASCE/SEI Table 11.6-1 based solely on S_{DS} provided all of the four conditions listed under IBC 1613.5.6.1 or ASCE/SEI 11.6 are satisfied.

Where S_1 is greater than or equal to 0.75, conditions under which SDC E and SDC F are to be assigned are also given in IBC 1613.5.6 and ASCE/SEI 11.6.

The SDC is a trigger mechanism for many seismic requirements, including

- permissible seismic force-resisting systems
- limitations on building height
- consideration of structural irregularities
- the need for additional special inspections, structural testing and structural observation for seismic resistance

6.2.4 Design Requirements for SDC A

Structures assigned to SDC A need only comply with the requirements of 11.7.[6] To ensure general structural integrity, the lateral force-resisting system must be proportioned to resist a lateral force at each floor level equal to 1 percent of the total dead load at that floor level, as depicted in Figure 6.1.

According to 11.7.2, the lateral forces are to be applied independently in each of two orthogonal directions.

Requirements for load path connections, connection to supports, and anchorage of concrete or masonry walls are given in 11.7.3, 11.7.4 and 11.7.5, respectively.

---

[6] From this point on in this chapter, referenced section, table and figure numbers are from ASCE/SEI 7-05 unless noted otherwise.

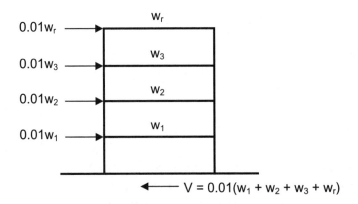

Figure 6.1 Design Seismic Force Distribution for Structures Assigned to SDC A

6.3 SEISMIC DESIGN REQUIREMENTS FOR BUILDING STRUCTURES

6.3.1 Basic Requirements

Basic requirements for seismic analysis and design of building structures are contained in 12.1. In general, the structure must have a complete lateral and vertical force-resisting system that is capable of providing adequate strength, stiffness and energy dissipation capacity when subjected to the design ground motion. Also, a continuous load path with adequate strength and stiffness must be provided to transfer all forces from the point of application to the final point of resistance.

Design seismic forces and their distribution over the height of a structure are to be established in accordance with one of the procedures in 12.6. A simplified design procedure may also be used, provided all of the limitations of 12.14 are satisfied. These methods are covered in subsequent sections of this publication.

General requirements for connections between members and connection to supports are given in 12.1.3 and 12.1.4, respectively. Basic requirements for foundation design are contained in 12.1.5.

6.3.2 Seismic Force-Resisting Systems

The basic seismic force-resisting systems are listed in Table 12.2-1. Included in the table are the response modification coefficients R to be used in determining the base shear V, the system overstrength factors Ω_O to be used in determining element design forces, and the deflection amplification factors C_d to be used in determining design story drift.

Table 12.2-1 also contains important information on system limitations with respect to SDC and height. Some systems, such as special steel moment frames and special reinforced concrete moment frames, can be utilized in structures assigned to any SDC with no height limitations. Other less ductile systems can be utilized with no height limitations in some SDCs while others are permitted in structures up to certain heights. The least ductile systems are not permitted in the higher SDCs under any circumstances.

Information on combinations of framing systems in different and in the same direction is given in 12.2.2 and 12.2.3, respectively. Different seismic force-resisting systems may be used along two orthogonal axes of the structure. The more stringent system limitations of Table 12.2-1 apply where different seismic force-resisting systems are used in the same direction. Occupancy Category I or II buildings that are two stories or less in height and that use light-frame construction or flexible diaphragms are permitted to be designed using the least value of R for the different structural systems in each independent line of resistance (12.2.3.2). Also, the value of R used for the design of the diaphragms in such structures must be less than or equal to the least value for any of the systems utilized in that same direction.

Specific requirements for dual systems, cantilever column systems, inverted pendulum-type structures, special moment frames and other systems are given in 12.2.5. In shear wall-frame interactive systems, which are permitted in structures assigned to SDC B only, the shear walls must be able to resist at least 75 percent of the design story shear at each story (12.2.5.10).

6.3.3 Diaphragm Flexibility, Configuration Irregularities, and Redundancy

Diaphragm Flexibility

The relative flexibility of diaphragms must be considered in the structural analysis (12.3.1). Floor and roof systems that can be idealized as flexible or rigid diaphragms are given in 12.3.1.1 and 12.3.1.2, respectively. Floor/roof systems that do not satisfy the conditions of these sections are permitted to be idealized as flexible diaphragms where the computed in-plane deflection of the diaphragm under lateral load is greater than two times the average story drift of adjoining vertical elements of the seismic force-resisting system (see 12.3.1.3 and Figure 12.3-1).

Irregular and Regular Classifications

Structures are classified as regular or irregular based on the criteria in 12.3.2. In general, buildings having irregular configurations in plan and/or elevation suffer greater damage than those having regular configurations when subjected to design earthquakes.

Table 12.3-1 contains the types and descriptions of horizontal structural irregularities that must be considered as a function of SDC. Similar information is provided in Table 12.3-2 for vertical structural irregularities. Tables 6.2 and 6.3 summarize and illustrate the information provided in Tables 12.3-1 and 12.3-2, respectively.

Limitations and additional requirements for systems with particular types of structural irregularities are given in 12.3.3 and are summarized in Table 6.4.

The equivalent lateral force procedure using elastic analysis is not capable of accurately predicting the earthquake effects on certain types of irregular buildings. As indicated in Section 6.3.6 of this publication, more comprehensive structural analyses must be performed for structures with specific types of horizontal and vertical irregularities.

Table 6.2 Horizontal Structural Irregularities (12.3.2.1)

	Irregularity Type	Description	Reference Section	SDC Application
1a	Torsional Irregularity	$\Delta_2 > 1.2\left(\dfrac{\Delta_1 + \Delta_2}{2}\right)$	12.3.3.4 12.8.4.3 12.7.3 12.12.1 Table 12.6-1 16.2.2	D, E and F C, D, E and F B, C, D, E and F C, D, E and F D, E and F B, C, D, E and F
1b	Extreme Torsional Irregularity	$\Delta_2 > 1.4\left(\dfrac{\Delta_1 + \Delta_2}{2}\right)$	12.3.3.1 12.3.3.4 12.7.3 12.8.4.3 12.12.1 Table 12.6-1 16.2.2	E and F D B, C and D C and D C and D D B, C and D
2	Re-entrant Corners	• Projection b > 0.15a and • Projection d > 0.15c	12.3.3.4 Table 12.6-1	D, E and F D, E and F

(continued)

Table 6.2 Horizontal Structural Irregularities (12.3.2.1) (continued)

	Irregularity Type	Description	Reference Section	SDC Application
3	Diaphragm Discontinuity	• Area of opening > 0.5ab or • Changes in effective diaphragm stiffness > 50% from one story to the next	12.3.3.4 Table 12.6-1	D, E and F D, E and F
4	Out-of-Plane Offsets	Discontinuities in lateral-force-resisting path, such as out-of-plane offsets of vertical elements	12.3.3.4 12.3.3.3 12.7.3 Table 12.6-1 16.2.2	D, E and F B, C, D, E and F B, C, D, E and F D, E and F B, C, D, E and F
5	Nonparallel Systems	Vertical lateral force-resisting elements are not parallel to or symmetric about the major orthogonal axes of the seismic force-resisting system	12.5.3 12.7.3 Table 12.6-1 16.2.2	C, D, E and F B, C, D, E and F D, E and F B, C, D, E and F

Table 6.3 Vertical Structural Irregularities (12.3.2.2)

	Irregularity Type		Description	Reference Section	SDC Application
1a	Stiffness Irregularity— Soft Story		Soft story stiffness < 70% (story stiffness above) or < 80% (average stiffness of 3 stories above)	Table 12.6-1	D, E and F
1b	Stiffness Irregularity— Extreme Soft Story		Soft story stiffness < 60% (story stiffness above) or < 70% (average stiffness of 3 stories above)	12.3.3.1 Table 12.6-1	E and F D, E and F
2	Weight (Mass) Irregularity		Story mass > 150% (adjacent story mass) (a roof that is lighter than the floor below need not be considered)	Table 12.6-1	D, E and F
3	Vertical Geometric Irregularity		Horizontal dimension of seismic force-resisting system in story > 130% of that in adjacent story	Table 12.6-1	D, E and F

(continued)

Table 6.3 Vertical Structural Irregularities (12.3.2.2) (continued)

	Irregularity Type	Description	Reference Section	SDC Application
4	In-Plane Discontinuity in Vertical Lateral Force-Resisting Elements	In-plane offset of lateral force-resisting elements > lengths of those elements, or reduction in stiffness of resisting elements in story below	12.3.3.3 12.3.3.4 Table 12.6-1	B, C, D, E and F D, E and F D, E and F
5a	Discontinuity in Capacity— Weak Story	• Weak story strength < 80% (story strength above) • Story strength = total strength of seismic-resisting elements sharing story shear for direction under consideration	12.3.3.1 Table 12.6-1	E and F D, E and F
5b	Discontinuity in Capacity— Extreme Weak Story	• Weak story strength < 65% (story strength above) • Story strength = total strength of seismic-resisting elements sharing story shear for direction under consideration	12.3.3.1 12.3.3.2 Table 12.6-1	D, E and F B and C D, E and F

Table 6.4 Limitations and Additional Requirements for Systems with Structural Irregularities (12.3.3)

SDC	Irregularity Type	Limitations/Additional Requirements
D	Vertical irregularity Type 5b	Not permitted
E or F	• Horizontal irregularity Type 1b • Vertical irregularities Type 1b, 5a or 5b	
B – F	Vertical irregularity Type 5b	Height limited to two stories or 30 ft*
B – F	Horizontal irregularity Type 4 or vertical irregularity Type 4	Design supporting members and their connections to resist load combinations with overstrength factor of 12.4.3.2
D – F	• Horizontal irregularities Type 1a, 1b, 2, 3 or 4 • Vertical irregularity Type 4	Design forces determined by 12.8.1 shall be increased by 1.25 for connections of diaphragms to vertical elements and to collectors and for connections of collectors to vertical elements.**

* The limit does not apply where the weak story is capable of resisting a total seismic force equal to Ω_o times the design force prescribed in 12.8.

** Collectors and their connections must also be designed for these increased forces unless they are designed for the load combinations with overstrength factor of 12.4.3.2 in accordance with 12.10.2.1.

Redundancy

The redundancy factor ρ is a measure of the redundancy inherent in the seismic force-resisting system. The value of this factor is either 1.0 or 1.3. Conditions where ρ is equal to 1.0 are given in 12.3.4. The redundancy factor is 1.0 for structures assigned to SDC B or C.

In essence, the redundancy factor has the effect of reducing the response modification coefficient R for less redundant structures, which, in turn, increases the seismic demand on the system.

6.3.4 Seismic Load Effects and Combinations

The seismic load effect E that is used in the load combinations defined in 2.3 and 2.4 consists of effects of horizontal ($E_h = \rho Q_E$) and vertical ($E_v = 0.2 S_{DS} D$) seismic forces where Q_E are the effects (bending moments, shear forces, axial forces) on the structural members obtained from the structural analysis where the base shear V is distributed over the height of the structure (12.4).

Basic load combinations for strength design and allowable stress design are summarized in 12.4.2.3 and Table 6.5. Additional information on these load combinations can be found in Chapter 2 of this publication.

Table 6.5 Seismic Load Combinations (12.4.2.3)

Combination No.	Combination
Strength Design*	
5	$(1.2D + 0.2S_{DS})D + \rho Q_E + L + 0.2S$
7	$(0.9 - 0.2S_{DS})D + \rho Q_E + 1.6H$
Allowable Stress Design	
5	$(1.0 + 0.14S_{DS})D + H + F + 0.7\rho Q_E$
6	$(1.0 + 0.105S_{DS})D + H + F + 0.525\rho Q_E + 0.75L + 0.75(L_r \text{ or } S \text{ or } R)$
8	$(0.6 - 0.14S_{DS})D + 0.7\rho Q_E + H$

* Load factor on L in combination 5 is permitted to equal 0.5 for all occupancies in which L_o in Table 4-1 is less than or equal to 100 psf, with the exception of garages or areas occupied as places of public assembly. The load factor on H in combination 7 shall be set equal to zero if the structural action due to H counteracts that due to E. Where lateral earth pressure provides resistance to structural actions from other forces, it shall not be included in H but shall be included in the design resistance.

The seismic load effect including overstrength factor E_m that is used in the load combinations defined in 12.4.3.2 consists of the horizontal seismic load effect with overstrength factor ($E_{mh} = \Omega_o Q_E$) and the effects of vertical ($E_v = 0.2S_{DS}D$) seismic forces where the overstrength factor Ω_o is given in Table 12.2-1 as a function of the seismic force-resisting system (12.4.3).

Basic combinations for strength design with overstrength factor and allowable stress design with overstrength factor are summarized in 12.4.3.2 and Table 6.6. These combinations pertain only to those structural elements that are listed in 12.3.3.3 (elements supporting discontinuous walls or frames in structures assigned to SDC B through F) and 12.10.2.1 (collector elements, splices and their connections to resisting elements in structures assigned to SDC C through F). See Chapter 2 of this publication for additional information on these load combinations.

6.3.5 Direction of Loading

According to 12.5.1, seismic forces must be applied to the structure in directions that produce the most critical load effects on the structural members. The requirements are based on the SDC and are summarized in Table 6.7.

Table 6.6 Seismic Load Combinations with Overstrength Factor (12.4.3.2)

Combination No.	Combination
Strength Design*	
5	$(1.2D + 0.2S_{DS})D + \Omega_o Q_E + L + 0.2S$
7	$(0.9 - 0.2S_{DS})D + \Omega_o Q_E + 1.6H$
Allowable Stress Design	
5	$(1.0 + 0.14S_{DS})D + H + F + 0.7\Omega_o Q_E$
6	$(1.0 + 0.105S_{DS})D + H + F + 0.525\Omega_o Q_E + 0.75L + 0.75(L_r \text{ or } S \text{ or } R)$
8	$(0.6 - 0.14S_{DS})D + 0.7\Omega_o Q_E + H$

* Load factor on L in combination 5 is permitted to equal 0.5 for all occupancies in which L_o in Table 4-1 is less than or equal to 100 psf, with the exception of garages or areas occupied as places of public assembly. The load factor on H in combination 7 shall be set equal to zero if the structural action due to H counteracts that due to E. Where lateral earth pressure provides resistance to structural actions from other forces, it shall not be included in H but shall be included in the design resistance.

Table 6.7 Direction of Loading Requirements (12.5)

SDC	Requirement
B	Design seismic forces applied independently in each of two orthogonal directions and orthogonal interaction effects are permitted to be neglected.
C	• Conform to requirements of SDC B • Structures with horizontal irregularity Type 5: – *Orthogonal combination procedure*: Apply 100 percent of the seismic forces for one direction plus 30 percent of the seismic forces for the perpendicular direction on the structure simultaneously where the forces are computed in accordance with 12.8 (equivalent lateral force analysis procedure), 12.9 (modal response spectrum analysis procedure) or 16.1 (linear response history procedure), or – *Simultaneous application of orthogonal ground motion*: Apply orthogonal pairs of ground motion acceleration histories simultaneously to the structure using 16.1 (linear response history procedure) or 16.2 (nonlinear response history procedure).
D – F	• Conform to requirements of SDC C • Any column or wall that forms part of two or more intersecting seismic force-resisting systems that is subjected to axial load due to seismic forces along either principal axis that is greater than or equal to 20 percent of the axial design strength of the column or wall must be designed for the most critical load effect due to application of seismic forces in any direction.*

* Either of the procedures of 12.5.3 a or b are permitted to be used to satisfy this requirement.

6.3.6 Analysis Procedure Selection

Requirements on the type of procedure that can be used to analyze the structure for seismic loads are given in 12.6 and are summarized in Table 12.6-1. As noted previously, the permitted analytical procedures depend on the SDC, the occupancy of the structure, characteristics of the structure (height and period) and the presence of any structural irregularities.

6.3.7 Modeling Criteria

Requirements pertaining to the construction of an adequate model for the purposes of seismic load analysis are given in 12.7. It is permitted to assume that the base of the structure is fixed. However, if the flexibility of the foundation is considered, the requirements of 12.13.3 (foundation load-deformation characteristics) or Chapter 19 (soil structure interaction for seismic design) must be satisfied.

A three-dimensional analysis is required for structures that have horizontal irregularity Types 1a, 1b, 4 or 5. In such cases, a minimum of three dynamic degrees of freedom (translation in two orthogonal directions and torsional rotation about the vertical axis) must be included at each level of the structure.

Cracked section properties must be used when analyzing concrete and masonry structures. In steel moment frames, the contribution of panel zone deformations to overall story drift must be included.

The definition of the effective seismic weight W that is used in determining the base shear V is given in 12.7.2. In addition to the dead load of the structure, a portion of the storage live load, the partition load, the weight of permanent operating equipment and a portion of the snow load, where applicable, must be included in W.

6.3.8 Equivalent Lateral Force Procedure

The provisions of the Equivalent Lateral Force Procedure are contained in 12.8. This analysis procedure can be used for all structures assigned to SDC B and C as well as some types of structures assigned to SDC D, E and F (see Table 12.6-1).

Seismic Base Shear, V

In short, the seismic base shear V is determined as a function of the design response accelerations S_{DS} and S_{D1}, the response modification coefficient R, the importance factor I, the fundamental period of the structure T and the effective seismic weight W. The design spectrum defined by the equations in 12.8 is depicted in Figure 6.2. The long-period transition period T_L is given in Figures 22-15 through 22-20.

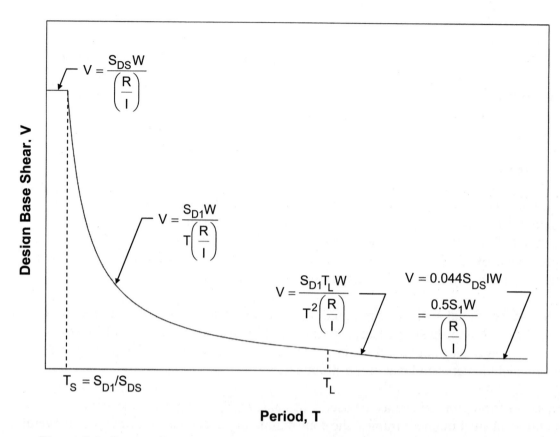

Figure 6.2 Design Response Spectrum According to the Equivalent Lateral Force Procedure (12.8)

Vertical Distribution of Seismic Forces

Once the seismic base shear V has been determined, it is distributed over the height of the building in accordance with 12.8.3. For structures with a fundamental period less than or equal to 0.5 sec, V is distributed linearly over the height, varying from zero at the base to a maximum value at the top (see Figure 6.3a). When T is greater than 2.5 sec, a parabolic distribution is to be used (see Figure 6.3b). For a period between these two values, a linear interpolation between a linear and parabolic distribution is permitted or a parabolic distribution may be utilized.

Horizontal Distribution of Forces

The seismic design story shear V_x in story x is the sum of the lateral forces acting at the floor or roof level supported by that story and all of the floor levels above, including the roof. The story shear is distributed to the vertical elements of the seismic force-resisting system in the story based on the lateral stiffness of the diaphragm.

For flexible diaphragms, V_x is distributed to the vertical elements of the seismic force-resisting system based on the area of the diaphragm tributary to each line of resistance.

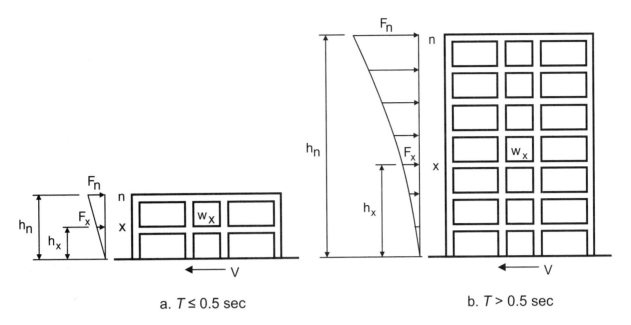

Figure 6.3 Vertical Distribution of Seismic Forces (12.8.3)

For diaphragms that are not flexible, V_x is distributed based on the relative stiffness of the vertical resisting elements and the diaphragm. Inherent and accidental torsion must be considered in the overall distribution (12.8.4.1 and 12.8.4.2). Where Type 1a or 1b irregularity is present in structures assigned to SDC C, D, E or F, the accidental torsional moment is to be amplified in accordance with 12.8.4.3 (see Figure 12.8-1).

Overturning

The structure must be designed to resist the overturning effects caused by the seismic forces. In such cases, the critical load combinations are typically those where the effects from gravity and seismic loads counteract.

Story Drift Determination

Design story drift Δ is determined in accordance with 12.8.6 and is computed as the difference of the deflections δ_x at the center of mass of the diaphragms at the top and bottom of the story under consideration (see Figure 12.8-2).

The deflections δ_x at each floor level are obtained by multiplying the deflections δ_{xe} (the deflections determined by an elastic analysis using the code-prescribed forces applied at each floor level) by the deflection amplification factor C_d in Table 12.2-1 and dividing by the importance factor I (see Eq. 12.8-15). Limits on the design story drifts are given in 12.12 and are covered in Section 6.3.12 of this publication.

P-Delta Effects

Member forces and story drifts induced by P-delta effects must be considered in member design and in the evaluation of overall stability of a structure where such effects are significant. Equation 12.8-16 can be used to evaluate the need to consider P-delta effects. Equation 12.8-17 is used to check if the structure is potentially unstable; this equation must be satisfied even where computer software is utilized to determine second-order effects.

6.3.9 Modal Response Spectral Analysis

Requirements on how to conduct a modal analysis of a structure are given in 12.9. Such an analysis is permitted for any structure assigned to SDC B through F and is required for regular and irregular structures where $T \geq 3.5\ T_S$ and certain types of irregular structures assigned to SDC D through F (see Table 12.6-1). The number of modes that need to be considered, methods on how to combine the results from the different modes and provisions on scaling the design values of the combined response are contained in 12.9.

6.3.10 Diaphragms, Chords, and Collectors

Diaphragm Design Forces

In structures assigned to SDC B and higher, floor and roof diaphragms must be designed to resist seismic forces from base shear determined from the structural analysis or the force F_{px} determined by Eq. 12.10-1, whichever is greater. Upper and lower limits of F_{px} are provided in 12.10.1.1.

An example of an offset in the vertical resisting elements of the seismic force-resisting system is illustrated in Figure 6.4. In such cases, the diaphragm that is required to transfer design forces from the vertical elements above the diaphragm to the vertical elements below the diaphragm must be designed for this additional force.

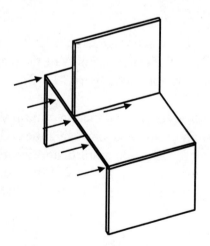

Figure 6.4 Example of Vertical Offsets in the Seismic Force-Resisting System (12.10.1)

The redundancy factor ρ applies to the design of diaphragms in structures assigned to SDC D, E and F; it is equal to 1.0 for inertial forces calculated by Eq 12.10-1. Where the diaphragm transfers design forces from the vertical elements above the diaphragm to the vertical elements below the diaphragm, the redundancy factor ρ for the structure applies to these forces. Also, the requirements of 12.3.3.4 must be satisfied for structures having horizontal or vertical irregularities indicated in that section.

Collector Elements

The provisions of 12.10.2.1 apply to collector elements in structures assigned to SDC C and higher. Collectors, which are also commonly referred to as drag struts, are elements in a structure that are used to transfer diaphragm loads from the diaphragm to the elements of the lateral force-resisting system where the lengths of the vertical elements in the lateral force-resisting system are less than the length of the diaphragm at that location. For example, the collector beams in Figure 6.5 collect the force from the diaphragm and distribute it to the shear wall.

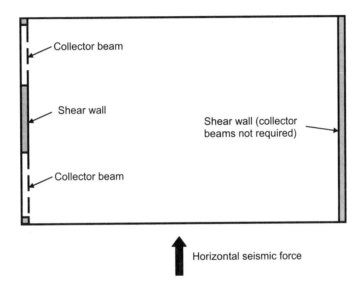

Figure 6.5 Example of Collector Beams and Shear Walls

In general, collector elements, splices and connections to resisting elements must all be designed to resist the load combinations with overstrength factor of 12.4.3.2 in addition to the applicable strength design or allowable stress design load combinations. Structures or portions of structures braced entirely by light-frame shear walls, collector elements, splices and connections need only be designed to resist the diaphragm forces prescribed in 12.10.1.1.

6.3.11 Structural Walls and Their Anchorage

Out-of-Plane Forces

In addition to forces in the plane of the wall, structural walls and their anchorage in structures assigned to SDC B and higher must be designed for an out-of-plane force equal to $0.4S_{DS}I$ times the weight of the structural wall or 0.10 times the weight of the structural wall, whichever is greater (12.11.1).

Anchorage of Concrete or Masonry Structural Walls

Concrete or masonry structural walls in structures assigned to SDC B and higher must be anchored to the roof and floor members that provide lateral support for the walls. The anchorage must be designed to resist the greater of three forces specified in 12.11.2. Where the anchor spacing is greater than 4 ft, the walls must be designed to resist bending between the anchors.

The requirements of 12.11.2.1 must be satisfied for concrete or masonry walls anchored to flexible diaphragms. Additional requirements for diaphragms in structures assigned to SDC C through F are given in 12.11.2.2.

6.3.12 Drift and Deformation

Once the design drifts have been determined in accordance with 12.8.6, they are compared to the allowable story drift Δ_a in Table 12.12-1. The drift limits depend on the occupancy category and are generally more restrictive for categories III and IV. These lower limits on drift are meant to provide a higher level of performance for more important occupancies. Drift limits also depend on the type of structure.

Provisions for moment frames in structures assigned to SDC D through F, diaphragm deflection, and building separation are contained in 12.12.1.1, 12.12.2 and 12.12.3, respectively.[7]

For structures assigned to SDC D through F, structural members that are not part of the seismic force-resisting system must be designed for deformation compatibility. These members must be designed to adequately resist applicable gravity load effects when subjected to the design story drift Δ determined in accordance with 12.8.6. The deformation compatibility requirements are given in 12.12.4.

6.3.13 Foundation Design

General requirements for the design of various types of foundation systems are given in 12.13. The special seismic requirements are driven by SDC.

---

[7] The requirements in ASCE/SEI 7 apply to structural separations. IBC 1613.6.7 provides requirements for separation between adjacent buildings on the same property and buildings adjacent to a property line not on a public way in SDC D, E or F. The minimum separation provisions were in the 1997 UBC and the 2000 and the 2003 editions of the IBC but not the 2006 IBC because it references ASCE/SEI 7 for seismic design requirements.

SEISMIC DESIGN REQUIREMENTS FOR BUILDING STRUCTURES 6-21

For other than cantilever column and inverted pendulum systems, overturning effects at the soil-foundation interface are permitted to be reduced by 25 percent where the equivalent lateral force procedure is used. A 10-percent reduction is permitted for foundations of structures designed in accordance with the modal analysis requirements of 12.9.

6.3.14 Simplified Alternative Structural Design Criteria for Simple Bearing Wall or Building Frame Systems

The simplified design procedure is entirely self-contained in 12.14. The simplified method is permitted to be used in lieu of other analytical procedures in Chapter 12 for the analysis and design of simple bearing wall or building frame systems provided the 12 limitations of 12.14.1.1 are satisfied.

The provisions of 12.14 are intended to apply to a defined set of essentially regularly-configured structures where a reduction in requirements is deemed to be warranted. Only those systems specifically listed in Table 12.14-1 are permitted to be used. Note that drift controlled structural systems such as moment resisting frames are not permitted. Some of the more noteworthy characteristics of this simplified method are as follows:

1. The simplified procedure applies to Occupancy Category I or II structures up to three stories in height, which are founded on Site Class A, B, C or D soils and which are assigned to SDC B, C, D or E.

2. The procedure is limited to bearing wall or building frame systems. Design coefficients for these structural systems are given in Table 12.14-1, which contains essentially the same information as Table 12.2-1, except that the system overstrength factor Ω_O, the deflection amplification factor C_d, and system height limitations have been omitted in Table 12.14-1, based on the limitations and requirements of the simplified method. For seismic load combinations involving overstrength, the system overstrength factor is 2.5 (12.14.3.2). Since the permissible types of lateral force-resisting systems are relatively stiff when compared to moment-resisting frames in structures within the prescribed limitations, design drift need not be calculated (12.14.8.5).

3. Given the prescriptive requirements for system configuration, definitions of and design provisions for system irregularities are not needed.

4. Design and detailing requirements are independent of the SDC.

5. The redundancy coefficient ρ does not apply.

6. The seismic base shear V is determined by Eq. 12.14-11 and is a function of the design spectral response acceleration at short periods S_{DS} (short period plateau), the number of stories in the structure and the response modification coefficient R. Determination of the period of the structure T is not needed.

7. Vertical distribution of the seismic base shear V is based on tributary weight (Eq. 12.14-12). Diaphragms must be designed to resist the design seismic forces calculated by the provisions of 12.14.8.2 (12.14.7.4).

8. For flexible diaphragms, horizontal distribution of seismic forces to the vertical elements of the seismic force-resisting system is based on tributary area. For diaphragms that are not flexible, distribution is based on the relative stiffness of the seismic force-resisting elements considering any torsional moment resulting from eccentricity between the center of mass and the center of rigidity (12.14.8.3). Accidental torsion and dynamic amplification of torsion need not be considered.

9. Calculations for P-delta effects need not be considered.

10. Load combinations are prescribed in 12.14.3 for strength design, allowable stress design and seismic load effect including a 2.5 overstrength factor.

11. Design seismic forces are permitted to be applied separately in each orthogonal direction and the combination of effects from the two directions need not be considered (12.14.6).

Additional requirements of the simplified method can be found in 12.14.

6.4 SEISMIC DESIGN REQUIREMENTS FOR NONSTRUCTURAL COMPONENTS

6.4.1 General

Chapter 13 establishes minimum design criteria for nonstructural components that are permanently attached to structures and for their supports and attachments. Included are provisions for architectural components, mechanical and electrical components, and anchorage.

Nonstructural components that weigh greater than or equal to 25 percent of the effective seismic weight W of the structure must be designed as nonbuilding structures in accordance with 15.3.2 (13.1.1). Nonstructural components are assigned to the same SDC as the structure that they occupy or to which they are attached (13.1.2).

The component importance factor I_p is equal to 1.0 except when the following conditions apply where it is equal to 1.5 (13.1.3):

1. The component is required to function for life-safety purposes after an earthquake (including fire protection sprinkler systems).

2. The component contains hazardous materials.

3. The component is in or attached to an Occupancy Category IV structure and is needed for continued operation of the facility or its failure could impair the continued operation of the facility.

A list of nonstructural components that are exempt from the requirements of this section is given in 13.1.4. Also see Table 13.2-1 for a summary of applicable requirements.

6.4.2 Seismic Demands on Nonstructural Components

The horizontal seismic design force that is to be applied to the center of gravity of the component F_p is determined by Eq. 13.3-1. This force is a function of the following:

a_p = component amplification factor given in Table 13.5-1 for architectural components and 13.6-1 for mechanical and electrical components.

S_{DS} = design spectral response acceleration at short periods determined in accordance with 11.4.4.

W_p = component operating weight.

R_p = component response modification factor given in Table 13.5-1 for architectural components and 13.6-1 for mechanical and electrical components.

I_p = component importance factor (1.0 or 1.5).

z = height in structure of point of attachment of component with respect to the base. For items at or below the base, $z = 0$. The value of z/h need not exceed 1.0.

h = average roof height of structure with respect to the base.

Component seismic forces F_p are to be applied independently in at least two orthogonal horizontal directions in combination with the service loads that are associated with the component. Requirements are also given for vertically cantilevered systems. Minimum and maximum values of F_p are determined by Eqs. 13.3-2 and 13.3-3, respectively.

Equation 13.3-4 can be used to calculate F_p based on accelerations determined by a modal analysis. Requirements for seismic relative displacements are given in 13.3.2.

6.4.3 Nonstructural Component Anchorage

Components and their supports are to be attached or anchored to the structure in accordance with the requirements of 13.4. Forces in the attachments are to be determined using the prescribed forces and displacements specified in 13.3.1 and 13.3.2. Anchors in concrete or masonry elements must satisfy the provisions of 13.4.2.

6.4.4 Architectural Components

General design and detailing requirements for architectural components are contained in 13.5. All architectural components and attachments must be designed for the seismic forces defined in 13.3.1. Specific requirements are stipulated for:

- Exterior nonstructural wall panels
- Glass
- Suspended ceilings

- Access floors
- Partitions
- Glass in glazed curtain walls, glazed storefronts and glazed partitions

6.4.5 Mechanical and Electrical Components

The requirements of 13.6 are to be satisfied for mechanical and electrical components and their supports.

Equation 13.6-1 can be used to determine the fundamental period T_p of the mechanical or electrical component.

Requirements are provided for the following systems:

- Utility and service lines
- HVAC ductwork
- Piping systems
- Boilers and pressure vessels
- Elevators and escalators

6.5 SEISMIC DESIGN REQUIREMENTS FOR NONBUILDING STRUCTURES

6.5.1 General

Nonbuilding structures supported by the ground or by other structures must be designed and detailed to resist the minimum seismic forces set forth in Chapter 15.

The selection of a structural analysis procedure for a nonbuilding structure is based on its similarity to buildings. Nonbuilding structures that are similar to buildings exhibit behavior similar to that of building structures; however, their function and performance are different. According to 15.1.3, structural analysis procedures for such buildings are to be selected in accordance with 12.6 and Table 12.6-1, which are applicable to building structures. Guidelines and recommendations on the use of these methods are given in C15.1.3. In short, the provisions for building structures need to be carefully examined before they are applied to nonbuilding structures.

Nonbuilding structures that are not similar to buildings exhibit behavior that is markedly different than that of building structures. Most of these types of structures have reference documents that address their unique structural performance and behavior. Such reference documents are permitted to be used to analyze the structure (15.1.3). In addition, the following procedures may be used: equivalent lateral force procedure (12.8), modal analysis procedure (12.9), linear response history analysis procedure (16.1) and nonlinear response history analysis procedure (16.2). In the case of nonbuilding structures similar to buildings, guidelines and recommendations on the proper analysis method to utilize for nonbuilding structures that are not similar to buildings are given in C15.1.3.

6.5.2 Reference Documents

As noted above, reference documents may be used to design nonbuilding structures for earthquake load effects. References that have seismic requirements based on the same force and displacement levels used in ASCE/SEI 7-05 are listed in Chapter 23 (15.2). The provisions in the reference documents are subject to the amendments given in 15.4.1. See C15.2 for additional references that cannot be referenced directly by ASCE/SEI 7-05.

It is important to note that the provisions of an industry standard or document must not be used unless the seismic ground accelerations and seismic coefficients are in conformance with the requirements of 11.4. The values for total lateral force and total base overturning moment from the reference documents must be taken greater than or equal to 80 percent of the corresponding values determined by the seismic provisions of ASCE/SEI 7-05.

6.5.3 Nonbuilding Structures Supported by Other Structures

Provisions are given in 15.3 for the nonbuilding structures in Table 15.4-2 (i.e., nonbuilding structures that are not similar to buildings) that are supported by other structures and that are not part of the primary seismic force-resisting system. The design method depends on the weight of the nonbuilding structure relative to the weight of the combined nonbuilding and supporting structure (see 15.3.1 and 15.3.2).

6.5.4 Structural Design Requirements

Specific design requirements for nonbuilding structures are given in 15.4. As noted previously, provisions in referenced documents are amended by the requirements of this section.

For nonbuilding structures that are similar to buildings, the permitted structural systems, design values and limitations are given in Table 15.4-1. Similar information is provided in Table 15.4-2 for nonbuilding structures that are not similar to buildings. Requirements on the determination of the base shear, vertical distribution of seismic forces, importance factor and load combinations are contained in 15.4.1.

Additional provisions for rigid buildings, loads, fundamental period, drift limitations, deflection limits and other requirements can be found in 15.4.2 through 15.4.8.

6.5.5 Nonbuilding Structures Similar to Buildings

Additional requirements are given in 15.5 for pipe racks, steel storage racks, electrical power generating facilities, structural towers for tanks and vessels, and piers and wharves.

6.5.6 Nonbuilding Structures Not Similar to Buildings

Additional requirements are given in 15.6 for earth-retaining structures, stacks and chimneys, amusement structures, special hydraulic structures, secondary containment systems and telecommunication towers.

6.5.7 Tanks and Vessels

Comprehensive seismic design requirements are given in 15.7 for tanks and vessels. As noted in C15.7, most, if not all, industry standards that contain seismic design requirements are based on earlier seismic codes. Many of the provisions of 15.7 show how to modify existing standards to get to the same force levels as ASCE/SEI 7-05.

6.6 FLOWCHARTS

A summary of the flowcharts provided in this chapter is given in Table 6.7. Included is a description of the content of each flowchart.

All referenced section numbers and equations in the flowcharts are from ASCE/SEI 7-05 unless noted otherwise.

Table 6.7 Summary of Flowcharts Provided in Chapter 6

Flowchart	Title	Description
Section 6.6.1 Seismic Design Criteria		
Flowchart 6.1	Consideration of Seismic Design Requirements	Summarizes conditions where the seismic requirements of ASCE/SEI 7-05 must be considered and need not be considered.
Flowchart 6.2	Seismic Ground Motion Values	Provides step-by-step procedure on how to determine the design spectral accelerations for a site in accordance with 11.4.
Flowchart 6.3	Site Classification Procedure for Seismic Design	Provides step-by-step procedure on how to determine the Site Class of a site in accordance with Chapter 20.
Flowchart 6.4	Seismic Design Category	Provides step-by-step procedure on how to determine the seismic design category of a structure.
Flowchart 6.5	Design Requirements for SDC A	Summarizes the seismic design requirements for structures assigned to SDC A.

(continued)

Table 6.7 Summary of Flowcharts Provided in Chapter 6 (continued)

Flowchart	Title	Description
Section 6.6.2 Seismic Design Requirements for Building Structures		
Flowchart 6.6	Diaphragm Flexibility	Provides methods on how to determine whether a diaphragm is flexible, rigid or semi-rigid.
Flowchart 6.7	Permitted Analytical Procedures	Summarizes analytical procedures that are permitted in determining design seismic forces for building structures.
Flowchart 6.8	Equivalent Lateral Force Procedure	Provides step-by-step procedure on how to determine the design seismic forces and their distribution based on the requirements of this procedure.
Flowchart 6.9	Alternate Simplified Design Procedure	Provides step-by-step procedure on how to determine the design seismic forces and their distribution based on the requirements of this procedure.
Section 6.6.3 Seismic Design Requirements for Nonstructural Components		
Flowchart 6.10	Seismic Demands on Nonstructural Components	Provides step-by-step procedure on how to determine design seismic forces on nonstructural components.
Section 6.6.4 Seismic Design Requirements for Nonbuilding Structures		
Flowchart 6.11	Seismic Design Requirements for Nonbuilding Structures	Provides step-by-step procedure on how to determine design seismic forces on nonbuilding structures that are similar to buildings and that are not similar to buildings.

6.6.1 Seismic Design Criteria

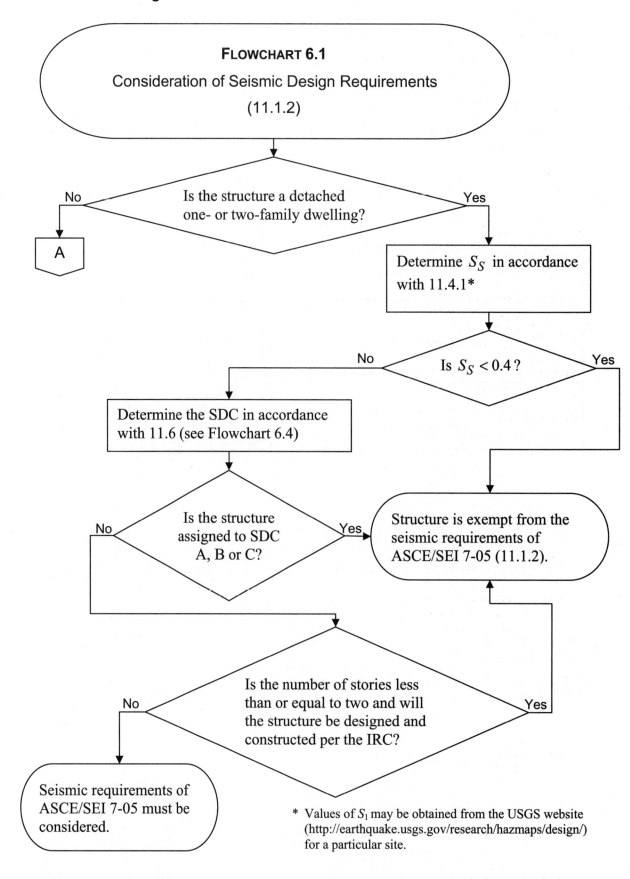

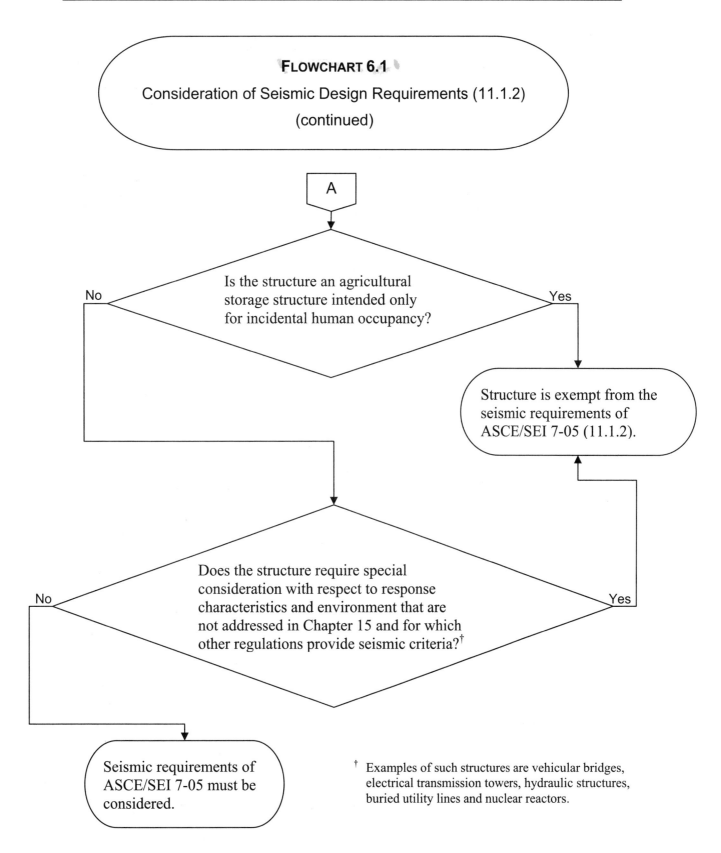

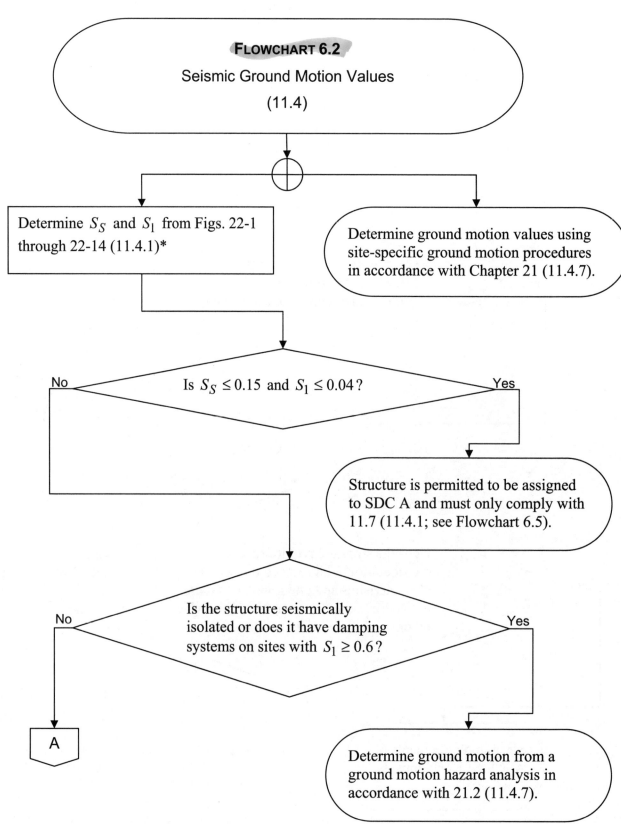

* Values of S_S and S_1 may be obtained from the USGS website (http://earthquake.usgs.gov/research/hazmaps/design/) for a particular site.

FLOWCHART 6.2

Seismic Ground Motion Values (11.4)

(continued)

Determine the site class of the soil in accordance with 11.4.2 and Chapter 20 (11.4.2; see Flowchart 6.3)

Is the site classified as Site Class F?

Yes → Determine ground motion values using a site response analysis in accordance with 21.1 (11.4.7).**

No ↓

Determine S_{MS} and S_{M1} by Eqs. 11.4-1 and 11.4-2, respectively:

$S_{MS} = F_a S_S$

$S_{M1} = F_v S_1$

where site coefficients F_a and F_v are given in Tables 11.4-1 and 11.4-2, respectively (11.4.3).†

Determine S_{DS} and S_{D1} by Eqs. 11.4-3 and 11.4-4, respectively:†

$S_{DS} = 2S_{MS}/3$

$S_{D1} = 2S_{M1}/3$ (11.4.4)

** A site response analysis in accordance with 21.1 is required for structures on Site Class F sites unless the exception in 20.3.1(1) is satisfied for structures with periods $T \leq 0.5$ sec.

† Where the simplified design procedure of 12.14 is used, only the values of F_a and S_{DS} must be determined in accordance with 12.14.8.1 (11.4.3, 11.4.4).

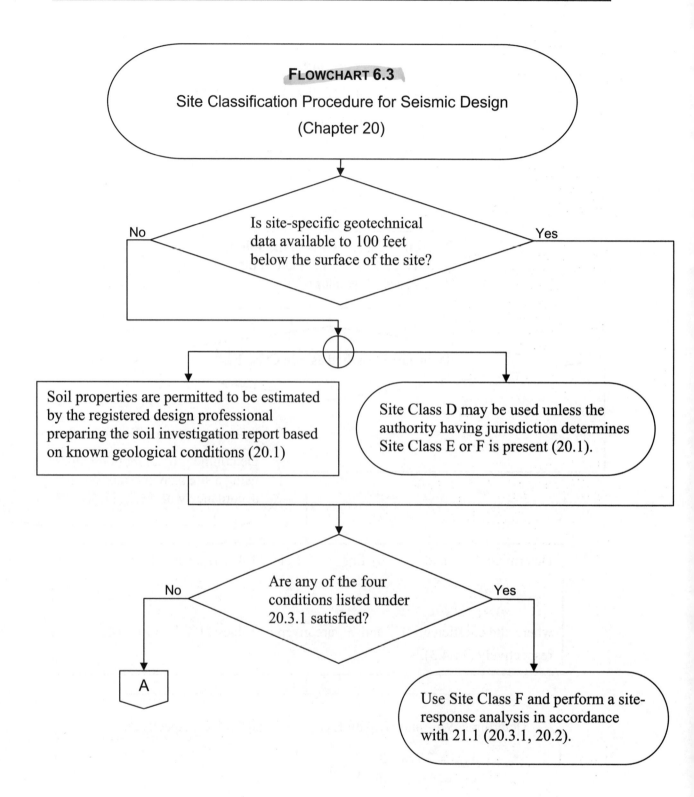

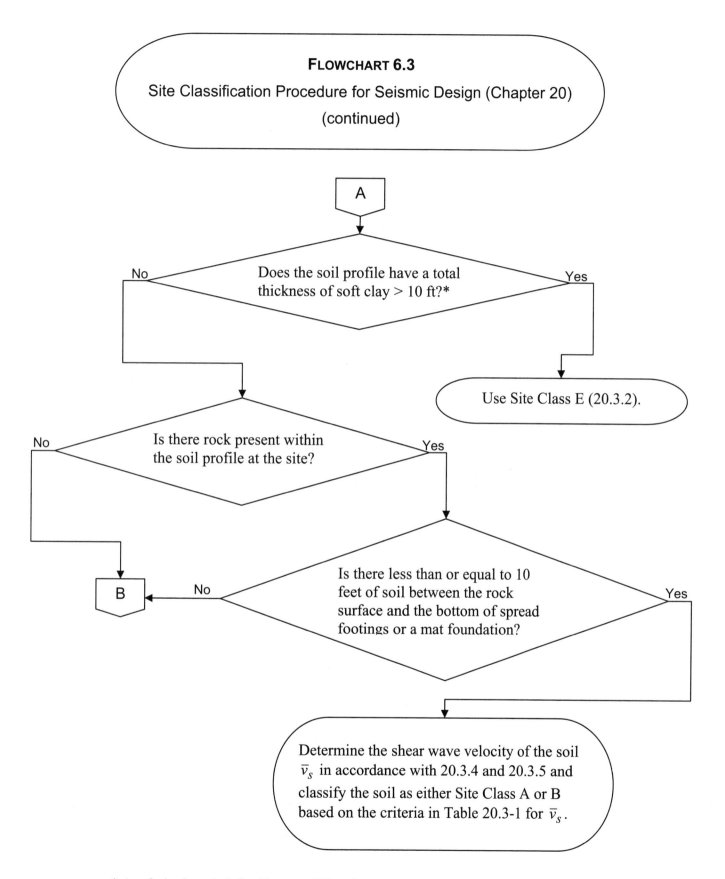

* A soft clay layer is defined by $s_u < 500$ psf, $w > 40$ percent, and $PI > 20$ (20.3.2).

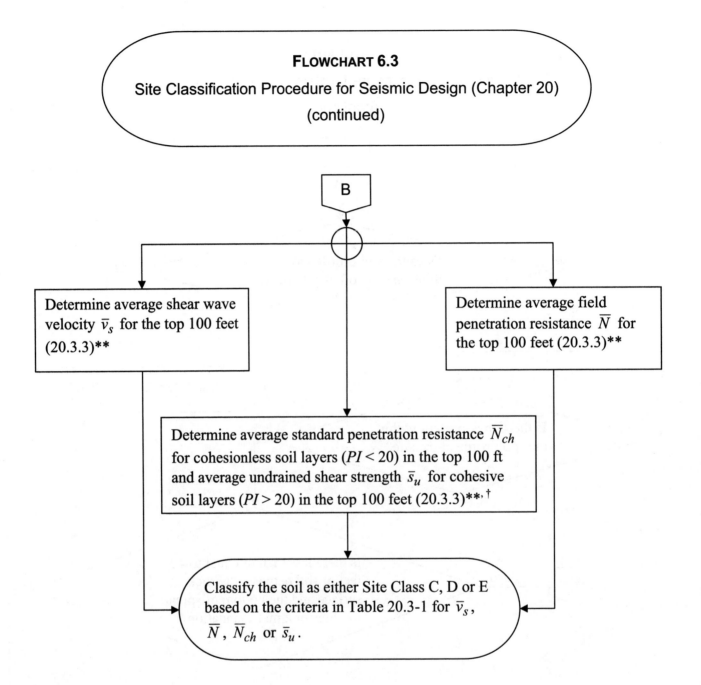

FLOWCHART 6.3

Site Classification Procedure for Seismic Design (Chapter 20) (continued)

** Values of $\bar{v}_s$, $\bar{N}$ and $\bar{s}_u$ are computed in accordance with 20.4 where soil profiles contain distinct soil and rock layers (20.3.3).

† Where the $\bar{N}_{ch}$ and $\bar{s}_u$ criteria differ, the site shall be assigned to the category with the softer soil [20.3.3(3)].

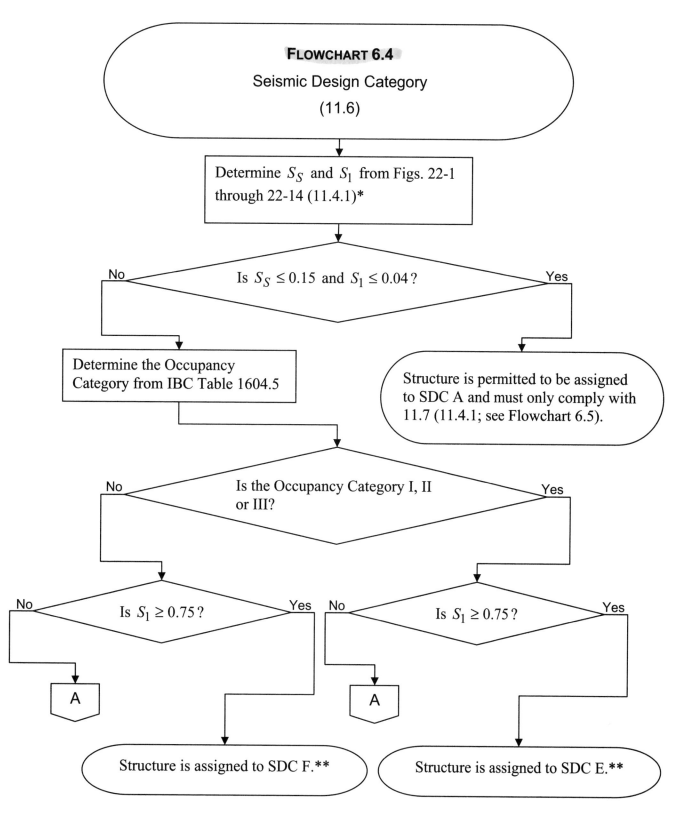

* Values of S_S and S_1 may be obtained from the USGS website (http://earthquake.usgs.gov/research/hazmaps/design/) for a particular site.

** A structure assigned to SDC E or F shall not be located where there is a known potential for an active fault to cause rupture of the ground surface at the structure (11.8).

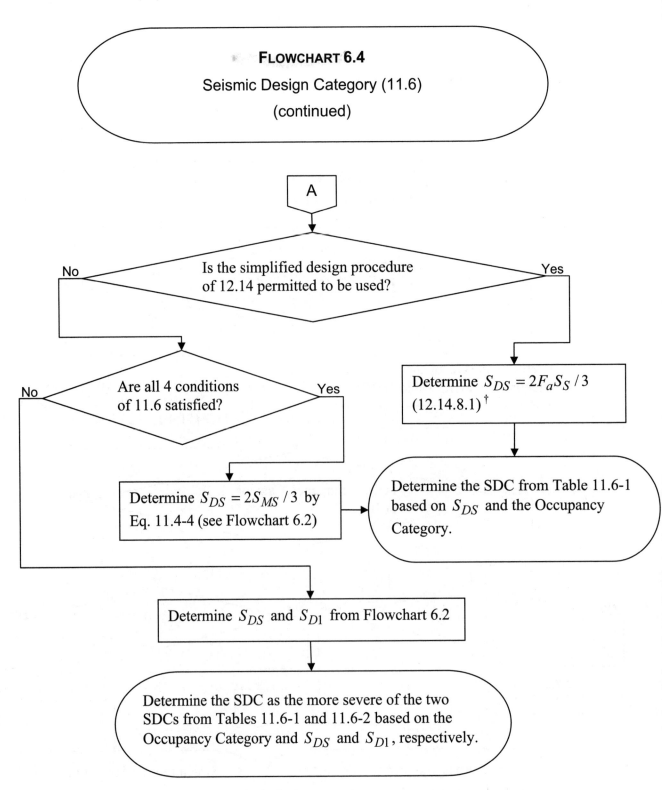

† Short-period site coefficient F_a is permitted to be taken as 1.0 for rock sites, 1.4 for soil sites or may be determined in accordance with 11.4.3. Rock sites have no more than 10 feet of soil between the rock surface and the bottom of spread footing or mat foundation. Mapped spectral response acceleration S_S is determined in accordance with 11.4.1 and need not be taken larger than 1.5 (12.14.8.1).

FLOWCHART 6.5

Design Requirements for SDC A
(11.7)

↓

Determine the seismic force F_x applied at each floor level by Eq. 11.7-1: $F_x = 0.01 w_x$ where w_x = portion of the total dead load of the structure located or assigned to level x (11.7.2)*

↓

Provide load path connections and connection to supports in accordance with 11.7.3 and 11.7.4, respectively.

↓

Anchor any concrete or masonry walls to roof and floors in accordance with 11.7.5.

---

* These forces are applied simultaneously at all levels in one direction. The structure is analyzed for the effects of these forces applied independently in each of two orthogonal directions (11.7.2). The effects from these forces on the structure and its components shall be taken as E and combined with the effects of other loads in accordance with 2.3 or 2.4 (11.7.1).

6.6.2 Seismic Design Requirements for Building Structures

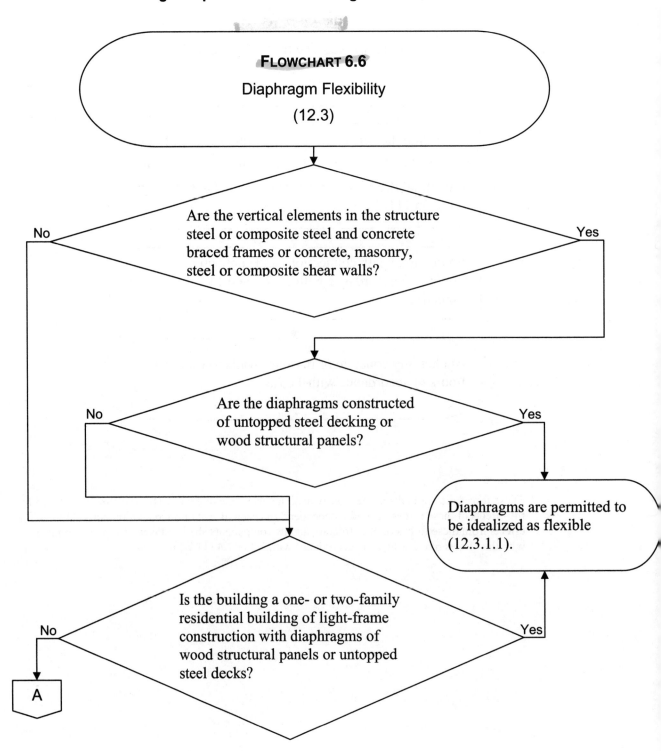

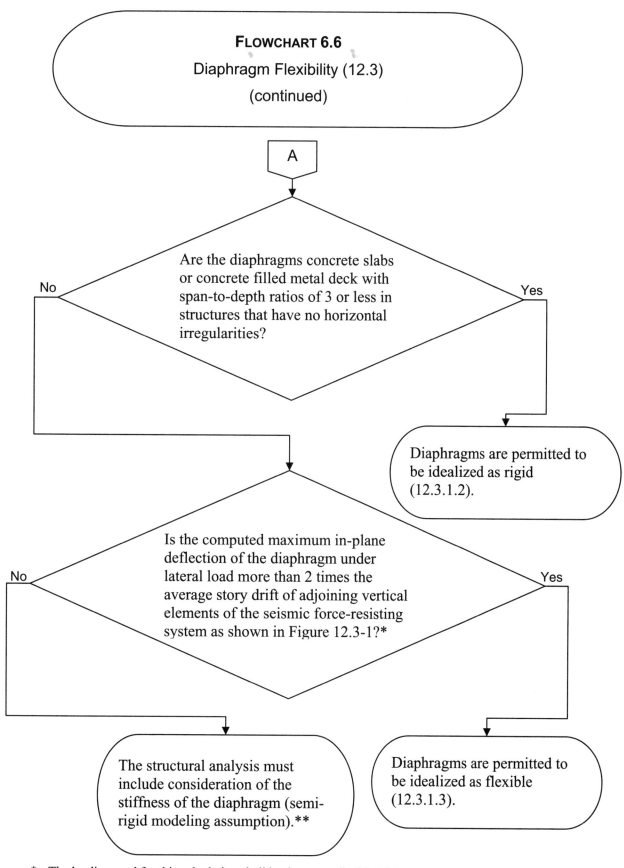

* The loading used for this calculation shall be that prescribed in 12.8.
** IBC 1602 provides definitions for rigid and flexible diaphragms only; semi-rigid diaphragms are not defined in the IBC. See IBC 1613.6.1 for modifications made to 12.3.1.

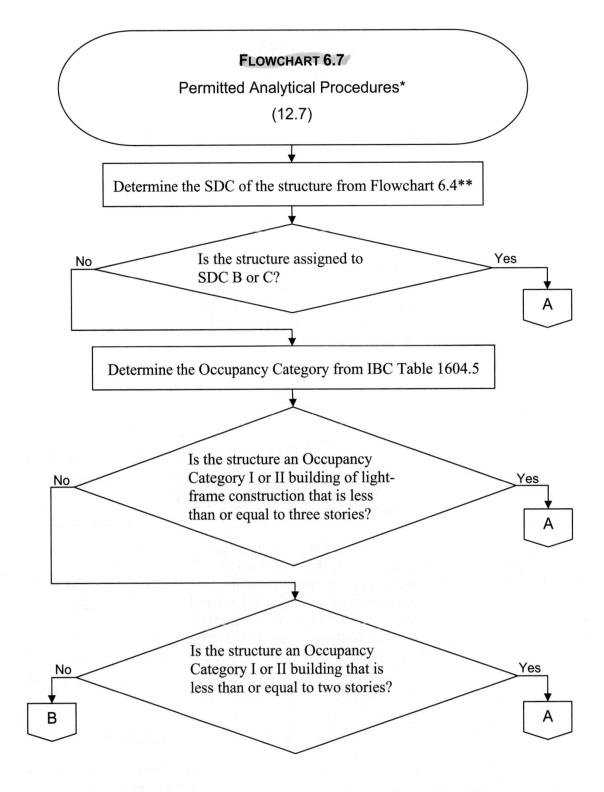

* The simplified alternative structural design method of 12.14 may be used for simple bearing wall or building frame systems that satisfy the 12 limitations in 12.14.1.1.

** This flowchart is applicable to buildings assigned to SDC B and higher. See Flowchart 6.5 for design requirements for SDC A.

FLOWCHART 6.7

Permitted Analytical Procedures (12.6)

(continued)

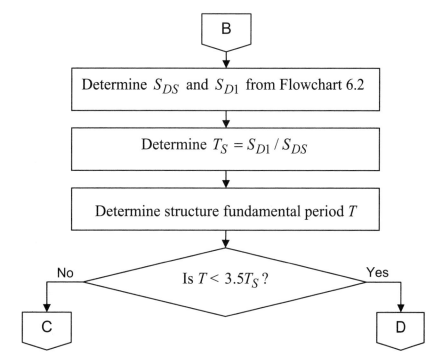

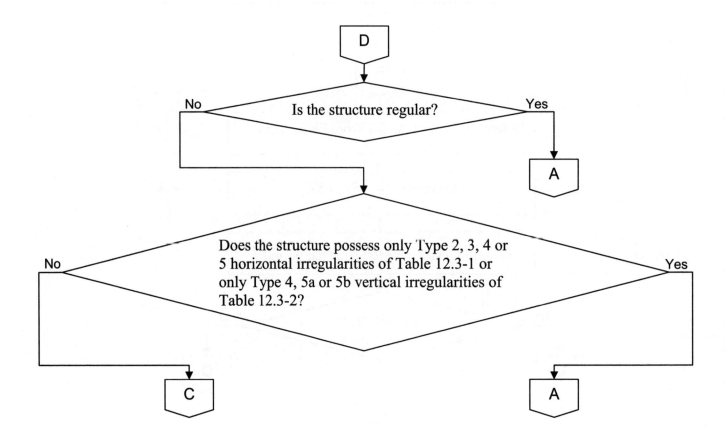

FLOWCHART 6.7
Permitted Analytical Procedures (12.6)
(continued)

FLOWCHART 6.8
Equivalent Lateral Force Procedure
(12.6)

↓

Determine S_S, S_1, S_{DS}, S_{D1} and the SDC from Flowchart 6.4

↓

Determine the response modification coefficient R from Table 12.2-1 for the appropriate structural system based on SDC

↓

Determine the importance factor I from Table 11.5-1 based on the Occupancy Category

↓

Left branch:

Determine the fundamental period of the structure T by a substantiated analysis that considers the structural properties and deformational characteristics of the structure

↓

Determine the approximate fundamental period of the structure T_a by Eq. 12.8-7: $T_a = C_t h_n^x$ where values of approximate period parameters C_t and x are given in Table 12.8-2*,**

↓

A

Right branch:

Determine the approximate fundamental period of the structure T_a by Eq. 12.8-7: $T_a = C_t h_n^x$ where values of approximate period parameters C_t and x are given in Table 12.8-2*,**

↓

B

---

* h_n height in feet above the base to the highest level of the structure.

** Alternate equations for T_a are given in 12.8.2.1 for concrete or steel moment resisting frames and masonry or concrete shear wall structures.

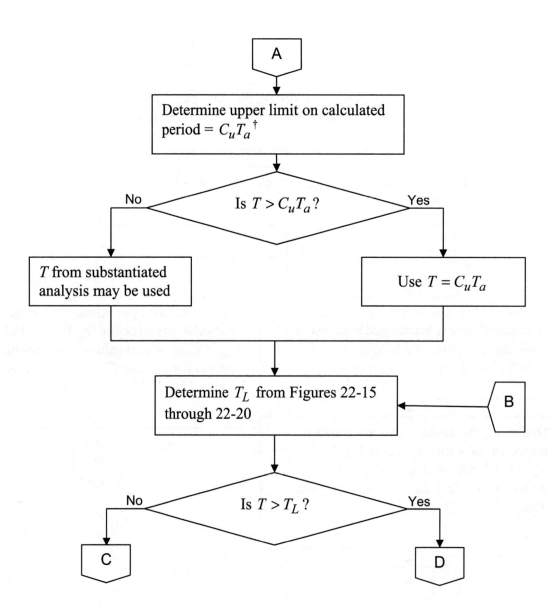

† C_u = coefficient for upper limit on calculated period given in Table 12.8-1.

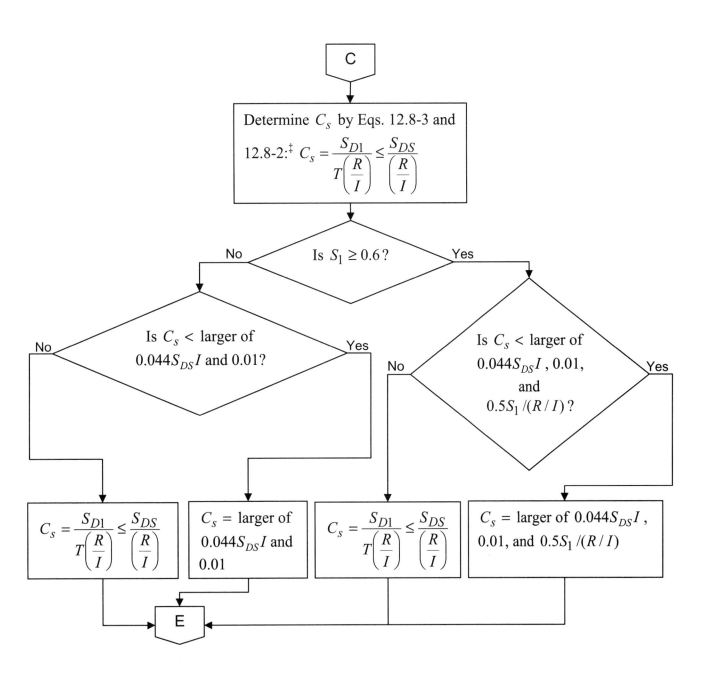

‡ For regular structures five stories or less in height and having a period T less than or equal to 0.5 sec, C_s is permitted to be calculated using a value of 1.5 for S_S (12.8.1.3).

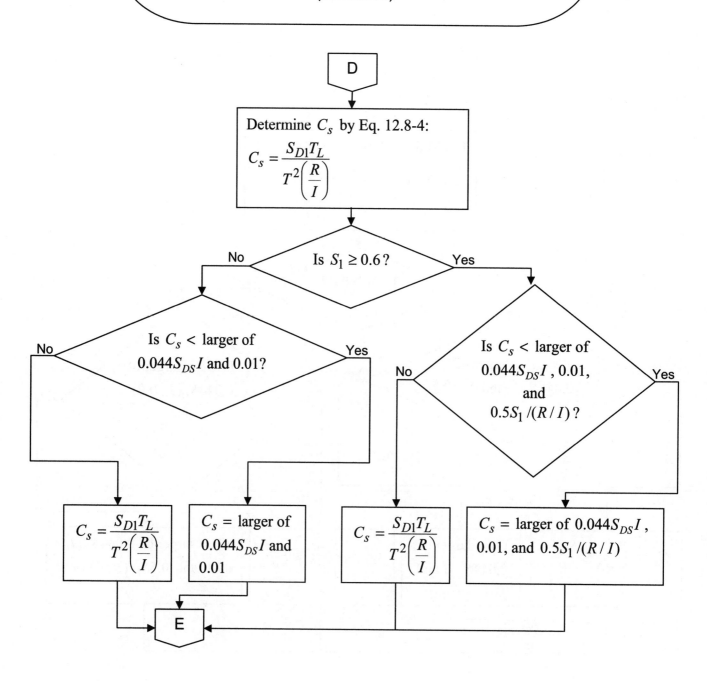

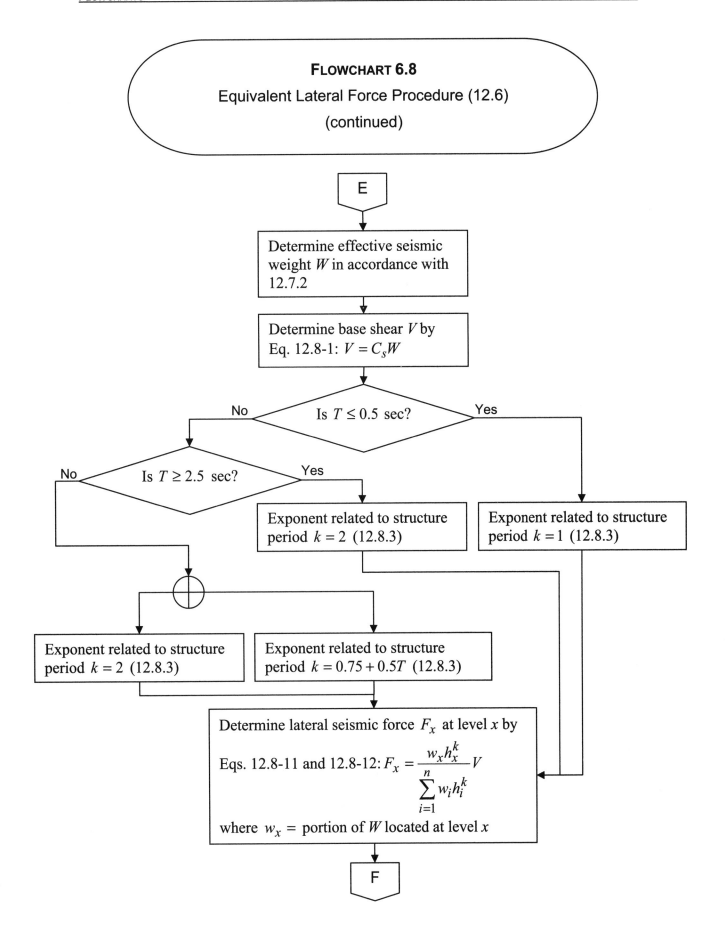

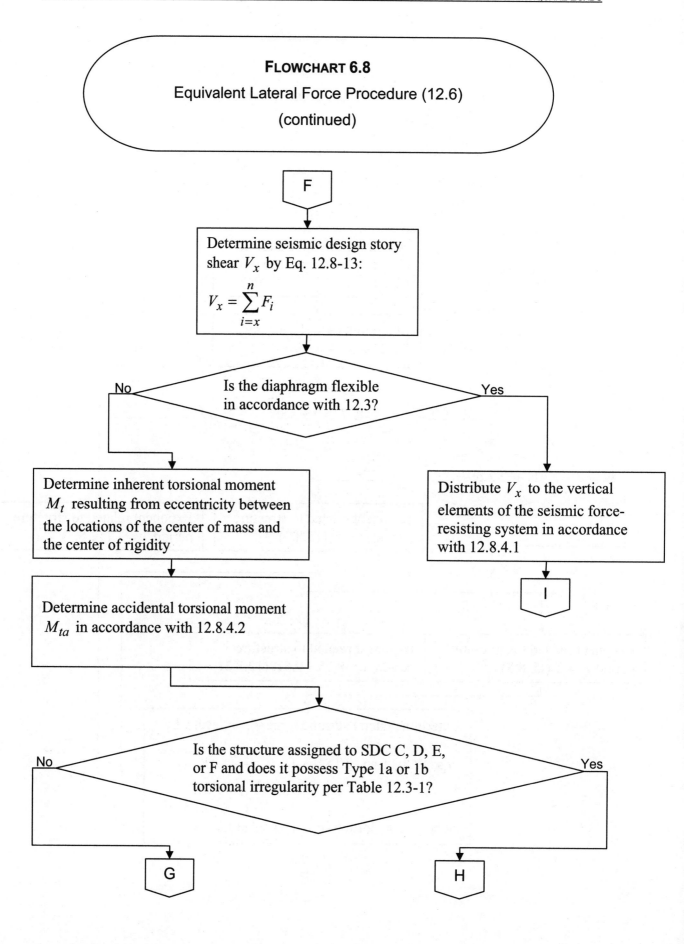

FLOWCHART 6.8

Equivalent Lateral Force Procedure (12.6)

(continued)

Distribute V_x to the vertical elements of the seismic force-resisting system considering the relative lateral stiffness of the vertical resisting elements and the diaphragm, including $M_t + M_{ta}$

Determine the torsional amplification factor A_x by Eq. 12.8-14: $A_x = \left(\dfrac{\delta_{max}}{1.2\delta_{avg}} \right)^2 \leq 3$

where δ_{max} and δ_{avg} are defined in 12.8.4.3

Determine $M'_{ta} = A_x M_{ta}$

Distribute V_x to the vertical elements of the seismic force-resisting system considering the relative lateral stiffness of the vertical resisting elements and the diaphragm, including $M_t + M'_{ta}$

Design structure to resist overturning effects caused by the seismic forces F_x (12.8.5)

Determine the deflection amplification factor C_d from Table 12.2-1

Determine the deflection δ_x at levels x by Eq. 12.8-15: $\delta_x = \dfrac{C_d \delta_{xe}}{I}$ where δ_{xe} are the deflections at level x based on an elastic analysis of the structure subjected to the seismic forces F_x [††]

[††] It is permitted to determine the δ_{xe} using seismic design forces based on the computed fundamental period of the structure without the upper limit $C_u T_a$ specified in 12.8.2 (12.8.6.2).

FLOWCHART 6.8
Equivalent Lateral Force Procedure (12.6)
(continued)

J

Determine the design story drift Δ as the difference of the deflections δ_x at the center of mass at the top and bottom of the story under consideration (see Figure 12.8-2)

↓

Check that the allowable story drifts Δ_a given in Table 12.12-1 are satisfied at each story

↓

Ensure that the other applicable drift and deformation requirements of 12.12 are satisfied

↓

Determine stability coefficient θ at each story by Eq. 12.8-16:
$\theta = \dfrac{P_x \Delta}{V_x h_{sx} C_d}$ where P_x = total vertical design load at and above level x (P_x is determined using load factors no greater than 1.0) and h_{sx} = story height below level x

↓

Is $\theta \leq 0.1$?

- **No** →
  Determine θ_{max} by Eq. 12.8-17:
  $\theta_{max} = \dfrac{0.5}{C_d \beta}$ where β = ratio of shear demand to shear capacity for the story‡‡

  ↓

  K

- **Yes** →
  P-delta effects need not be considered on the structure (12.8.7).

‡‡ β can conservatively be taken as 1.0. Where P-delta effects are included in an automated analysis, the value of θ computed by Eq. 12.8-16 is permitted to be divided by $(1 + \theta)$ before checking Eq. 12.8-17 (12.8.7).

FLOWCHART 6.8
Equivalent Lateral Force Procedure (12.6)
(continued)

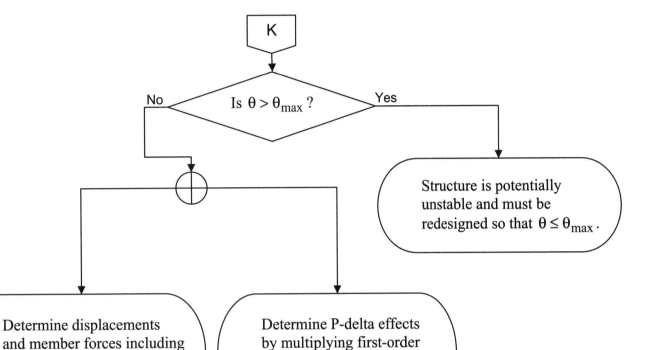

FLOWCHART 6.9

Alternate Simplified Design Procedure (12.14)

Are the 12 limitations of 12.14.1.1 satisfied?

- **No** → Use another analytical procedure in Chapter 12.
- **Yes** ↓

Determine S_S, S_{DS} and the SDC from Flowchart 6.4

↓

Determine the response modification coefficient R from Table 12.14-1 for the appropriate structural system based on SDC

↓

Determine effective seismic weight W in accordance with 12.14.8.1

↓

Determine base shear V by Eq. 12.14-11: $V = \dfrac{FS_{DS}W}{R}$

where

F = 1.0 for one-story buildings
  = 1.1 for two-story buildings
  = 1.2 for three-story buildings

$S_{DS} = 2F_a S_S / 3$

F_a = 1.0 for rock sites
  = 1.4 for soil sites, or
  = value determined in accordance with 11.4.3

↓

A

FLOWCHART 6.9
Alternate Simplified Design Procedure (12.14)
(continued)

A

Determine lateral seismic force F_x at level x by Eq. 12.14-12: $F_x = \dfrac{w_x}{W} V$ where w_x = portion of W located at level x

Determine seismic design story shear V_x by Eq. 12.14-13: $V_x = \sum_{i=x}^{n} F_i$

Is the diaphragm flexible in accordance with 12.14.5?

No: Determine inherent torsional moment M_t resulting from eccentricity between the locations of the center of mass and the center of rigidity (12.14.8.3.2.1)

Distribute V_x to the vertical elements of the seismic force-resisting system based on relative stiffness of the vertical elements and the diaphragm, including M_t

Yes: Distribute V_x to the vertical elements of the seismic force-resisting system based on tributary area (12.14.8.3.1)

B

FLOWCHART 6.9

Alternate Simplified Design Procedure (12.14)

(continued)

Design structure to resist overturning effects caused by the seismic forces F_x (12.14.8.4)

Foundations of structures shall be designed for not less than 75 percent of the foundation overturning design moment.

Structural drift need not be calculated. If drift is required for other design requirements, it shall be taken as 1 percent of building height unless it is computed to be less (12.14.8.5).

6.6.3 Seismic Design Requirements for Nonstructural Components

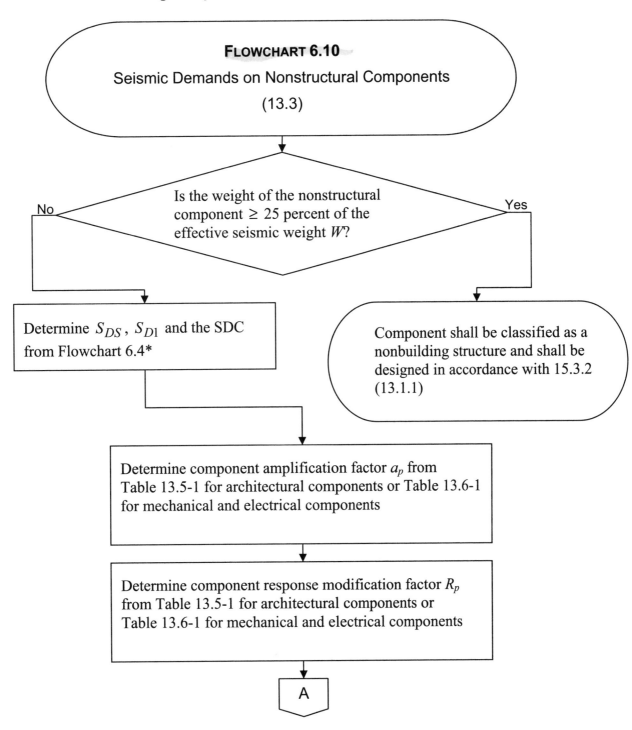

* Nonstructural components shall be assigned to the same SDC as the structure that they occupy or to which they are attached (13.1.2).

FLOWCHART 6.10
Seismic Demands on Nonstructural Components (13.3)
(continued)

A

Determine component importance factor I_p in accordance with 13.1.3

Determine horizontal seismic design force F_p applied at component's center of gravity by Eqs. 13.3-1, 13.3-2 and 13.3-3:**

$$0.3 S_{DS} I_p W_p \leq F_p = \frac{0.4 a_p S_{DS} W_p}{\left(\frac{R_p}{I_p}\right)}\left(1 + 2\frac{z}{h}\right) \leq 1.6 S_{DS} I_p W_p$$

where W_p = component operating weight

z = height in structure of point of attachment of component with respect to the base.†

h = average roof height of structure with respect to the base

---

** F_p shall be applied independently in at least two orthogonal horizontal directions in combination with service loads. For vertically cantilevered systems, F_p shall be assumed to act in any horizontal direction, and the component shall be designed for a concurrent vertical force $\pm 0.2 S_{DS} W_p$. Redundancy factor ρ is permitted to be taken equal to 1.0 and the overstrength factor Ω_o does not apply (13.3.1).

† For items at or below the base, z shall be taken as zero. The value of z/h need not exceed 1.0.

6.6.4 Seismic Design Requirements for Nonbuilding Structures

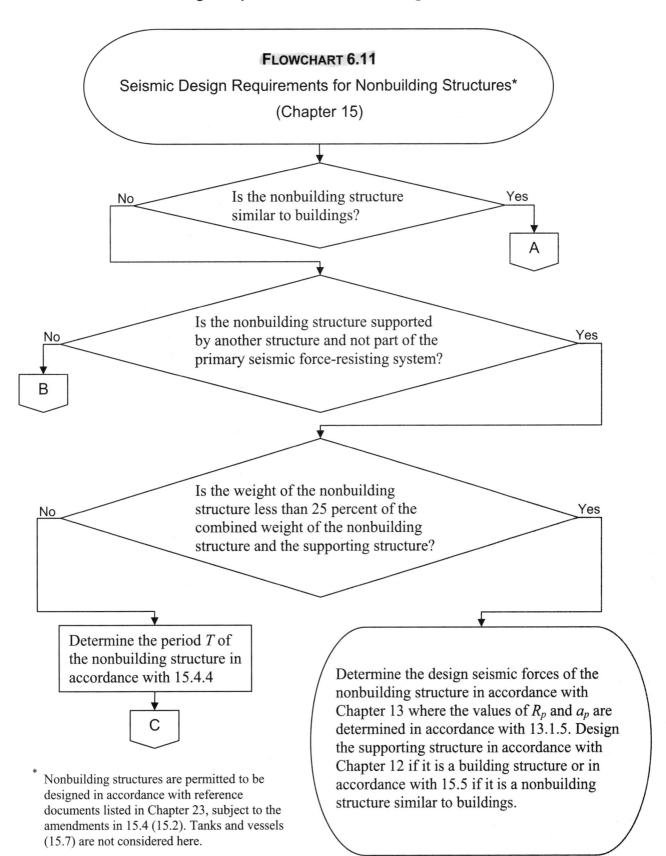

*Nonbuilding structures are permitted to be designed in accordance with reference documents listed in Chapter 23, subject to the amendments in 15.4 (15.2). Tanks and vessels (15.7) are not considered here.

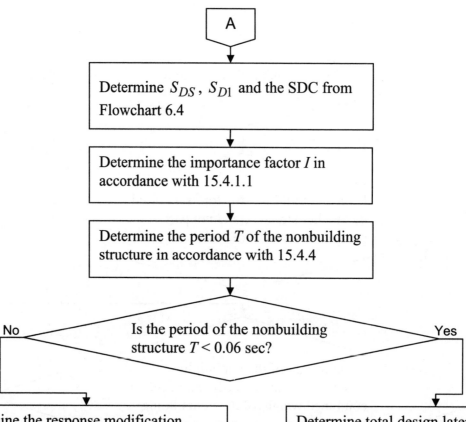

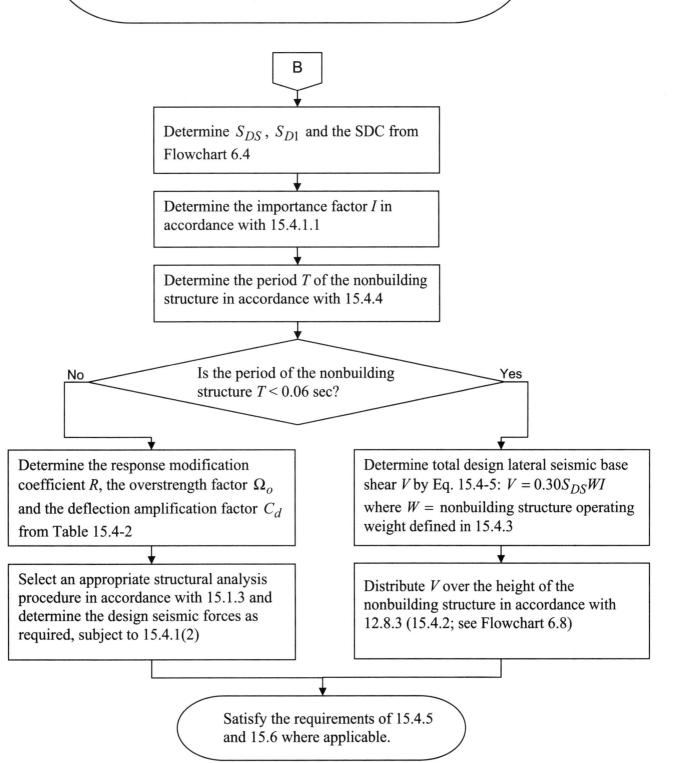

FLOWCHART 6.11
Seismic Design Requirements for Nonbuilding Structures*
(continued)

C

Is the period of the nonbuilding structure $T < 0.06$ sec?

No: Determine the design seismic forces of the nonbuilding structure based on a combined model of the nonbuilding structure and supporting structure. Design the combined structure in accordance with 15.5 using an R value of the combined system as the lesser R value of the nonbuilding structure or the supporting structure.

Yes: Determine the design seismic forces of the nonbuilding structure in accordance with Chapter 13 where the value of R_p shall be taken as the R value of the nonbuilding structure in accordance with Table 15.4-2 and a_p shall be taken as 1.0. Design the supporting structure in accordance with Chapter 12 if it is a building structure or in accordance with 15.5 if it is a nonbuilding structure similar to buildings. The R-value of the combined system is permitted to be taken as the R-value of the supporting system.

6.7 EXAMPLES

The following examples illustrate the IBC and ASCE/SEI 7 requirements for seismic design loads.

6.7.1 Example 6.1 – Residential Building, Seismic Design Category

A typical floor plan and elevation of a 12-story residential building is depicted in Figure 6.6. Given the design data below, determine the Seismic Design Category (SDC).

DESIGN DATA	
Location:	Charleston, SC (Latitude: 32.74°, Longitude: -79.93°)
Soil classification:	Site Class D
Occupancy:	Residential occupancy where less than 300 people congregate in one area
Material:	Cast-in-place reinforced concrete
Structural system:	Building frame system

SOLUTION

- Step 1: Determine the seismic ground motion values from Flowchart 6.2.

  1. Determine the mapped accelerations S_S and S_1.

     In lieu of using Figures 22-1 and 22-2, the mapped accelerations are determined by inputting the latitude and longitude of the site into the USGS Ground Motion Parameter Calculator. The output is as follows: $S_S = 1.37$ and $S_1 = 0.34$.

  2. Determine the site class of the soil.

     The site class of the soil is given in the design data as Site Class D.

  3. Determine soil-modified accelerations S_{MS} and S_{M1}.

     Site coefficients F_a and F_v are determined from Tables 11.4-1 and 11.4-2, respectively:

     For Site Class D and $S_S > 1.25$: $F_a = 1.0$

     For Site Class D and $0.3 < S_1 < 0.4$: $F_v = 1.72$ from linear interpolation

6-62 CHAPTER 6 EARTHQUAKE LOADS

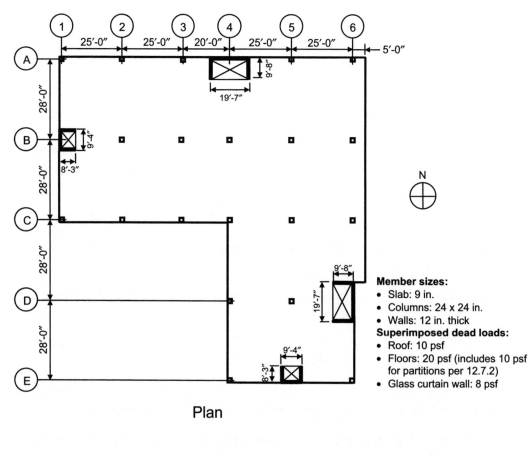

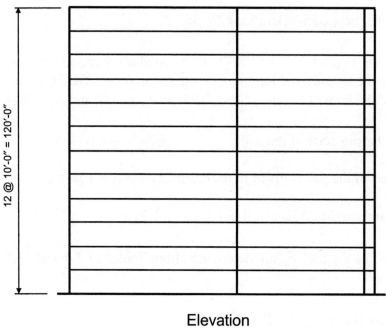

Figure 6.6 Typical Floor Plan and Elevation of 12-story Residential Building

Thus,

$$S_{MS} = 1.0 \times 1.37 = 1.37$$

$$S_{M1} = 1.72 \times 0.34 = 0.59$$

4. Determine design accelerations S_{DS} and S_{D1}.

   From Eqs. 11.4-3 and 11.4-4:

   $$S_{DS} = \frac{2}{3} \times 1.37 = 0.91$$

   $$S_{D1} = \frac{2}{3} \times 0.59 = 0.39$$

- Step 2: Determine the SDC from Flowchart 6.4.

   1. Determine if the building can be assigned to SDC A in accordance with 11.4.1.

      Since $S_S = 1.37 > 0.15$ and $S_1 = 0.34 > 0.04$, the building cannot be automatically assigned to SDC A.

   2. Determine the Occupancy Category from IBC Table 1604.5.

      For a residential occupancy where less than 300 people congregate in one area, the Occupancy Category is II.

   3. Since $S_1 < 0.75$, the building is not assigned to SDC E or F.

   4. Check if all four conditions of 11.6 are satisfied.

      Check if the approximate period T_a is less than $0.8T_S$.

      Use Eq. 12.8-7 with approximate period parameters for "other structural systems":

      $$T_a = C_t h_n^x = 0.02(120)^{0.75} = 0.73 \text{ sec}$$

      where C_t and x are given in Table 12.8-2.

      $$T_S = S_{D1}/S_{DS} = 0.39/0.91 = 0.43 \text{ sec}$$

$0.73 \text{ sec} > 0.8 \times 0.43 = 0.34 \text{ sec}$

Since this condition is not satisfied, the SDC cannot be determined by Table 11.6-1 alone (11.6).

5. Determine the SDC from Tables 11.6-1 and 11.6-2.

   From Table 11.6-1, with $S_{DS} > 0.50$ and Occupancy Category II, the SDC is D.

   From Table 11.6-2, with $S_{D1} > 0.20$ and Occupancy Category II, the SDC is D.

Therefore, the SDC is D for this building.

6.7.2   Example 6.2 – Residential Building, Permitted Analytical Procedure

For the 12-story residential building in Example 6.1, determine the analytical procedure that can be used in calculating the seismic forces.

SOLUTION

Use Flowchart 6.7 to determine the permitted analytical procedure.

1. Determine the SDC from Flowchart 6.4.

   It was determined in Step 2 of Example 6.1 that the SDC is D.

2. Determine S_{DS} and S_{D1} from Flowchart 6.2.

   The design accelerations were determined in Step 1, item 4 of Example 6.1: $S_{DS} = 0.91$ and $S_{D1} = 0.39$.

3. Determine T_S.

   $T_S = S_{D1} / S_{DS}$ was determined in Step 2, item 4 of Example 6.1 as 0.43 sec.

4. Determine the fundamental period of the building T.

   It was determined in Step 2, item 4 of Example 6.1 that $T = T_a = 0.73$ sec.

5. Check if $T < 3.5T_S$.

   $T = 0.73 \text{ sec} < 3.5T_S = 3.5 \times 0.43 = 1.5 \text{ sec}$

6. Determine if the structure is regular or not.

   a. Determine if the structure has any horizontal structural irregularities in accordance with Table 12.3-1.

      i. Torsional irregularity

         In accordance with Table 12.3-1, Type 1a torsional irregularity and Type 1b extreme torsional irregularity for rigid or semirigid diaphragms exist where the ratio of the maximum story drift at one end of a structure to the average story drifts at two ends of the structure exceeds 1.2 and 1.4, respectively. The story drifts are to be determined using code-prescribed forces, including accidental torsion. In this example, the floors and roof are cast-in-place reinforced concrete slabs, which are considered to be rigid diaphragms (12.3.1.2).

         At this point in the analysis, it is obviously not evident which method is required to be used to determine the prescribed seismic forces. In lieu of using a more complicated higher order analysis, the equivalent lateral force procedure may be used to determine the lateral seismic forces. These forces are applied to the building and the subsequent analysis yields the story drifts Δ, which are used in determining whether Type 1a or 1b torsional irregularity exists. The results from the equivalent lateral force procedure will be needed if it is subsequently determined that a modal analysis is required (see 12.9.4).

         Use Flowchart 6.8 to determine the lateral seismic forces from the equivalent lateral force procedure.

         a) The design accelerations and the SDC have been determined in Example 6.1.

         b) Determine the response modification coefficient R from Table 12.2-1.

            The walls in this building frame system must be special reinforced concrete shear walls, since the building is assigned to SDC D (system B5 in Table 12.2-1). In this case, $R = 6$. Note that height of the building, which is 120 ft, is less than the limiting height of 160 ft for this type of system in SDC D.[8]

         c) Determine the importance factor I from Table 11.5-1.

            For Occupancy Category II, $I = 1.0$.

         d) Determine the period of the structure T.

---

[8] The increased building height limit of 12.2.5.4 is not considered.

It was determined in Step 2, item 4 of Example 6.1 that the approximate period of the structure T_a, which is permitted to be used in the equivalent lateral force procedure, is equal to 0.73 sec.

e) Determine long-period transition period T_L from Figure 22-15.

For Charleston, SC, $T_L = 8$ sec $> T_a = 0.73$ sec.

f) Determine seismic response coefficient C_s.

The seismic response coefficient C_s is determined by Eq. 12.8-3:

$$C_s = \frac{S_{D1}}{T(R/I)} = \frac{0.39}{0.73(6/1.0)} = 0.09$$

The value of C_s need not exceed that from Eq. 12.8-2:

$$C_s = \frac{S_{DS}}{R/I} = \frac{0.91}{6/1.0} = 0.15$$

Also, C_s must not be less than the larger of $0.044 S_{DS} I = 0.04$ (governs) and 0.01 (Eq. 12.8-5).

Thus, the value of C_s from Eq. 12.8-3 governs.

g) Determine effective seismic weight W in accordance with 12.7.2.

The member sizes and superimposed dead loads are given in Figure 6.6 and the effective weights at each floor level are given in Table 6.8. The total weight W is the summation of the effective dead loads at each level.

h) Determine seismic base shear V.

Seismic base shear is determined by Eq. 12.8-1:

$$V = C_s W = 0.09 \times 19{,}920 = 1{,}793 \text{ kips}$$

i) Determine exponent related to structure period k.

Since 0.5 sec $< T = 0.73$ sec < 2.5 sec, k is determined as follows:

$$k = 0.75 + 0.5T = 1.12$$

Table 6.8 Seismic Forces and Story Shears

Level	Story weight, w_x (kips)	Height, h_x (ft)	$w_x h_x^k$	Lateral force, F_x (kips)	Story Shear, V_x (kips)
R	1,308	120	278,799	233	233
11	1,692	110	327,160	274	507
10	1,692	100	294,036	246	753
9	1,692	90	261,308	219	972
8	1,692	80	229,014	192	1,164
7	1,692	70	197,202	165	1,329
6	1,692	60	165,932	139	1,468
5	1,692	50	135,284	113	1,581
4	1,692	40	105,368	88	1,669
3	1,692	30	76,344	64	1,733
2	1,692	20	48,479	41	1,774
1	1,692	10	22,305	19	1,793
Σ	19,920		2,141,231	1,793	

j) Determine lateral seismic force F_x at each level x.

F_x is determined by Eqs. 12.8-11 and 12.8-12. A summary of the lateral forces F_x and the story shears V_x is given in Table 6.8.

Three-dimensional analyses were performed independently in the N-S and E-W directions for the seismic forces in Table 6.8 using a commercial computer program. In the model, rigid diaphragms were assigned at each floor level. The stiffness properties of the shear walls were input assuming cracked sections (12.7.3): $I_{eff} = 0.5 I_g$ where I_g is the gross moment of inertia of the section. In accordance with 12.8.4.2, the center of mass was displaced each way from its actual location a distance equal to 5 percent of the building dimension perpendicular to the applied forces to account for accidental torsion in seismic design.

A summary of the elastic displacements δ_{xe} at each end of the building in both the N-S and E-W directions due to the code-prescribed forces in Table 6.8 is given in Table 6.9 at all floor levels. Also provided in the table are the story drifts Δ at each end of the building in both directions.

According to Table 12.3-1, a torsional irregularity occurs where maximum story drift at one of the structure is greater than 1.2 times the average of the story drifts at the two ends of the structure. The average story drift Δ_{avg} and the ratio of the maximum story drift to the average story drift $\Delta_{max} / \Delta_{avg}$ are also provided in Table 6.9.

Table 6.9 Lateral Displacements and Story Drifts due to Seismic Forces

Story	N-S Direction						E-W Direction					
	$(\delta_{xe})_1$ (in.)	Δ_1 (in.)	$(\delta_{xe})_2$ (in.)	Δ_2 (in.)	Δ_{avg} (in.)	$\frac{\Delta_{max}}{\Delta_{avg}}$	$(\delta_{xe})_1$ (in.)	Δ_1 (in.)	$(\delta_{xe})_2$ (in.)	Δ_2 (in.)	Δ_{avg} (in.)	$\frac{\Delta_{max}}{\Delta_{avg}}$
12	10.79	1.16	5.72	0.60	0.88	1.32	7.29	0.75	6.20	0.65	0.70	1.07
11	9.63	1.18	5.12	0.60	0.89	1.33	6.54	0.77	5.55	0.66	0.72	1.07
10	8.45	1.17	4.52	0.61	0.89	1.32	5.77	0.76	4.89	0.66	0.71	1.07
9	7.28	1.17	3.91	0.61	0.89	1.32	5.01	0.77	4.23	0.66	0.72	1.07
8	6.11	1.13	3.30	0.59	0.86	1.31	4.24	0.76	3.57	0.65	0.71	1.07
7	4.98	1.09	2.71	0.58	0.84	1.30	3.48	0.74	2.92	0.62	0.68	1.09
6	3.89	1.02	2.13	0.54	0.78	1.31	2.74	0.69	2.30	0.58	0.64	1.08
5	2.87	0.90	1.59	0.49	0.70	1.29	2.05	0.63	1.72	0.53	0.58	1.09
4	1.97	0.78	1.10	0.42	0.60	1.30	1.42	0.56	1.19	0.46	0.51	1.10
3	1.19	0.61	0.68	0.34	0.48	1.27	0.86	0.43	0.73	0.40	0.42	1.04
2	0.58	0.41	0.34	0.24	0.33	1.24	0.43	0.31	0.37	0.26	0.29	1.07
1	0.17	0.17	0.10	0.10	0.14	1.21	0.12	0.12	0.11	0.11	0.12	1.00

For example, at the 12th story in the N-S direction:

$$\Delta_1 = 10.79 - 9.63 = 1.16 \text{ in.}$$

$$\Delta_2 = 5.72 - 5.12 = 0.60 \text{ in.}$$

$$\Delta_{avg} = \frac{1.16 + 0.60}{2} = 0.88 \text{ in.}$$

$$\frac{\Delta_{max}}{\Delta_{avg}} = \frac{1.16}{0.88} = 1.32 > 1.2$$

Therefore, a Type 1a torsional irregularity exists at all floor levels in the N-S direction.[9]

According to 12.8.4.3, where torsional irregularity exists at floor level x, the accidental torsional moments M_{ta} must be increased by the torsional amplification factor A_x given by Eq. 12.8-14:

$$A_x = \left(\frac{\delta_{max}}{1.2\delta_{avg}}\right)^2$$

---

[9] A Type 1b extreme torsional irregularity does not exist since the ratio of maximum drift to average drift is less than 1.4 (see Table 12.3-1).

For example, at the 12th story in the N-S direction:

$$A_{12} = \left[\frac{10.79}{1.2\left(\frac{10.79 + 5.72}{2}\right)}\right]^2 = 1.19 > 1.0$$

ii. Re-entrant corner irregularity

According to Table 12.3-1, a re-entrant corner irregularity exists where both plan projections of the structure beyond a re-entrant corner are greater than 15 percent of the plan dimension in a given direction.

Using Table 6.2:

Projection $b = 56$ ft $> 0.15a = 0.15 \times 112 = 16.8$ ft

Projection $d = 70$ ft $> 0.15c = 0.15 \times 120 = 18.0$ ft

Therefore, a Type 2 re-entrant corner irregularity exists.

iii. Diaphragm discontinuity irregularity

This type of irregularity does not exist, since the area of any of the openings is much less than 50 percent of the area of the diaphragm. Also, the diaphragm has the same effective stiffness on all of the floor levels.

iv. Out-of-plane offsets irregularity

There are no out-of-plane offsets of the shear walls, so this irregularity does not exist.

v. Nonparallel systems irregularity

This discontinuity does not exist, since all of the shear walls are parallel to a major orthogonal axis of the building.

b. Determine if the structure has any vertical structural irregularities in accordance with Table 12.3-2.

By inspection, none of the vertical irregularities defined in Table 12.3-2 exist for this building (also see Table 6.3).

In summary, the building is not regular and has the following horizontal irregularities: Type 1a torsional irregularity and Type 2 re-entrant corner irregularity.

7. Determine the permitted analytical procedure from Table 12.6-1.

The structure is irregular with $T < 3.5\, T_S$. If the structure had only a Type 2 re-entrant corner irregularity, the equivalent lateral force procedure could be used to analyze the structure. However, since a Type 1a torsional irregularity also exists, the equivalent lateral force procedure is not permitted; either a modal response spectrum analysis (12.9) or a seismic response history procedure (Chapter 16) must be utilized.[10]

6.7.3 Example 6.3 – Office Building, Seismic Design Category

Typical floor plans and elevations of a seven-story office building are depicted in Figure 6.7. Given the design data below, determine the Seismic Design Category (SDC).

DESIGN DATA

Location:	Memphis, TN (Latitude: 35.13°, Longitude: -90.05°)
Soil classification:	Site Class D
Occupancy:	Office occupancy where less than 300 people congregate in one area
Material:	Structural steel
Structural system:	Moment-resisting frame and building frame systems

SOLUTION

- Step 1: Determine the seismic ground motion values from Flowchart 6.2.

    1. Determine the mapped accelerations S_S and S_1.

        In lieu of using Figures 22-1 and 22-2, the mapped accelerations are determined by inputting the latitude and longitude of the site into the USGS Ground Motion Parameter Calculator. The output is as follows: $S_S = 1.35$ and $S_1 = 0.37$.

    2. Determine the site class of the soil.

        The site class of the soil is given in the design data as Site Class D.

---

[10] See the reference sections in Table 12.3-1 that must be satisfied for these types of irregularities in structures assigned to SDC D. Design forces shall be increased 25 percent for connections of diaphragms to vertical elements and to collectors and for connection of collectors to the vertical elements (12.3.3.4). Members that are not part of the seismic force-resisting system must satisfy the deformational compatibility requirements of 12.12.4.

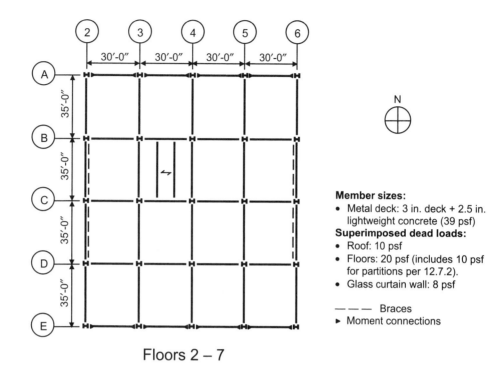

Floors 2 – 7

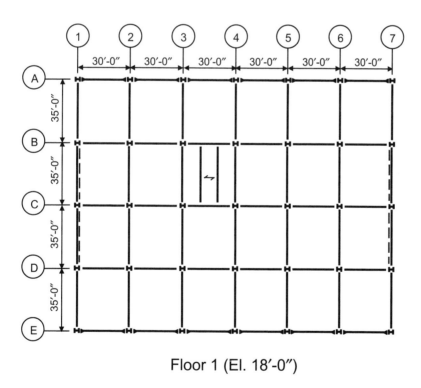

Floor 1 (El. 18'-0")

Figure 6.7 Typical Floor Plans and Elevations of Seven-story Office Building

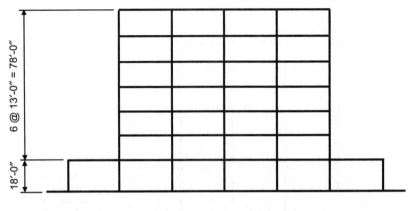

North or South Elevation

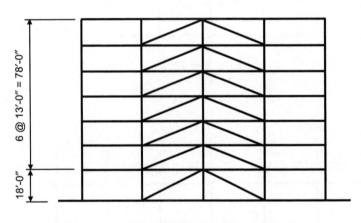

East or West Elevation

Figure 6.7 Typical Floor Plans and Elevations of Seven-story Office Building (continued)

3. Determine soil-modified accelerations S_{MS} and S_{M1}.

   Site coefficients F_a and F_v are determined from Tables 11.4-1 and 11.4-2, respectively:

   For Site Class D and $S_S > 1.25$: $F_a = 1.0$

   For Site Class D and $0.3 < S_1 < 0.4$: $F_v = 1.66$ from linear interpolation

   Thus,

$$S_{MS} = 1.0 \times 1.35 = 1.35$$

$$S_{M1} = 1.66 \times 0.37 = 0.61$$

4. Determine design accelerations S_{DS} and S_{D1}.

   From Eqs. 11.4-3 and 11.4-4:

   $$S_{DS} = \frac{2}{3} \times 1.35 = 0.90$$

   $$S_{D1} = \frac{2}{3} \times 0.61 = 0.41$$

- Step 2: Determine the SDC from Flowchart 6.4.

  1. Determine if the building can be assigned to SDC A in accordance with 11.4.1.

     Since $S_S = 1.35 > 0.15$ and $S_1 = 0.37 > 0.04$, the building cannot be automatically assigned to SDC A.

  2. Determine the Occupancy Category from IBC Table 1604.5.

     For a business occupancy where less than 300 people congregate in one area, the Occupancy Category is II.

  3. Since $S_1 < 0.75$, the building is not assigned to SDC E or F.

  4. Check if all four conditions of 11.6 are satisfied.

     Check if the approximate period T_a is less than $0.8T_S$.

     In the N-S direction, the concentrically braced steel frames fall under "other structural systems" in Table 12.8-2. Using Eq. 12.8-7, T_a is determined as follows:

     $$T_a = C_t h_n^x = 0.02(96)^{0.75} = 0.61 \text{ sec}$$

     In the E-W direction, values of C_t and x for steel moment-resisting frames are used from Table 12.8-2:

     $$T_a = C_t h_n^x = 0.028(96)^{0.8} = 1.1 \text{ sec}$$

$T_S = S_{D1}/S_{DS} = 0.41/0.90 = 0.46$ sec

$0.8 \times 0.46 = 0.37$ sec is less than the approximate periods in both the N-S and E-W directions.

Since this condition is not satisfied, the SDC cannot be determined by Table 11.6-1 alone (11.6).

5. Determine the SDC from Tables 11.6-1 and 11.6-2.

   From Table 11.6-1, with $S_{DS} > 0.50$ and Occupancy Category II, the SDC is D.

   From Table 11.6-2, with $S_{D1} > 0.20$ and Occupancy Category II, the SDC is D.

Therefore the SDC is D for this building.

6.7.4 Example 6.4 – Office Building, Permitted Analytical Procedure

For the seven-story office building in Example 6.3, determine the analytical procedure that can be used in calculating the seismic forces.

SOLUTION

Use Flowchart 6.6 to determine the permitted analytical procedure.

1. Determine the SDC from Flowchart 6.4.

   It was determined in Step 2 of Example 6.3 that the SDC is D.

2. Determine S_{DS} and S_{D1} from Flowchart 6.2.

   The design accelerations were determined in Step 1, item 4 of Example 6.3: $S_{DS} = 0.90$ and $S_{D1} = 0.41$.

3. Determine T_S.

   It was determined in Step 2, item 4 of Example 6.3 that $T_S = S_{D1}/S_{DS} = 0.46$ sec.

4. Determine the fundamental period of the building T.

   The periods T were determined in Step 2, item 4 of Example 6.3 as 0.61 sec in the N-S direction and 1.1 sec in the E-W direction.

5. Check if $T < 3.5 T_S$.

   $3.5 T_S = 3.5 \times 0.46 = 1.6$ sec, which is greater than the periods in both directions.

6. Determine if the structure is regular or not.

   a. Determine if the structure has any horizontal structural irregularities in accordance with Table 12.3-1.

      i. Torsional irregularity

         In accordance with Table 12.3-1, Type 1a torsional irregularity and Type 1b extreme torsional irregularity for rigid or semirigid diaphragms exist where the ratio of the maximum story drift at one end of a structure to the average story drifts at two ends of the structure exceeds 1.2 and 1.4, respectively. The story drifts are to be determined using code-prescribed forces, including accidental torsion. In this example, the floor and roof are metal deck with concrete, which is considered to be a rigid diaphragm (12.3.1.2).

         At this point in the analysis, it is obviously not evident which method is required to be used to determine the prescribed seismic forces. In lieu of using a more complicated higher order analysis, the equivalent lateral force procedure may be used to determine the lateral seismic forces. These forces are applied to the building and the subsequent analysis yields the story drifts Δ, which are used in determining whether Type 1a or 1b torsional irregularity exists. The results from the equivalent lateral force procedure will be needed if it is subsequently determined that a modal analysis is required (see 12.9.4).

         Use Flowchart 6.8 to determine the lateral seismic forces from the equivalent lateral force procedure:

         a) The design accelerations and the SDC have been determined in Example 6.3.

         b) Determine the response modification coefficient R from Table 12.2-1.

            In the N-S direction, special steel concentrically braced frames are required, since the building is assigned to SDC D (system B3 in Table 12.2-1). In this case, $R = 6$. Note that height of the building, which is 96 ft, is less than the limiting height of 160 ft for this type of system in SDC D.[11]

---

[11] The increased building height limit of 12.2.5.4 is not considered.

In the E-W direction, special steel moment frames are required (system C1 in Table 12.2-1). In this case, $R = 8$, and there is no height limit.

c) Determine the importance factor I from Table 11.5-1.

For Occupancy Category II, $I = 1.0$.

d) Determine the period of the structure T.

It was determined in Step 2, item 4 of Example 6.3 that the approximate period of the structure T_a, which is permitted to be used in the equivalent lateral force procedure, is 0.61 sec in the N-S direction and 1.1 sec in the E-W direction.

e) Determine long-period transition period T_L from Figure 22-15.

For Memphis, TN, $T_L = 12$ sec, which is greater than the periods in both directions.

f) Determine seismic response coefficients C_s in both directions.

- N-S direction:

The seismic response coefficient C_s is determined by Eq. 12.8-3:

$$C_s = \frac{S_{D1}}{T(R/I)} = \frac{0.41}{0.61(6/1.0)} = 0.11$$

The value of C_s need not exceed that from Eq. 12.8-2:

$$C_s = \frac{S_{DS}}{R/I} = \frac{0.90}{6/1.0} = 0.15$$

Also, C_s must not be less than the larger of $0.044 S_{DS} I = 0.04$ (governs) and 0.01 (Eq. 12.8-5).

Thus, the value of C_s from Eq. 12.8-3 governs in the N-S direction.

- E-W direction:

$$C_s = \frac{S_{D1}}{T(R/I)} = \frac{0.41}{1.1(8/1.0)} = 0.05$$

The value of C_s need not exceed that from Eq. 12.8-2:

$$C_s = \frac{S_{DS}}{R/I} = \frac{0.90}{8/1.0} = 0.11$$

Also, C_s must not be less than the larger of $0.044S_{DS}I = 0.04$ (governs) and 0.01 (Eq. 12.8-5).

Thus, the value of C_s from Eq. 12.8-3 governs in the E-W direction.

g) Determine effective seismic weight W in accordance with 12.7.2.

The member sizes and superimposed dead loads are given in Figure 6.7 and the effective weights at each floor level are given in Tables 6.10 and 6.11. The total weight W is the summation of the effective dead loads at each level.

h) Determine seismic base shear V.

Seismic base shear is determined by Eq. 12.8-1 in both the N-S and E-W directions.

- N-S direction: $V = C_s W = 0.11 \times 9,960 = 1,096$ kips

- E-W direction: $V = C_s W = 0.05 \times 9,960 = 498$ kips

i) Determine exponent related to structure period k in both directions.

Since $0.5 \text{ sec} < T < 2.5 \text{ sec}$ in both directions, k is determined as follows:

- N-S direction: $k = 0.75 + 0.5T = 1.06$

- E-W direction: $k = 0.75 + 0.5T = 1.30$

j) Determine lateral seismic force F_x at each level x.

Lateral forces F_x are determined by Eqs. 12.8-11 and 12.8-12.

A summary of the lateral forces F_x and the story shears V_x are given in Table 6.10 and 6.11 for the N-S and E-W directions, respectively.

Table 6.10 Seismic Forces and Story Shears in the N-S Direction

Level	Story weight, w_x (kips)	Height, h_x (ft)	$w_x h_x^k$	Lateral force, F_x (kips)	Story Shear, V_x (kips)
R	1,018	96	128,515	209	209
6	1,381	83	149,423	242	451
5	1,381	70	124,738	202	653
4	1,381	57	100,328	163	816
3	1,381	44	76,252	124	940
2	1,381	31	52,606	85	1,025
1	2,037	18	43,609	71	1,096
Σ	9,960		675,471	1,096	

Table 6.11 Seismic Forces and Story Shears in the E-W Direction

Level	Story weight, w_x (kips)	Height, h_x (ft)	$w_x h_x^k$	Lateral force, F_x (kips)	Story Shear, V_x (kips)
R	1,018	96	384,327	105	105
6	1,381	83	431,514	118	223
5	1,381	70	345,797	94	317
4	1,381	57	264,747	72	389
3	1,381	44	189,096	52	441
2	1,381	31	119,940	33	474
1	2,037	18	87,266	24	498
Σ	9,960		1,822,687	498	

Three-dimensional analyses were performed independently in the N-S and E-W directions for the seismic forces in Table 6.10 and 6.11 using a commercial computer program. In the model, rigid diaphragms were assigned at each floor level. In accordance with 12.8.4.2, the center of mass was displaced each way from its actual location a distance equal to 5 percent of the building dimension perpendicular to the applied forces to account for accidental torsion in seismic design.

A summary of the elastic displacements δ_{xe} at each end of the building in both the N-S and E-W directions due to the code-prescribed forces in Tables 6.10 and 6.11 is given in Table 6.12 at all floor levels.

According to Table 12.3-1, a torsional irregularity occurs where maximum story drift at one of the structure is greater than 1.2 times the average of the story drifts at the two ends of the structure. The average story drift Δ_{avg} and the ratio of the maximum story drift to the average story drift $\Delta_{max}/\Delta_{avg}$ are also provided in Table 6.12.

Table 6.12 Lateral Displacements and Story Drifts due to Seismic Forces

Story	N-S Direction						E-W Direction					
	$(\delta_{xe})_1$ (in.)	Δ_1 (in.)	$(\delta_{xe})_2$ (in.)	Δ_2 (in.)	Δ_{avg} (in.)	$\dfrac{\Delta_{max}}{\Delta_{avg}}$	$(\delta_{xe})_1$ (in.)	Δ_1 (in.)	$(\delta_{xe})_2$ (in.)	Δ_2 (in.)	Δ_{avg} (in.)	$\dfrac{\Delta_{max}}{\Delta_{avg}}$
7	1.59	0.14	1.23	0.12	0.13	1.08	5.36	0.36	5.22	0.34	0.35	1.03
6	1.45	0.17	1.11	0.15	0.16	1.06	5.00	0.49	4.88	0.47	0.48	1.02
5	1.28	0.22	0.96	0.20	0.21	1.05	4.51	0.71	4.41	0.71	0.71	1.00
4	1.06	0.21	0.76	0.19	0.20	1.05	3.80	1.10	3.70	1.06	1.08	1.02
3	0.85	0.22	0.57	0.17	0.19	1.16	2.70	1.19	2.64	1.16	1.17	1.02
2	0.63	0.17	0.40	0.22	0.19	1.16	1.51	0.89	1.48	0.87	0.88	1.01
1	0.46	0.46	0.18	0.18	0.32	1.44	0.62	0.62	0.61	0.61	0.61	1.02

For example, at the first story in the N-S direction:

$\Delta_1 = 0.46$ in.

$\Delta_2 = 0.18$ in.

$$\Delta_{avg} = \frac{0.46 + 0.18}{2} = 0.32 \text{ in.}$$

$$\frac{\Delta_{max}}{\Delta_{avg}} = \frac{0.46}{0.32} = 1.44 > 1.4$$

Therefore, a Type 1b extreme torsional irregularity exists at the first story in the N-S direction.

According to 12.8.4.3, where torsional irregularity exists at floor level x, the accidental torsional moments M_{ta} must be increased by the torsional amplification factor A_x given by Eq. 12.8-14:

$$A_x = \left(\frac{\delta_{max}}{1.2\delta_{avg}}\right)^2$$

At the first story in the N-S direction:

$$A_1 = \left[\frac{0.46}{1.2\left(\dfrac{0.46 + 0.18}{2}\right)}\right]^2 = 1.44 > 1.0$$

Therefore, the accidental torsional moment at the first story is:[12]

$$(M_{ta})_1 = A_1 F_1 e$$
$$= 1.44 \times 71 \times (0.05 \times 180) = 920 \text{ ft-kips}$$

    ii. Re-entrant corner irregularity

    By inspection, this irregularity does not exist.

    iii. Diaphragm discontinuity irregularity

    This irregularity does not exist in this building when opening sizes for typical elevators and stairs are present.

    iv. Out-of-plane offsets irregularity

    In the first story, the seismic force-resisting system consists of special steel concentrically braced frames along column lines 1 and 7. Above the first floor, there is a 30-ft offset of the braced frames, which occur along column lines 2 and 6.

    Therefore, a Type 4 out-of-plane offset irregularity exists.

    Note that the forces from the braced frames along column lines 2 and 6 must be transferred through the structure to the braced frames along column lines 1 and 7, respectively.

    v. Nonparallel systems irregularity

    This discontinuity does not exist, since all of the braced frames and moment-resisting frames are parallel to a major orthogonal axis of the building.

b. Determine if the structure has any vertical structural irregularities in accordance with Table 12.3-2.

    i. Stiffness–Soft Story Irregularity

    A soft story is defined in Table 12.3-2 based on the relative lateral stiffness of stories in a building. In general, it is not practical to determine story stiffness. Instead, this type of irregularity can be evaluated using drift ratios due to the code-prescribed lateral forces.[13] A soft story exists when one of the following conditions are satisfied:

---

[12] It is assumed that the center of mass and center of rigidity are at the same location in this building.
[13] Story displacements based on the code-prescribed lateral forces can be used to evaluate soft stories when the story heights are equal.

Soft story irregularity: $0.7 \dfrac{\delta_{1e}}{h_1} > \dfrac{\delta_{2e} - \delta_{1e}}{h_2}$ or

$$0.8 \dfrac{\delta_{1e}}{h_1} > \dfrac{1}{3} \left[\dfrac{\delta_{2e} - \delta_{1e}}{h_2} + \dfrac{\delta_{3e} - \delta_{2e}}{h_3} + \dfrac{\delta_{4e} - \delta_{3e}}{h_4} \right]$$

Extreme soft story irregularity: $0.6 \dfrac{\delta_{1e}}{h_1} > \dfrac{\delta_{2e} - \delta_{1e}}{h_2}$ or

$$0.7 \dfrac{\delta_{1e}}{h_1} > \dfrac{1}{3} \left[\dfrac{\delta_{2e} - \delta_{1e}}{h_2} + \dfrac{\delta_{3e} - \delta_{2e}}{h_3} + \dfrac{\delta_{4e} - \delta_{3e}}{h_4} \right]$$

Check for a soft story in the first story:

In the N-S direction, the displacements of the center of mass at the first, second, third and fourth floors are $\delta_{1e} = 0.31$ in., $\delta_{2e} = 0.51$ in., $\delta_{3e} = 0.69$ in., and $\delta_{4e} = 0.91$ in.

$$0.7 \dfrac{\delta_{1e}}{h_1} = 0.7 \dfrac{0.31}{18 \times 12} = 0.0010 < \dfrac{\delta_{2e} - \delta_{1e}}{h_2} = \dfrac{0.51 - 0.31}{13 \times 12} = 0.0013$$

$$0.8 \dfrac{\delta_{1e}}{h_1} = 0.8 \dfrac{0.31}{18 \times 12} = 0.0011 < \dfrac{1}{3} \left[\dfrac{\delta_{2e} - \delta_{1e}}{h_2} + \dfrac{\delta_{3e} - \delta_{2e}}{h_3} + \dfrac{\delta_{4e} - \delta_{3e}}{h_4} \right]$$

$$= \dfrac{1}{3} \left[\dfrac{0.51 - 0.31}{13 \times 12} + \dfrac{0.69 - 0.51}{13 \times 12} + \dfrac{0.91 - 0.69}{13 \times 12} \right] = 0.0013$$

In the E-W direction, the displacements of the center of mass at the first, second, third and fourth floors are $\delta_{1e} = 0.61$ in., $\delta_{2e} = 1.49$ in., $\delta_{3e} = 2.68$ in., and $\delta_{4e} = 3.75$ in.

$$0.7 \dfrac{\delta_{1e}}{h_1} = 0.7 \dfrac{0.61}{18 \times 12} = 0.0020 < \dfrac{\delta_{2e} - \delta_{1e}}{h_2} = \dfrac{1.49 - 0.61}{13 \times 12} = 0.0056$$

$$0.8 \dfrac{\delta_{1e}}{h_1} = 0.8 \dfrac{0.61}{18 \times 12} = 0.0023 < \dfrac{1}{3} \left[\dfrac{\delta_{2e} - \delta_{1e}}{h_2} + \dfrac{\delta_{3e} - \delta_{2e}}{h_3} + \dfrac{\delta_{4e} - \delta_{3e}}{h_4} \right]$$

$$= \dfrac{1}{3} \left[\dfrac{1.49 - 0.61}{13 \times 12} + \dfrac{2.68 - 1.49}{13 \times 12} + \dfrac{3.75 - 2.68}{13 \times 12} \right] = 0.0067$$

Therefore, a soft story irregularity does not exist in the first story.

ii. Weight (mass) irregularity.

Check the weight ratio of the first and second stories: 2,037/1,381 = 1.48 < 1.50. Thus, this irregularity is not present.

iii. Vertical geometric irregularity.

A vertical geometric irregularity is considered to exist where the horizontal dimension of the seismic force-resisting system in any story is 1.3 times that in an adjacent story.

In this case, the setbacks at the first floor level must be investigated for the moment-resisting frames along column lines A and E:

Width of floor 1/Width of floor 2 = 180/120 = 1.5 > 1.3

Thus, a Type 3 vertical geometric irregularity exists.

iv. In-plane discontinuity in vertical lateral force-resisting element irregularity.

There are no in-plane offsets of this type, so this irregularity does not exist.

v. Discontinuity in lateral strength-weak story irregularity.

This type of irregularity exists where a story lateral strength is less than 80 percent of that in the story above. The story strength is considered to be the total strength of all seismic-resisting elements that share the story shear for the direction under consideration.

<u>E-W Direction</u>

Determine whether a weak story exists in the first story in the E-W direction. In this case, the story strength is equal to the sum of the column shears in the moment-resisting frames in that story when the member moment capacity is developed by lateral loading. It is assumed in this example that the same column and beam sections are used in the moment-resisting frames in the first and second stories.

Assume the following nominal flexural strengths:[14]

Columns: $M_{nc} = 550$ ft-kips

Beams: $M_{nb} = 525$ ft-kips

---

[14] The assumed nominal flexural strengths of the columns and beams are based on preliminary member sizes and are provided for illustration purposes only.

EXAMPLES

- First story shear strength

    Corner columns A1/E1 and A7/E7 are checked for strong column-weak beam considerations:

    $$2M_{nc} = 2 \times 550 = 1,100 \text{ ft-kips} > M_{nb} = 525 \text{ ft-kips}$$

    The maximum shear force that can develop in each exterior column is based on the moment capacity of the beam (525/2 = 263 ft-kips), since it is less than the moment capacity of the column (550 ft-kips) at the top of the column:[15]

    $$V_1 = V_7 = \frac{263 + 550}{18} = 45 \text{ kips}$$

    Interior columns A2 through A6/E2 through E6 are checked for strong column-weak beam considerations:

    $$2M_{nc} = 2 \times 550 = 1,100 \text{ ft-kips} > 2M_{nb} = 1,050 \text{ ft-kips}$$

    The maximum shear force that can develop in each interior column is based on the moment capacity of the beam (525 ft-kips), since it is less than the moment capacity of the column (550 ft-kips) at the top of the column:

    $$V_2 = V_3 = V_4 = V_5 = V_6 = \frac{525 + 550}{18} = 60 \text{ kips}$$

    Total first story strength = $2(V_1 + V_2 + V_3 + V_4 + V_5 + V_6 + V_7) = 780$ kips

- Second story shear strength

    $$V_1 = V_7 = \frac{263 + 263}{13} = 41 \text{ kips}$$

    $$V_2 = V_3 = V_4 = V_5 = V_6 = \frac{525 + 525}{13} = 81 \text{ kips}$$

    Total second story strength = $2(V_1 + V_2 + V_3 + V_4 + V_5 + V_6 + V_7) =$ 974 kips

    780 kips > $0.80 \times 974 = 779$ kips

---

[15] At the bottom of the column, it is assumed that the full moment capacity of the column can be developed.

Therefore, a weak story irregularity does not exist in the first story in the E-W direction.

N-S Direction

Assuming the same beam, column and brace sizes in the first and second floors, the shear strengths of these floors are essentially the same, and no weak story irregularity exists in the N-S direction.

In summary, the building is not regular and has the following irregularities: horizontal Type 1b extreme torsional irregularity, horizontal Type 4 out-of-plane offsets irregularity and vertical Type 3 vertical geometric irregularity.

7. Determine the permitted analytical procedure from Table 12.6-1.

The structure is irregular with $T < 3.5T_S$. If the structure had only a Type 4 out-of-plane offsets irregularity, the equivalent lateral force procedure could be used to analyze the structure. However, since a Type 1b extreme torsional irregularity and a Type 3 vertical geometric irregularity also exist, the equivalent lateral force procedure is not permitted; either a modal response spectrum analysis (12.9) or a seismic response history procedure (Chapter 16) must be utilized.[16]

6.7.5 Example 6.5 – Office Building, Allowable Story Drift

For the seven-story office building in Example 6.3, check the story drift limits in both the N-S and E-W directions. For illustration purposes, use the lateral deflections determined by the equivalent lateral force procedure.

SOLUTION
---

1. Drift limits in N-S direction.

   To check drift limits, the deflections determined by Eq. 12.8-15 must be used:

   $$\delta_x = \frac{C_d \delta_{xe}}{I}$$

---

[16] See the reference sections in Tables 12.3-1 and 12.3-2 that must be satisfied for these types of irregularities in structures assigned to SDC D. Columns B2, C2, D2, B6, C6 and D6 must be designed to resist the load combinations with overstrength factor of 12.4.3.2, since they support the discontinued braced frames along column lines 2 and 6 (12.3.3.2). Design forces shall be increased 25 percent for connections of diaphragms to vertical elements and to collectors and for connection of collectors to the vertical elements (12.3.3.4). Members that are not part of the seismic force-resisting system must satisfy the deformational compatibility requirements of 12.12.4.

The maximum story displacements δ_{xe} in the N-S direction are summarized in Table 6.12. For special steel concentrically braced frames, the deflection amplification factor C_d is equal to 5 from Table 12.2-1.

A summary of the displacements at each floor level in the N-S direction is given in Table 6.13.

The interstory drifts Δ computed from the δ_x are also given in the table. The drift at story level x is determined by subtracting the design earthquake displacement at the bottom of the story from the design earthquake displacement at the top of the story:

$$\Delta = \delta_x - \delta_{x-1}$$

Table 6.13 Lateral Displacements and Story Drifts due to Seismic Forces in the N-S Direction

Story	δ_{xe} (in.)	δ_x (in.)	Δ (in.)
7	1.59	7.95	0.70
6	1.45	7.25	0.85
5	1.28	6.40	1.10
4	1.06	5.30	1.05
3	0.85	4.25	1.10
2	0.63	3.15	0.85
1	0.46	2.30	2.30

The design story drifts Δ shall not exceed the allowable story drift Δ_a given in Table 12.12-1. For Occupancy Category II and "all other structures," $\Delta_a = 0.020 h_{sx}$ where h_{sx} is the story height below level x.

For the 18-ft story height, $\Delta_a = 0.020 \times 18 \times 12 = 4.32$ in. > 2.30 in.

For the 13-ft story heights, $\Delta_a = 0.020 \times 13 \times 12 = 3.12$ in., which is greater than the values of Δ at floor levels 2 through 7.

Thus, drift limits are satisfied in the N-S direction.

2. Drift limits in the E-W direction.

   The maximum story displacements δ_{xe} in the N-S direction are summarized in Table 6.12. For special steel moment frames, the deflection amplification factor C_d is equal to 5.5 from Table 12.2-1.

A summary of the displacements at each floor level in the E-W direction is given in Table 6.14. The interstory drifts Δ computed from the δ_x are also given in the table.

Table 6.14 Lateral Displacements and Story Drifts due to Seismic Forces in the E-W Direction

Story	δ_{xe} (in.)	δ_x (in.)	Δ (in.)
7	5.36	29.48	1.98
6	5.00	27.50	2.69
5	4.51	24.81	3.91
4	3.80	20.90	6.05
3	2.70	14.85	6.54
2	1.51	8.31	4.90
1	0.62	3.41	3.41

In accordance with 12.12.1.1, design story drifts Δ must not exceed Δ_a/ρ for seismic force-resisting systems comprised solely of moment frames in structures assigned to SDC D, E or F where ρ is the redundancy factor determined in accordance with 12.3.4.2.

Due to the Type 1b extreme torsional irregularity, ρ must be equal to 1.3. Therefore, for the 18-ft story height, $\Delta_a/\rho = 0.020 \times 18 \times 12/1.3 = 3.32$ in. < 3.41 in.

For the 13-ft story heights, $\Delta_a/\rho = 0.020 \times 13 \times 12/1.3 = 2.40$ in., which is less than the design drifts at stories 2 through 6.

Thus, drift limits are not satisfied in the N-S direction. Increasing member sizes may not be sufficient to reduce the design drift; including additional members in the seismic force-resisting system may be needed to control drift. This, in turn, may help reduce the torsional effects.

6.7.6   Example 6.6 – Office Building, P-delta Effects

For the seven-story office building in Example 6.3, determine the P-delta effects in both the N-S and E-W directions.

For illustration purposes, use the lateral deflections determined by the equivalent lateral force procedure.

Assume a 10 psf live load on the roof and a 50 psf live load on the floors.

EXAMPLES 6-87

SOLUTION

1. P-delta effects in the N-S direction.

   In lieu of automatically considering P-delta effects in a computer analysis, the following procedure can be used to determine whether P-delta effects need to be considered in accordance with 12.8.7.

   P-delta effects need not be considered when the stability coefficient θ determined by Eq. 12.8-16 is less than or equal to 0.10:

   $$\theta = \frac{P_x \Delta}{V_x h_{sx} C_d}$$

   where P_x = total unfactored vertical design load at and above level x
   Δ = design story drift occurring simultaneously with V_x
   V_x = seismic shear force acting between level x and x–1
   h_{sx} = story height below level x
   C_d = deflection amplification factor in Table 12.2-1

   The stability coefficient θ must not exceed θ_{max} determined by Eq. 12.8-17:

   $$\theta_{max} = \frac{0.5}{\beta C_d} \leq 0.25$$

   Where β is the ratio of shear demand to shear capacity between level x and x–1, which may be taken equal to 1.0 when it is not calculated.

   The P-delta calculations for the N-S direction are shown in Table 6.15. It is clear that P-delta effects need not be considered at any of the floor levels. Note that θ_{max} is equal to 0.1000 in the N-S direction using $\beta = 1.0$.

   Table 6.15 P-delta Effects in the N-S Direction

Level	h_{sx} (ft)	P_x (kips)	V_x (kips)	Δ (in.)	θ
7	13	1,186	209	0.70	0.0051
6	13	3,407	451	0.85	0.0082
5	13	5,628	653	1.10	0.0122
4	13	7,849	816	1.05	0.0129
3	13	10,070	940	1.10	0.0151
2	13	12,291	1,025	0.85	0.0131
1	18	15,588	1,096	2.30	0.0303

2. P-delta effects in the E-W direction.

The P-delta calculations for the E-W direction are shown in Table 6.16. Note that θ_{max} is equal to 0.0909 in the E-W direction using $\beta = 1.0$, and, since θ is greater than θ_{max} at levels 2 through 4, the structure is potentially unstable and needs to be redesigned. It was determined in Example 6.4 that the shear capacity of the second floor is equal to 780 kips. Thus, $\beta = 474/780 = 0.61$, and $\theta_{max} = 0.5/(0.61 \times 5.5) = 0.15$. Assuming the same shear capacities at levels 3 and 4, $\theta_{max} = 0.16$ at level 3 and $\theta_{max} = 0.18$ at level 4. Therefore, the structure is still potentially unstable.

Table 6.16 P-delta Effects in the E-W Direction

Level	h_{sx} (ft)	P_x (kips)	V_x (kips)	Δ (in.)	θ
7	13	1,186	105	1.98	0.0261
6	13	3,407	223	2.69	0.0479
5	13	5,628	317	3.91	0.0809
4	13	7,849	389	6.05	0.1423
3	13	10,070	441	6.54	0.1741
2	13	12,291	474	4.90	0.1481
1	18	15,588	498	3.41	0.0898

6.7.7 Example 6.7 – Health Care Facility, Diaphragm Design Forces

Determine the diaphragm design forces for the three-story health care facility depicted in Figure 6.8 given the design data below.

DESIGN DATA

Location:	St. Louis, MO (Latitude: 38.63°, Longitude: -90.20°)
Soil classification:	Site Class C
Occupancy:	Health care facility without surgery or emergency treatment facilities
Material:	Cast-in-place concrete
Structural system:	Moment-resisting frames in both directions

---
SOLUTION

In order to determine the diaphragm design forces in accordance with 12.10.1.1, the design seismic forces must be determined at each floor level.

Assuming that the building is regular, the equivalent lateral force procedure can be used to determine the design seismic forces (see Table 12.6-1).

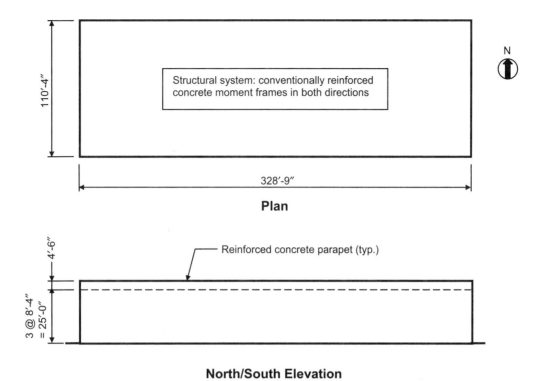

Figure 6.8 Plan and Elevation of Health Care Facility

- Step 1: Determine the seismic ground motion values from Flowchart 6.2.

  1. Determine the mapped accelerations S_S and S_1.

     In lieu of using Figures 22-1 and 22-2, the mapped accelerations are determined by inputting the latitude and longitude of the site into the USGS Ground Motion Parameter Calculator. The output is as follows: $S_S = 0.58$ and $S_1 = 0.17$.

  2. Determine the site class of the soil.

     The site class of the soil is given in the design data as Site Class C.

  3. Determine soil-modified accelerations S_{MS} and S_{M1}.

     Site coefficients F_a and F_v are determined from Tables 11.4-1 and 11.4-2, respectively:

     For Site Class C and $0.5 < S_S < 0.75$: $F_a = 1.17$ from linear interpolation

For Site Class C and $0.1 < S_1 < 0.2$: $F_v = 1.63$ from linear interpolation
Thus,

$$S_{MS} = 1.17 \times 0.58 = 0.68$$

$$S_{M1} = 1.63 \times 0.17 = 0.28$$

4. Determine design accelerations S_{DS} and S_{D1}.

From Eqs. 11.4-3 and 11.4-4:

$$S_{DS} = \frac{2}{3} \times 0.68 = 0.45$$

$$S_{D1} = \frac{2}{3} \times 0.28 = 0.19$$

- Step 2: Determine the SDC from Flowchart 6.4.

  1. Determine if the building can be assigned to SDC A in accordance with 11.4.1.

     Since $S_S = 0.58 > 0.15$ and $S_1 = 0.17 > 0.04$, the building cannot be automatically assigned to SDC A.

  2. Determine the Occupancy Category from IBC Table 1604.5.

     For a health care facility, the Occupancy Category is III.

  3. Since $S_1 < 0.75$, the building is not assigned to SDC E or F.

  4. Check if all four conditions of 11.6 are satisfied.

     - Check if the approximate period T_a is less than $0.8 T_S$.

       From Eq. 12.8-7 for a concrete moment-resisting frame:

       $$T_a = C_t h_n^x = 0.016(25.0)^{0.9} = 0.29 \text{ sec}$$

       where C_t and x are given in Table 12.8-2.

       $$T_S = S_{D1} / S_{DS} = 0.19 / 0.45 = 0.42 \text{ sec}$$

$$0.29 \text{ sec} < 0.8 \times 0.42 = 0.34 \text{ sec}$$

- The fundamental period used to calculate the design drift is taken as 0.29 sec, which is less than $T_S = 0.42$ sec.

- Equation 12.8-2 will be used to determine the seismic response coefficient C_s.

- Since the roof and the floors are cast-in-place concrete, the diaphragms are considered to be rigid.

Since all four conditions are satisfied, the SDC can be determined by Table 11.6-1 alone (11.6).

5. Determine the SDC from Table 11.6-1.

   From Table 11.6-1, with $0.33 < S_{DS} = 0.45 < 0.50$ and Occupancy Category III, the SDC is C.[17]

- Step 3: Determine the design seismic forces of the equivalent lateral force procedure from Flowchart 6.8.

  1. The design accelerations and the SDC have been determined in Steps 1 and 2 above.

  2. Determine the response modification coefficient R from Table 12.2-1.

     The moment-resisting frames in this building must be intermediate reinforced concrete moment frames, since the building is assigned to SDC C (system C6 in Table 12.2-1). In this case, $R = 5$. There is no height limit for this system in SDC C.

  3. Determine the importance factor I from Table 11.5-1.

     For Occupancy Category III, $I = 1.25$.

  4. Determine the period of the structure T.

     It was determined in Step 2, item 4 above that the approximate period of the structure T_a, which is permitted to be used in the equivalent lateral force procedure, is equal to 0.29 sec.

---

[17] If Table 11.6-2 were also used, the SDC would also be C.

5. Determine long-period transition period T_L from Figure 22-15.

   For St. Louis, MO, $T_L = 12$ sec $> T_a = 0.29$ sec

6. Determine seismic response coefficient C_s.

   The value of C_s must be determined by Eq. 12.8-2 (see Step 2, item 4):

   $$C_s = \frac{S_{DS}}{R/I} = \frac{0.45}{5/1.25} = 0.11$$

7. Determine effective seismic weight W in accordance with 12.7.2.

   The effective weights at each floor level are given in Table 6.17. The total weight W is the summation of the effective dead loads at each level.

8. Determine seismic base shear V.

   Seismic base shear is determined by Eq. 12.8-1:

   $$V = C_s W = 0.11 \times 16,320 = 1,795 \text{ kips}$$

9. Determine exponent related to structure period k.

   Since $T < 0.5$ sec, $k = 1.0$.

10. Determine lateral seismic force F_x at each level x.

    F_x is determined by Eqs. 12.8-11 and 12.8-12. A summary of the lateral forces F_x and the story shears V_x is given in Table 6.17.

Table 6.17 Seismic Forces and Story Shears

Level	Story weight, w_x (kips)	Height, h_x (ft)	$w_x h_x^k$	Lateral force, F_x (kips)	Story Shear, V_x (kips)
3	4,958	25.00	123,950	837	837
2	5,681	16.67	94,702	639	1,476
1	5,681	8.33	47,323	319	1,795
Σ	16,320		265,975	1,795	

- Step 4: Determine the diaphragm design seismic forces using Eq. 12.10-1.

Diaphragm design force $F_{px} = \dfrac{\sum_{i=x}^{n} F_i}{\sum_{i=x}^{n} w_i} w_{px}$

where w_i = weight tributary to level i and w_{px} = weight tributary to the diaphragm at level x.

Minimum $F_{px} = 0.2 S_{DS} I w_{px} = 0.2 \times 0.45 \times 1.25 \times w_{px} = 0.1125 w_{px}$

Maximum $F_{px} = 0.4 S_{DS} I w_{px} = 0.2250 w_{px}$

Assuming that the exterior walls are primarily glass, which weigh significantly less than the diaphragm weight at each level, the weight that is tributary to each diaphragm is identical to the weight of the structure at that level (i.e., $w_{px} = w_x$).

A summary of the diaphragm forces is given in Table 6.18.

Table 6.18 Design Seismic Diaphragm Forces

Level	w_x (kips)	Σw_x (kips)	F_x (kips)	ΣF_x (kips)	$\Sigma F_x / \Sigma w_x$	w_{px} (kips)	F_{px} (kips)
3	4,958	4,958	837	837	0.1688	4,958	837
2	5,681	10,639	639	1,476	0.1387	5,681	788
1	5,681	16,320	319	1,795	0.1125*	5,681	639

* Minimum value governs.

6.7.8 Example 6.8 – Health Care Facility, Nonstructural Component

Determine the design seismic force on the parapet of the health care facility in Example 6.7.

SOLUTION

Use Flowchart 6.10 to determine the seismic force on the parapet.

1. Determine S_{DS}, S_{D1} and the SDC.

   The design accelerations and the SDC are determined in Example 6.7.

   The parapet is assigned to SDC C, which is the same SDC as the building to which it is attached (13.1.2).

2. Determine the component amplification factor a_p and the component response modification factor R_p from Table 13.5-1 for architectural components.

   Assuming that the parapet is not braced, $a_p = 2.5$ and $R_p = 2.5$ from Table 13.5-1.

3. Determine component importance factor I_p in accordance with 13.1.3.

   Since the parapet does not meet any of the three criteria that require $I_p = 1.5$, then $I_p = 1.0$.

4. Determine the horizontal seismic design force F_p by Eq. 13.3-1.

$$F_p = \frac{0.4 a_p S_{DS} W_p}{\left(\frac{R_p}{I_p}\right)}\left(1 + 2\frac{z}{h}\right)$$

Assuming that the thickness of the parapet is 8 in. and that normal weight concrete is utilized, $W_p = 8 \times 150/12 = 100$ psf.

Since the parapet is attached to the top of the structure, $z/h = 1$.

Thus,

$$F_p = \frac{0.4 \times 2.5 \times 0.45 \times 100}{\left(\frac{2.5}{1.0}\right)}(1+2) = 54 \text{ psf}$$

Minimum $F_p = 0.3 S_{DS} I_p W_p = 13.5$ psf

Maximum $F_p = 1.6 S_{DS} I_p W_p = 72.0$ psf

The 54 psf seismic load is applied to the parapet as shown in Figure 6.9.

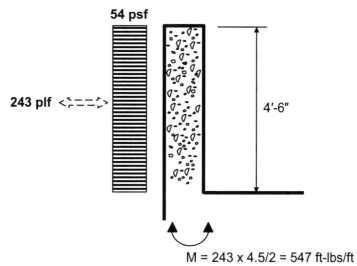

M = 243 × 4.5/2 = 547 ft-lbs/ft

Figure 6.9 Design Seismic Force on Parapet

6.7.9 Example 6.9 – Residential Building, Vertical Combination of Structural Systems

Determine the design seismic forces on the residential building depicted in Figure 6.10 given the design data below.

DESIGN DATA	
Location:	Philadelphia, PA (Latitude: 39.92°, Longitude: -75.23°)
Soil classification:	Site Class D
Occupancy:	Residential occupancy where less than 300 people congregate in one area
Structural systems:	Light-frame wood bearing walls with shear panels rated for shear resistance and cast-in-place reinforced concrete building frame system with ordinary reinforced concrete shear walls

SOLUTION

- Step 1: Determine the seismic ground motion values from Flowchart 6.2.

    Using the USGS Ground Motion Parameter Calculator, $S_S = 0.27$ and $S_1 = 0.06$.

    Using Tables 11.4-1 and 11.4-2, the soil-modified accelerations are $S_{MS} = 0.44$ and $S_{M1} = 0.15$.

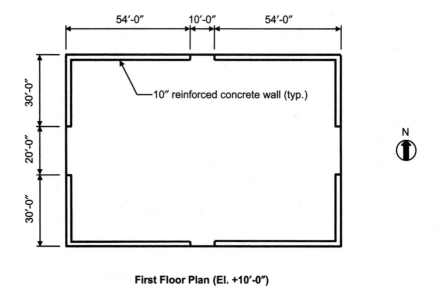

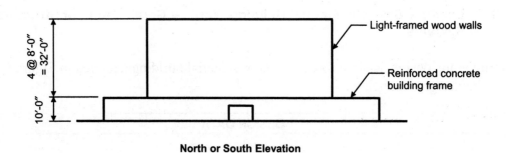

Figure 6.10 First Floor Plan and Elevation of Residential Building

Design accelerations: $S_{DS} = \frac{2}{3} \times 0.44 = 0.29$ and $S_{D1} = \frac{2}{3} \times 0.15 = 0.10$

- Step 2: Determine the SDC from Flowchart 6.4.

From IBC Table 1604.5, the Occupancy Category is II.

From Table 11.6-1 with $0.167 < S_{DS} < 0.33$ and Occupancy II, the SDC is B.

From Table 11.6-2 with $0.067 < S_{D1} < 0.133$ and Occupancy Category II, the SDC is B.

Therefore, the SDC is B for this building.

EXAMPLES   6-97

- Step 3: Determine the response modification coefficients R in accordance with 12.2.3.1 for vertical combinations of structural systems.

  The vertical combination of structural systems in this building is a flexible wood frame upper portion above a rigid concrete lower portion. Thus, a two-stage equivalent lateral force procedure is permitted to be used provided the design of the structure complies with the four criteria listed in 12.2.3.1:

  a. The stiffness of lower portion must be at least 10 times the stiffness of the upper portion.

     It can be shown that the stiffness of the lower portion of this structure is more than 10 times that of the upper portion. **O.K.**

  b. The period of the entire structure shall not be greater than 1.1 times the period of the upper portion considered as a separate structure fixed at the base.

     Determine the period of the upper portion by Eq. 12.8-7 using the approximate period coefficients in Table 12.8-2 for "all other structural systems":

     $$T_a = C_t h_n^x = 0.02(32)^{0.75} = 0.27 \text{ sec}$$

     Determine the period of the lower portion in the N-S direction by Eqs. 12.8-9 and 12.8-10 for concrete shear wall structures:

     $$T_a = \frac{0.0019}{\sqrt{C_W}} h_n$$

     $$C_W = \frac{100}{A_B} \sum_{i=1}^{x} \left(\frac{h_n}{h_i}\right)^2 \frac{A_i}{\left[1 + 0.83\left(\frac{h_i}{D_i}\right)^2\right]}$$

     where

     A_B = area of base of structure = 118×80 = 9,440 sq ft

     h_n = height of lower portion of building = 10 ft

     h_i = height of shear wall i = 10 ft

     A_i = area of shear wall i = $\frac{10}{12} \times 30$ = 25 sq ft

D_i = length of shear wall i = 30 ft

Thus,

$$C_W = \frac{4 \times 100}{9{,}440} \frac{25}{\left[1 + 0.83\left(\dfrac{10}{30}\right)^2\right]} = 0.97$$

$$T_a = \frac{0.0019}{\sqrt{0.97}} \times 10 = 0.02 \text{ sec}$$

The period of the lower portion in the E-W direction is equal to 0.01 sec.

The period of the combined structure is approximately 0.28 sec.[18]

0.28 sec < 1.1×0.27 = 0.30 sec **O.K.**

c. The flexible upper portion shall be designed as a separate structure using the appropriate values of R and ρ.

From Table 12.2-1 for a bearing wall system with light-framed walls with wood structural panels rated for shear resistance (system A13), R = 6.5 with no height limitation for SDC B.

For SDC B, $\rho = 1.0$ (12.3.4.1).

d. The rigid lower portion shall be designed as a separate structure using the appropriate values of R and ρ. Amplified reactions from the upper portion are applied to the lower portion where the amplification factor is equal to the ratio of the R/ρ of the upper portion divided by R/ρ of the lower portion and must be greater than or equal to one.

From Table 12.2-1 for a building frame system with ordinary reinforced concrete shear walls (system B6), R = 5 with no height limitation for SDC B.

For SDC B, $\rho = 1.0$ (12.3.4.1).

Amplification factor = (6.5/1)/(5/1) = 1.3.

Therefore, a two-stage equivalent lateral force procedure is permitted to be used.

---

[18] The period of the combined structure was obtained from a commercial computer program.

EXAMPLES   6-99

- Step 4: Determine the design seismic forces on the upper and lower portions of the structure using the equivalent lateral force procedure.

    1. Use Flowchart 6.8 to determine the lateral seismic forces on the flexible upper portion of the structure.

        a. The design accelerations and the SDC have been determined in Steps 1 and 2 above, respectively.

        b. Determine the response modification coefficient R from Table 12.2-1.

            The response modification coefficient was determined in Step 3 as 6.5.

        c. Determine the importance factor I from Table 11.5-1.

            For Occupancy Category II, $I = 1.0$.

        d. Determine the period of the structure T.

            It was determined in Step 3 that the approximate period of the structure $T_a = 0.27$ sec.

        e. Determine long-period transition period T_L from Figure 22-15.

            For Philadelphia, PA, $T_L = 6$ sec $> T_a = 0.27$ sec.

        f. Determine seismic response coefficient C_s.

            The seismic response coefficient C_s is determined by Eq. 12.8-3:

            $$C_s = \frac{S_{D1}}{T(R/I)} = \frac{0.10}{0.27(6.5/1.0)} = 0.06$$

            The value of C_s need not exceed that from Eq. 12.8-2:

            $$C_s = \frac{S_{DS}}{R/I} = \frac{0.29}{6.5/1.0} = 0.05$$

            Also, C_s must not be less than the larger of $0.044 S_{DS} I = 0.013$ (governs) and 0.01 (Eq. 12.8-5).

            Thus, the value of C_s from Eq. 12.8-2 governs.

g. Determine effective seismic weight W in accordance with 12.7.2.

The effective weights at each floor level are given in Table 6.19. The total weight W is the summation of the effective dead loads at each level.

h. Determine seismic base shear V.

Seismic base shear is determined by Eq. 12.8-1:

$$V = C_s W = 0.05 \times 761 = 38 \text{ kips}$$

i. Determine exponent related to structure period k.

Since $T = 0.27$ sec < 0.5 sec, $k = 1.0$.

Table 6.19 Seismic Forces and Story Shears on Flexible Upper Portion

Level	Story weight, w_x (kips)	Height, h_x (ft)	$w_x h_x^k$	Lateral force, F_x (kips)	Story Shear, V_x (kips)
4	185	32	5,920	15	15
3	192	24	4,608	11	26
2	192	16	3,072	8	34
1	192	8	1,536	4	38
Σ	761		15,136	38	

j. Determine lateral seismic force F_x at each level x.

F_x is determined by Eqs. 12.8-11 and 12.8-12. A summary of the lateral forces F_x and the story shears V_x is given in Table 6.19.

2. Use Flowchart 6.8 to determine the lateral seismic forces on the rigid lower portion of the structure in the N-S direction.

   a. The design accelerations and the SDC have been determined in Steps 1 and 2 above, respectively.

   b. Determine the response modification coefficient R from Table 12.2-1.

   It was determined in Step 3 that the response modification coefficient $R = 5$.

   c. Determine the importance factor I from Table 11.5-1.

   For Occupancy Category II, $I = 1.0$.

EXAMPLES

d. Determine the period of the structure T.

It was determined in Step 3 that the approximate period of the structure $T_a = 0.02$ sec in the N-S direction.

e. Determine long-period transition period T_L from Figure 22-15.

For Philadelphia, PA, $T_L = 6$ sec $> T_a = 0.02$ sec.

f. Determine seismic response coefficient C_S.

The seismic response coefficient C_S is determined by Eq. 12.8-3:

$$C_S = \frac{S_{D1}}{T(R/I)} = \frac{0.10}{0.02(5/1.0)} = 1.0$$

The value of C_S need not exceed that from Eq. 12.8-2:

$$C_S = \frac{S_{DS}}{R/I} = \frac{0.29}{5/1.0} = 0.06$$

Also, C_S must not be less than the larger of $0.044 S_{DS} I = 0.013$ (governs) and 0.01 (Eq. 12.8-5).

Thus, the value of C_S from Eq. 12.8-2 governs.

g. Determine effective seismic weight W in accordance with 12.7.2.

The effective weight W is equal to 1,458 kips for the lower portion.

h. Determine seismic base shear V.

Seismic base shear is determined by Eq. 12.8-1:

$$V = C_S W = 0.06 \times 1{,}458 = 88 \text{ kips}$$

i. Determine total lateral seismic forces on the lower portion.

For the one-story lower portion, the total seismic force is equal to the lateral force due to the base shear of the lower portion plus the amplified seismic force from the upper portion:

$$V_{total} = 88 + (1.3 \times 38) = 137 \text{ kips}$$

Since the base shear is independent of the period, the total seismic force in the E-W direction is also equal to 137 kips.

The distribution of the lateral seismic forces in the upper and lower parts of the structure is shown in Figure 6.11.

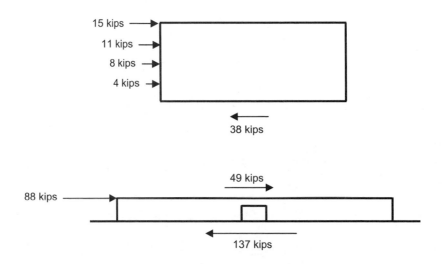

Figure 6.11 Distribution of Lateral Seismic Forces in the Upper and Lower Portions of the Structure

6.7.10 Example 6.10 – Warehouse Building, Design of Roof Diaphragm, Collectors, and Wall Panels

For the one-story warehouse illustrated in Figure 6.12, determine (1) design seismic forces on the diaphragm, including diaphragm shear forces in both directions, (2) design seismic forces on the steel collector beam in the N-S direction, and (3) out-of-plane design seismic forces on the precast concrete wall panels, given the design data below.

DESIGN DATA	
Location:	San Francisco, CA (Latitude: 37.75°, Longitude: -122.43°)
Soil classification:	Site Class D
Occupancy:	Less than 300 people congregate in one area and the building is not used to store hazardous or toxic materials
Structural system:	Building frame system with intermediate precast concrete shear walls

EXAMPLES   6-103

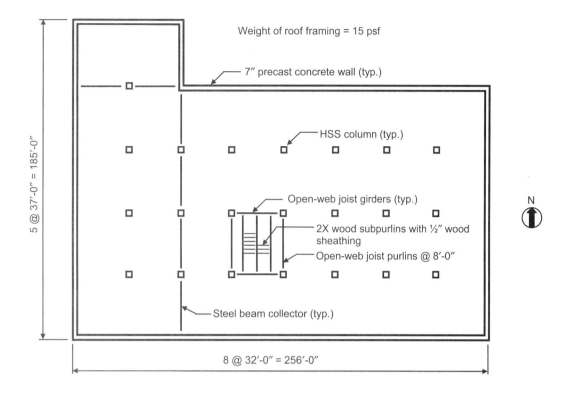

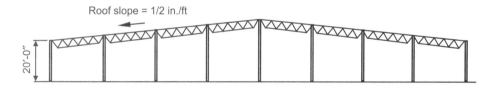

Figure 6.12 Plan and Elevation of Warehouse Building

SOLUTION

Part 1: Determine design seismic forces on the diaphragm

- Step 1: Determine the seismic ground motion values from Flowchart 6.2.

    Using the USGS Ground Motion Parameter Calculator, $S_S = 1.51$ and $S_1 = 0.76$.

    Using Tables 11.4-1 and 11.4-2, the soil-modified accelerations are $S_{MS} = 1.51$ and $S_{M1} = 1.14$.

Design accelerations: $S_{DS} = \frac{2}{3} \times 1.51 = 1.01$ and $S_{D1} = \frac{2}{3} \times 1.14 = 0.76$

- Step 2: Determine the SDC from Flowchart 6.4.

  From IBC Table 1604.5, the Occupancy Category is II.

  Since $S_1 > 0.75$, the SDC is E for this building (11.6).

- Step 3: Use Flowchart 6.8 to determine the seismic base shear using the equivalent lateral force procedure.

  1. Check if equivalent lateral force procedure can be used (see Flowchart 6.6).

     a. The building has a Type 2 re-entrant corner irregularity, since 37.0 ft > $0.15 \times 185.0 = 27.8$ ft and 192.0 ft > $0.15 \times 256.0 = 38.4$ ft (see Table 12.3-1 and Table 6.2).

     b. Determine the period in the N-S direction by Eqs. 12.8-9 and 12.8-10 for concrete shear wall structures:

     $$T_a = \frac{0.0019}{\sqrt{C_W}} h_n$$

     $$C_W = \frac{100}{A_B} \sum_{i=1}^{x} \left(\frac{h_n}{h_i}\right)^2 \frac{A_i}{\left[1 + 0.83\left(\frac{h_i}{D_i}\right)^2\right]}$$

     where

     A_B = area of base of structure = 40,256 sq ft

     h_n = average height of building = 22.67 ft

     h_i = average height of shear wall i = 22.67 ft

     A_i = area of shear wall i: $A_1 = \frac{7}{12} \times 185 = 107.9$ sq ft, $A_2 = \frac{7}{12} \times 37 = 21.6$ sq ft, $A_3 = \frac{7}{12} \times 148 = 86.3$ sq ft

     D_i = length of shear wall i: $D_1 = 185$ ft, $D_2 = 37$ ft, $D_3 = 148$ ft

Thus,

$$C_W = \frac{100}{40{,}256}\left\{\frac{107.9}{\left[1+0.83\left(\frac{22.67}{185}\right)^2\right]} + \frac{21.6}{\left[1+0.83\left(\frac{22.67}{37}\right)^2\right]} + \frac{86.3}{\left[1+0.83\left(\frac{22.67}{148}\right)^2\right]}\right\}$$

$$= \frac{100}{40{,}256}(106.6+16.5+84.7) = 0.52$$

$$T_a = \frac{0.0019}{\sqrt{0.52}} \times 22.67 = 0.06 \text{ sec}$$

c. Determine the period in the E-W direction.

$$C_W = \frac{100}{40{,}256}\left\{\frac{37.3}{\left[1+0.83\left(\frac{22.67}{64}\right)^2\right]} + \frac{112.0}{\left[1+0.83\left(\frac{22.67}{192}\right)^2\right]} + \frac{149.3}{\left[1+0.83\left(\frac{22.67}{256}\right)^2\right]}\right\}$$

$$= \frac{100}{40{,}256}(33.8+110.7+148.3) = 0.73$$

$$T_a = \frac{0.0019}{\sqrt{0.73}} \times 22.67 = 0.05 \text{ sec}$$

$3.5 T_s = 3.5 \times 0.76/1.01 = 2.63$ sec > period in both directions

In accordance with Table 12.6-1, the equivalent lateral force procedure is permitted to be used for this irregular structure with a Type 2 re-entrant corner irregularity and with $T < 3.5 T_s$.

2. Use Flowchart 6.8 to determine the lateral seismic forces in the N-S and E-W directions.

   a. The design accelerations and the SDC have been determined in Steps 1 and 2 above, respectively.

b. Determine the response modification coefficient R from Table 12.2-1.

For a building frame system with intermediate precast concrete shear walls (system B9), $R = 5$. Note that the average building height of 22.67 ft is less than the 45 ft height limit for this system assigned to SDC E (see footnote k in Table 12.2-1).

c. Determine the importance factor I from Table 11.5-1.

For Occupancy Category II, $I = 1.0$.

d. Determine the period of the structure T.

It was determined above that $T_a = 0.06$ sec in the N-S direction and $T_a = 0.05$ sec in the E-W direction.

e. Determine long-period transition period T_L from Figure 22-15.

For San Francisco, CA, $T_L = 12$ sec $> T_a$ in both directions.

f. Determine seismic response coefficient C_s.

The value of C_s from Eq. 12.8-2 is:

$$C_s = \frac{S_{DS}}{R/I} = \frac{1.01}{5/1.0} = 0.20$$

Also, C_s must not be less than the larger of $0.044 S_{DS} I = 0.013$ and 0.044 (Eq. 12.8-5). or the value obtained by Eq. 12.8-6 (governs), since $S_1 > 0.6$:

$$C_s = \frac{0.5 S_1}{\left(\dfrac{R}{I}\right)} = \frac{0.5 \times 0.76}{\left(\dfrac{5}{1}\right)} = 0.08$$

Thus, the value of C_s from Eq. 12.8-2 governs.

g. Determine effective seismic weight W in accordance with 12.7.2.

The effective weight W is equal to the weight of the roof framing plus the weight of the walls tributary to the roof:[19]

---

[19] The weight of the walls is conservatively calculated assuming that there are no openings in the walls.

EXAMPLES

Weight of roof framing = $0.015 \times 40,256 = 604$ kips

Weight of walls = $\frac{7}{12} \times 0.15 \times \frac{22.67}{2} \times [2(256 + 185)] = 875$ kips

$W = 604 + 875 = 1,479$ kips

h. Determine seismic base shear V.

Seismic base shear is determined by Eq. 12.8-1 and is the same in both the N-S and E-W directions, since the governing C_s is independent of the period:

$V = C_s W = 0.20 \times 1,479 = 296$ kips

3. Determine the design seismic forces in the diaphragm in both directions by Eq. 12.10-1.

Diaphragm design force $F_{px} = \dfrac{\sum_{i=x}^{n} F_i}{\sum_{i=x}^{n} w_i} w_{px}$

where w_i = weight tributary to level i and w_{px} = weight tributary to the diaphragm at level x.

Since this is a one-story building, Eq. 12.10-1 reduces to $F_{px} = 0.20 w_{px}$

Minimum $F_{px} = 0.2 S_{DS} I w_{px} = 0.2 \times 1.01 \times 1.0 \times w_{px} = 0.20 w_{px}$

Maximum $F_{px} = 0.4 S_{DS} I w_{px} = 0.40 w_{px}$

The wood sheathing is permitted to be idealized as a flexible diaphragm in accordance with 12.3.1.1. Seismic forces are computed from the tributary weight of the roof and the walls oriented perpendicular to the direction of analysis.[20]

<u>N-S direction</u>

Uniform diaphragm loads w_{N1} and w_{N2} are computed as follows (see Figure 6.13):

---

[20] Walls parallel to the direction of the seismic forces are typically not considered in the tributary weight, since these walls do not obtain support from the diaphragm in the direction of the seismic force.

$$w_{N1} = 0.20 \times 15 \times 185 + 0.20 \times 87.5 \times 2 \times \frac{22.67}{2} = 952 \text{ plf}$$

$$w_{N2} = 0.20 \times 15 \times 148 + 0.20 \times 87.5 \times 2 \times \frac{22.67}{2} = 841 \text{ plf}$$

Also shown in Figure 6.13 is the shear diagram for the diaphragm.

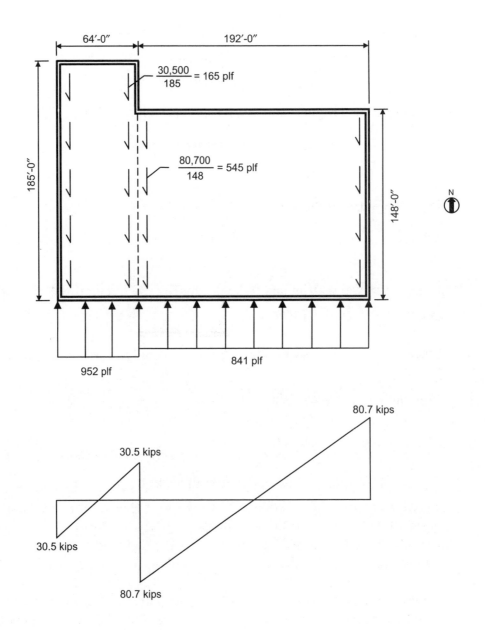

Figure 6.13 Design Seismic Forces and Shear Forces in the Diaphragm in the N-S Direction

E-W direction

Uniform diaphragm loads w_{E1} and w_{E2} are computed as follows (see Figure 6.14):

$$w_{E1} = 0.20 \times 15 \times 64 + 0.20 \times 87.5 \times 2 \times \frac{22.67}{2} = 589 \text{ plf}$$

$$w_{E2} = 0.20 \times 15 \times 256 + 0.20 \times 87.5 \times 2 \times \frac{22.67}{2} = 1{,}165 \text{ plf}$$

Also shown in Figure 6.14 is the shear diagram for the diaphragm.

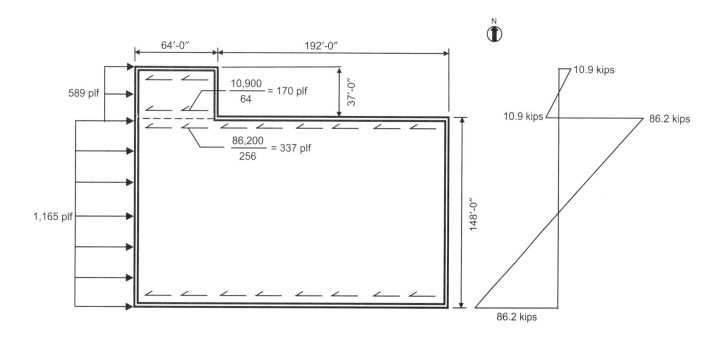

Figure 6.14 Design Seismic Forces and Shear Forces in the Diaphragm in the E-W Direction

4. Determine connection forces between the diaphragm and the shear walls.

Since the building has a Type 2 horizontal irregularity, the diaphragm connection design forces must be increased by 25 percent in accordance with 12.3.3.4. Thus, $F_{px} = 1.25 \times 0.20 w_{px} = 0.25 w_{px}$. This force increase applies to the row of diaphragm nailing that transfers the above diaphragm shears directly to the shear walls (and to the collectors) and to the bolts between the ledger beams and the shear walls.

Part 2: Determine design seismic forces on the steel collector beam in the N-S direction

From the diaphragm shear diagram in Figure 6.13, the maximum collector load is equal to 30.5(148/185) + 80.7 = 105.1 kips tension or compression.

The uniform axial load can be approximated by dividing the maximum load by the length of the collector: 105,100/148 = 710 plf.

The uniform axial load can be used to determine the axial force in any of the beams at any point along their length. For example, at the midspan of the northernmost collector beam, the axial force is equal to $710 \times (148 - 37/2)/1,000 = 92$ kips tension or compression. In accordance with 12.3.3.4, this force must be increased by 25 percent unless the collector is designed for the load combinations with overstrength factor of 12.4.3.2 (see 12.10.2.1).

The collector beams are subsequently designed for the combined effects of gravity and axial loads due to the design seismic forces.

Part 3: Determine out-of-plane design seismic forces on the precast concrete wall panels

1. Solid wall panels

    According to 12.11.1, structural walls shall be designed for a force normal to the surface equal to $0.4S_{DS}I$ times the weight of the wall. The minimum normal force is 10 percent of the weight of the wall.

    For a solid precast concrete wall panel:

    Weight $W_p = (7/12) \times 0.15 \times 22.67 = 2.0$ kips/ft

    $0.1 \times 2.0 = 0.2$ kips/ft

    $0.4 \times 1.01 \times 1.0 \times 2.0 = 0.8$ kips/ft (governs)

    This force governs over the three minimum anchorage forces specified in 12.11.2 as well.

    Distributed load = 0.8/22.67 = 0.04 kips/ft/ft width of wall

    This uniformly distributed load is applied normal to the wall in either direction.

    Anchorage of the precast walls to the flexible diaphragms must develop the out-of-plane force given by Eq. 12.11-1:

$F_p = 0.8 S_{DS} I W_p = 1.6$ kips/ft

Note that the 25 percent increase in the design force for diaphragm connections is not applied to out-of-plane wall anchorage force to the diaphragm (12.3.3.4).

2. Wall panels with openings

A typical wall panel on the east or west faces of the building is shown in Figure 6.15.

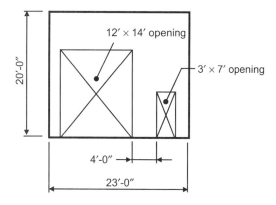

Figure 6.15 Typical Precast Wall Panel on East and West Faces

In lieu of a more rigorous analysis, the pier width between the two openings is commonly defined as a design strip. The total weight used in determining the out-of-plane design seismic force is taken as the weight of the design strip plus the weight of the wall tributary to the design strip above each adjacent opening (see Figure 6.16):

$$W_{p1} = \frac{7}{12} \times 150 \times 4 = 350 \text{ plf}$$

$$W_{p2} = \frac{7}{12} \times 150 \times 6 = 525 \text{ plf}$$

$$W_{p3} = \frac{7}{12} \times 150 \times 1.5 = 131 \text{ plf}$$

The out-of-plane seismic forces are determined by 12.11.1:

$$F_{p1} = 0.4 S_{DS} I W_{p1} = 0.4 \times 1.01 \times 1.0 \times 350 = 141.4 \text{ plf from 0 ft to 20 ft}$$

$$F_{p2} = 0.4 S_{DS} I W_{p1} = 0.4 \times 1.01 \times 1.0 \times 525 = 212.1 \text{ plf from 14 ft to 20 ft}$$

$F_{p3} = 0.4 S_{DS} I W_{p1} = 0.4 \times 1.01 \times 1.0 \times 131 = 52.9$ plf from 7 ft to 20 ft

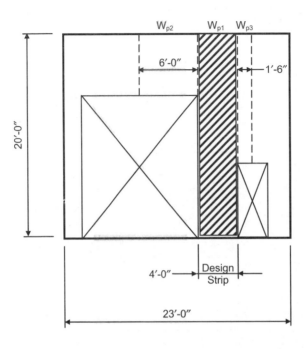

Figure 6.16 *Design Strip and Tributary Weights*

The wall is designed for the combination of axial load from the gravity forces and bending and shear from the out-of-plane seismic forces.

6.7.11 Example 6.11 – Retail Building, Simplified Design Method

For the one-story retail building illustrated in Figure 6.17, determine the seismic base shear using the simplified alternative structural design criteria of 12.14.

DESIGN DATA	
Location:	Seattle, WA (Latitude: 47.60°, Longitude: -122.33°)
Soil classification:	Site Class C
Occupancy:	Business occupancy
Structural system:	Bearing wall system with special reinforced masonry shear walls

SOLUTION

- Step 1: Determine the seismic ground motion values from Flowchart 6.2.

Using the USGS Ground Motion Parameter Calculator, $S_S = 1.47$ and $S_1 = 0.50$.

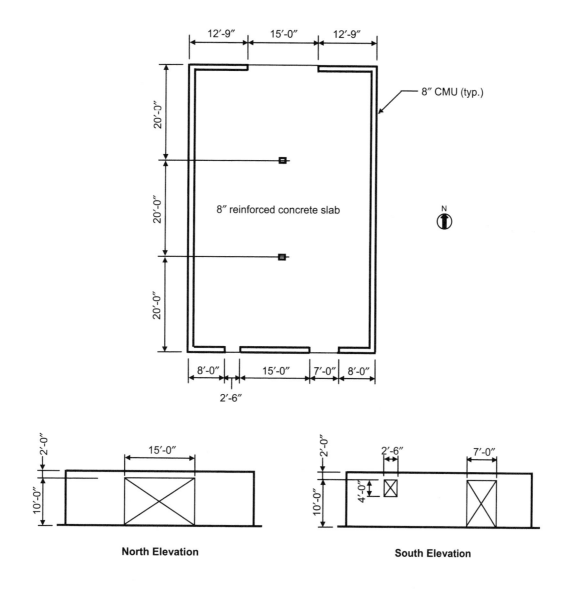

Figure 6.17 Plan and Elevations of One-story Retail Building

Using Tables 11.4-1 and 11.4-2, the soil-modified accelerations are $S_{MS} = 1.47$ and $S_{M1} = 0.65$.

Design accelerations: $S_{DS} = \frac{2}{3} \times 1.47 = 0.98$ and $S_{D1} = \frac{2}{3} \times 0.65 = 0.43$

- Step 2: Determine if the simplified method of 12.14 can be used for this building.

  The simplified method is permitted to be used if the following 12 limitations are met:

  1. The structure shall qualify for Occupancy I or II in accordance with Table 1-1.

From Table 1-1, the Occupancy Category is II. **O.K.**

2. The Site Class shall not be E or F.

   The Site Class is C in accordance with the design data. **O.K.**

3. The structure shall not exceed three stories in height.

   The structure is one story. **O.K.**

4. The seismic force-resisting system shall be either a bearing wall system or building frame system in accordance with Table 12.14-1.

   The seismic force-resisting system is a bearing wall system. **O.K.**

5. The structure has at least two lines of lateral resistance in each of the two major axis directions.

   Masonry shear walls are provided along two lines at the perimeter in both directions. **O.K.**

6. At least one line of resistance shall be provided on each side of the center of mass in each direction.

   The center of mass is approximately located at the geometric center of the building and walls are provided on all four sides at the perimeter. **O.K.**

7. For structures with flexible diaphragms, overhangs beyond the outside line of shear walls or braced frames shall satisfy: $a \leq d/5$.

   The diaphragm in this building is rigid, so this limitation is not applicable.

8. For buildings with diaphragms that are not flexible, the distance between the center of rigidity and the center of mass parallel to each major axis shall not exceed 15 percent of the greatest width of the diaphragm parallel to that axis.

   Assume that the center of mass is at the geometric center of this building.[21] This limitation is satisfied with respect to the east-west direction, since the center of rigidity and center of mass are on the same line due to the symmetrical layout of the walls on the east and west elevations.

   The center of rigidity must be located in the north-south direction, since the north and south walls are not identical. By inspection, the center of rigidity is located

---

[21] The exact location of the center of mass should be computed where it is anticipated to be offset from the geometric center of the building.

closer to the south wall, since the stiffness of that wall is greater than the stiffness of the north wall.

To locate the center of rigidity, the stiffnesses of the north and south walls must be determined. Assuming that the piers are fixed at the top and bottom ends, the stiffnesses (or rigidities) of the walls and piers can be determined by the following:

Total displacement of pier or wall i: $\delta_i = \delta_{fi} + \delta_{vi}$

$$\delta_{fi} = \text{displacement due to bending} = \frac{\left(\dfrac{h_i}{\ell_i}\right)^3}{Et}$$

$$\delta_{vi} = \text{displacement due to shear} = \frac{3\left(\dfrac{h_i}{\ell_i}\right)}{Et}$$

where h_i = height of pier or wall

ℓ_i = length of pier or wall

t = thickness of pier or wall

E = modulus of elasticity of pier or wall

Stiffness of pier or wall $k_i = \dfrac{1}{\delta_i}$

In lieu of a more rigorous analysis, the stiffness of a wall with openings is determined as follows: first, the deflection of the wall is obtained as though it were a solid wall with no openings. Next, the deflection of a solid strip of wall that contains the openings is subtracted from the total deflection. Finally, the deflection of each pier surrounded by the openings is added back.

Table 6.20 contains a summary of the stiffness calculations for the north wall. Similar calculations for the south wall are given in Table 6.21. The pier designations are provided in Figure 6.18.

North wall stiffness = $0.635Et$

South wall stiffness = $0.818Et$

East and west wall stiffness = $1.65Et$.

Table 6.20 Stiffness Calculations for North Wall

Pier/Wall	h_i (ft)	ℓ_i (ft)	$\delta_{fi}Et = (h_i/\ell_i)^3$	$\delta_{vi}Et = 3(h_i/\ell_i)$	$\delta_i Et$	k_i/Et
1 + 2 + 3 + 4	12	40.5	0.026	0.889	0.915	---
1 + 2 + 4	10	40.5	-0.015	-0.741	-0.756	---
1	10	12.75	0.483	2.353	---	0.353
2	10	12.75	0.483	2.353	---	0.353
1 + 2	---	---	---	---	1.416	0.706

1.575 → 0.635

Table 6.21 Stiffness Calculations for South Wall

Pier/Wall	h_i (ft)	ℓ_i (ft)	$\delta_{fi}Et = (h_i/\ell_i)^3$	$\delta_{vi}Et = 3(h_i/\ell_i)$	$\delta_i Et$	k_i/Et	$\delta_i Et$
1 + 2 + 3 + 4 + 5 + 6 + 7	12	40.5	0.026	0.889	0.915	---	---
1 + 2 + 3 + 4 + 6 + 7	10	40.5	-0.015	-0.741	-0.756	---	---
2 + 3 + 4 + 6	10	25.5	0.060	1.177	---	---	1.237
2 + 3 + 6	4	25.5	-0.004	-0.471	---	---	-0.475
2	4	15.0	0.019	0.800	---	1.221	---
3	4	8.0	0.125	1.500	---	0.615	---
2 + 3	---	---	---	---	---	1.836	0.545
2 + 3 + 4	---	---	---	---	---	0.765	1.307
1	10	8.0	1.953	3.750	---	0.175	---
1 + 2 + 3 + 4	---	---	---	---	1.064	0.940	---

1.223 → 0.818

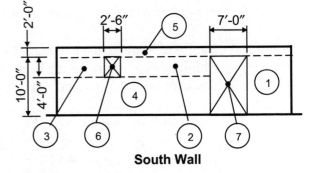

Figure 6.18 Pier Designations for Stiffness Calculations

The location of the center of rigidity in the north-south direction can be determined from the following equation:

$$\bar{y}_r = \frac{\sum k_i y_i}{\sum k_i}$$

where y_i is the distance from a reference point to wall i.

Using the centerline of the south wall as the reference line (see Figure 6.19),

$$\bar{y}_r = \frac{0.635 Et \left(60 - \dfrac{7.625}{12}\right)}{0.635 Et + 0.818 Et} = 25.9 \text{ ft}$$

$$e_1 = \left(30 - \frac{7.625}{2 \times 12}\right) - 25.9 = 3.8 \text{ ft}$$

$0.15 \times 60 = 9.0 \text{ ft} > 3.8 \text{ ft}$ **O.K.**

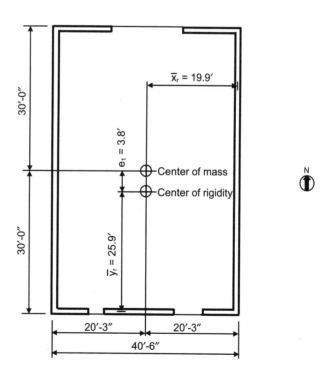

Figure 6.19 *Locations of Center of Mass and Center of Rigidity*

In addition, Eqs. 12.14-2A and 12.14-2B must be satisfied:

$$\sum_{i=1}^{m} k_{1i} d_{1i}^2 + \sum_{j=1}^{n} k_{2j} d_{2j}^2 \geq 2.5\left(0.05 + \frac{e_1}{b_1}\right) b_1^2 \sum_{i=1}^{m} k_{1i}$$

$$\sum_{i=1}^{m} k_{1i} d_{1i}^2 + \sum_{j=1}^{n} k_{2j} d_{2j}^2 \geq 2.5\left(0.05 + \frac{e_2}{b_2}\right) b_2^2 \sum_{j=1}^{n} k_{2j}$$

where the notation is defined in Figure 12.14-1 and 12.14.1.1.

$$\sum_{i=1}^{m} k_{1i} d_{1i}^2 + \sum_{j=1}^{n} k_{2j} d_{2j}^2 = (0.635 Et \times 33.5^2) + (0.818 Et \times 25.9^2) + (2 \times 1.65 Et \times 19.9^2)$$

$$= 2,568.2 Et$$

$$2.5\left(0.05 + \frac{e_1}{b_1}\right) b_1^2 \sum_{i=1}^{m} k_{1i} = 2.5\left(0.05 + \frac{3.8}{60}\right) \times 60^2 \times (0.635 Et + 0.818 Et)$$

$$= 1,482.1 Et < 2,568.2 Et \quad \textbf{O.K.}$$

$$2.5\left(0.05 + \frac{e_2}{b_2}\right) b_2^2 \sum_{j=1}^{n} k_{2j} = 2.5\left(0.05 + \frac{0}{40.5}\right) \times 40.5^2 \times (2 \times 1.65 Et)$$

$$= 676.6 Et < 2,568.2 Et \quad \textbf{O.K.}$$

Thus, all conditions of the eighth limitation are satisfied.

Note that Eqs. 12.14-2A and 12.14-2B need not checked where the following three conditions are met:

1. The arrangement of walls is symmetric about each major axis.

2. The distance between the two most separated wall lines is at least 90 percent of the structure dimension perpendicular to that axis direction.

3. The stiffness along each of the lines of resistance considered in item 2 above is at least 33 percent of the total stiffness in that direction.

In this example, only the second and third conditions are met.

9. Lines of resistance of the lateral force-resisting shall be oriented at angles of no more than 15 degrees from alignment with the major orthogonal axes of the building.

EXAMPLES 6-119

The shear walls in both directions are parallel to the major axes. **O.K.**

10. The simplified design procedure shall be used for each major orthogonal horizontal axis direction of the building.

    The simplified design procedure is used in both directions (see Step 3). **O.K.**

11. System irregularities caused by in-plane or out-of-plane offsets of lateral force-resisting elements shall not be permitted.

    This building does not have any irregularities. **O.K.**

12. The lateral load resistance of any story shall not be less than 80 percent of the story above.

    Since this is a one-story building, this limitation is not applicable. **O.K.**

Since all 12 limitations of 12.14.1.1 are satisfied, the simplified procedure may be used.

- Step 3: Determine the seismic base shear from Flowchart 6.9.

    1. Determine S_S, S_{DS} and the SDC from Flowchart 6.4.

       From Step 1, $S_S = 1.47$ and $S_{DS} = 0.98$.

       According to 11.6, the SDC is permitted to be determined from Table 11.6-1 alone where the simplified design procedure is used.

       For $S_{DS} > 0.50$ and Occupancy Category II, the SDC is D.

    2. Determine the response modification factor R from Table 12.14-1.

       For SDC D, a bearing wall system with special reinforced masonry shear walls is required (system A7). For this system, $R = 5$.

    3. Determine the effective seismic weight W in accordance with 12.14.8.1.

       Conservatively assume that the masonry walls are fully grouted and neglect any wall openings. Also assume a 10 psf superimposed dead load on the roof.

       Weight of masonry walls tributary to roof diaphragm
       $= 0.081 \times \dfrac{12}{2} \times 2(60.0 + 40.5) = 98$ kips

Weight of roof slab $= \dfrac{8}{12} \times 0.150 \times 60 \times 40.5 = 243$ kips

Superimposed dead load $= 0.010 \times 60 \times 40.5 = 24$ kips

$W = 98 + 243 + 24 = 365$ kips

4. Determine base shear V by Eq. 12.14-11.

$$V = \dfrac{FS_{DS}}{R} W = \dfrac{1 \times 0.98 \times 365}{5} = 72 \text{ kips}$$

where $F = 1$ for a one-story building (see 12.14.8.1).

Since this a one-story building, story shear $V_x = V$.

5. Distribute story shear to the shear walls.

Since the building has a rigid diaphragm, the design story shear is distributed to the shear walls based on the relative stiffness of the walls, including the effects from the torsional moment M_t resulting from eccentricity between the locations of the center of mass and the center of rigidity. Note that the simplified procedure does not require accidental torsion (12.14.8.3.2.1).

For lateral forces in the N-S direction, there is no torsional moment, since there is no eccentricity between the center of mass and center of rigidity in that direction. The east and west walls have the same stiffness, so each wall must resist 72/2 = 36 kips.

For lateral forces in the E-W direction, the torsional moment is equal to $72 \times 3.8 = 274$ ft-kips.

The total lateral force to be resisted by the north and south shear walls can be determined from the following equation:

$$V_{1i} = \dfrac{k_{1i}}{\sum k_{1i}} V_x + \dfrac{d_{1i} k_{1i}}{\sum\limits_{i=1}^{m} k_{1i} d_{1i}^2 + \sum\limits_{j=1}^{n} k_{2j} d_{2j}^2} M_t$$

For the north shear wall:

$$V_{11} = \frac{0.635Et}{0.635Et + 0.818Et} \times 72 + \frac{33.5 \times 0.635Et}{2{,}568.2Et} \times 274$$

$$= (0.437 \times 72) + (0.0083 \times 274)$$

$$= 31.5 + 2.3 = 33.8 \text{ kips}$$

For the south shear wall:

$$V_{12} = \frac{0.818Et}{0.635Et + 0.818Et} \times 72 - \frac{25.9 \times 0.818Et}{2{,}568.2Et} \times 274$$

$$= (0.563 \times 72) - (0.0082 \times 274)$$

$$= 40.5 - 2.3 = 38.2 \text{ kips}$$

The east and west shear walls are subjected to a shear force due to the torsional moment for lateral forces in the east-west direction, but that force is less than the 36 kip force that is required for lateral forces in the N-S direction.

6.7.12 Example 6.12 – Nonbuilding Structure

Determine the seismic base shear for the nonbuilding illustrated in Figure 6.20 using (1) 2L4x4x1/2 braces and (2) 2L4x4x1/4 braces, given the design data below.

DESIGN DATA	
Location:	Phoenix, AZ (Latitude: 33.42°, Longitude: -112.05°)
Soil classification:	Site Class D
Structural system:	Ordinary steel concentrically braced frame

SOLUTION

Part 1: Determine seismic base shear using 2L4x4x1/2 braces

Determine the seismic base shear from Flowchart 6.11.

This nonbuilding structure is similar to buildings and the appropriate design requirements from Chapter 15 are used to determine the seismic base shear.

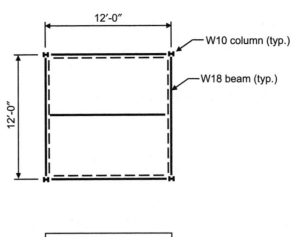

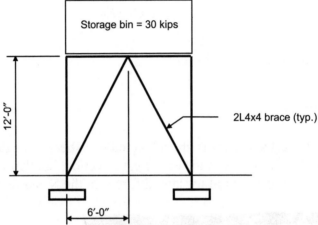

Figure 6.20 Plan and Elevation of Nonbuilding Structure

1. Determine S_{DS}, S_{D1} and the SDC from Flowchart 6.4.

   Using the USGS Ground Motion Parameter Calculator, $S_S = 0.18$ and $S_1 = 0.06$.

   Using Tables 11.4-1 and 11.4-2, the soil-modified accelerations are $S_{MS} = 0.28$ and $S_{M1} = 0.15$.

   Design accelerations: $S_{DS} = \frac{2}{3} \times 0.28 = 0.19$ and $S_{D1} = \frac{2}{3} \times 0.15 = 0.10$

   From IBC Table 1604.5, the Occupancy Category is I, assuming that the contents of the storage bin are not hazardous and that the structure represents a low hazard to human life in the event of failure.

   From Table 11.6-1, for $0.167 < S_{DS} < 0.33$, the SDC is B.

   From Table 11.6-2, for $0.067 < S_{D1} < 0.133$, the SDC is B.

EXAMPLES   6-123

Therefore, the SDC is B for this nonbuilding structure.

2. Determine the importance factor I in accordance with 15.4.1.1.

   Based on Occupancy Category I, the importance factor I is equal to 1.0 from Table 11.5-1.

3. Determine the period T in accordance with 15.4.4.[22]

   In lieu of a more rigorous analysis, Eq. 15.4-6 is used to determine the period T:

   $$T = 2\pi \sqrt{\frac{\sum_{i=1}^{n} w_i \delta_i^2}{g \sum_{i=1}^{n} f_i \delta_i}}$$

   where δ_i are the elastic deflections due to the forces f_i, which represent any lateral force distribution in accordance with the principles of structural mechanics.

   For this one-story nonbuilding structure, this equation reduces to

   $$T = 2\pi \sqrt{\frac{w}{gk}}$$

   where k is the lateral stiffness of the structure.

   The stiffness can be obtained by applying a unit horizontal load to the top of the frame. This load does not produce any forces in the columns. Assuming that the elastic shortening of the beams is negligible, only the braces in a given direction contribute to the stiffness of the frame.

   From statics, the force in one of the four braces due to a horizontal load of 1 applied to the top of the frame is equal to 0.5592. The horizontal deflection δ due to this unit load can be obtained from the following equation from the virtual work method:

   $$\delta = \sum u^2 L / AE$$

   where u = force in a brace due to the virtual (unit) load = 0.5592

   L = length of a brace = $\sqrt{6^2 + 12^2}$ = 13.4 ft = 161 in.

---

[22] The approximate fundamental period equations in 12.8.2.1 are not permitted to be used to determine the period of a nonbuilding structure (15.4.4).

A = area of a 2L4x4x1/2 brace = 7.49 sq in.

E = modulus of elasticity = 29,000 ksi

Thus,

$$\delta = \frac{4 \times 0.5592^2 \times 161}{7.49 \times 29,000} = 0.0009 \text{ in.}$$

The stiffness $k = \dfrac{1}{\delta} = 1,079$ kips/in.

Therefore, the period T is

$$T = 2\pi \sqrt{\frac{30}{386 \times 1,079}} = 0.05 \text{ sec}[23]$$

4. Determine the base shear V.

   Since the period is less than 0.06 sec, use Eq. 15.4-5 to determine V:

   $$V = 0.30 S_{DS} WI = 0.30 \times 0.19 \times 30 \times 1.0 = 1.7 \text{ kips}$$

Part 2: Determine seismic base shear using 2L4x4x1/4 braces

The calculations are similar to those in Part 1, except the stiffness and the period of the structure are different due to the use of lighter braces.

$$\delta = \frac{4 \times 0.5592^2 \times 161}{3.87 \times 29,000} = 0.0018 \text{ in.}$$

Stiffness $k = \dfrac{1}{\delta} = 556$ kips/in.

$$T = 2\pi \sqrt{\frac{30}{386 \times 556}} = 0.07 \text{ sec} > 0.06 \text{ sec}$$

Therefore, the base shear V can be determined by the equivalent lateral force procedure (15.1.3).

---

[23] The weight of the steel framing is negligible.

Determine seismic response coefficient C_s.

The value of C_s from Eq. 12.8-2 is:

$$C_s = \frac{S_{DS}}{R/I} = \frac{0.19}{1.5/1.0} = 0.13$$

where the seismic response coefficient $R = 1.5$ from Table 15.4-1 for an ordinary steel concentrically braced frame with unlimited height, which is permitted to be designed by AISC 360, Specification for Structural Steel Buildings (i.e., without any special seismic detailing).

Also, C_s must not be less than the larger of $0.044 S_{DS} I = 0.008$ and 0.01 (governs) (Eq. 12.8-5).

Thus, the value of C_s from Eq. 12.8-2 governs.

$$V = C_s W = 0.13 \times 30 = 3.9 \text{ kips}$$

CHAPTER 7   *FLOOD LOADS*

7.1 INTRODUCTION

All structures and portions of structures located in flood hazard areas must be designed and constructed to resist the effects of flood hazards and flood loads (IBC 1612.1). Flood hazards may include erosion and scour whereas flood loads include flotation, lateral hydrostatic pressures, hydrodynamic pressures (due to moving water), wave impact and debris impact.

In cases where a building or structure is located in more than one flood zone or is partially located in a flood zone, the entire building or structure must be designed and constructed according to the requirements of the more restrictive flood zone.

The following sections address the hazards and loads that need to be considered for structures located in flood hazard areas.

7.2 FLOOD HAZARD AREAS

By definition, a flood hazard area is the greater of the following two areas:

1. The area within a floodplain subject to a 1-percent or greater chance of flooding in any year.

2. The area designated as a flood hazard area on a community's flood hazard map, or otherwise legally designated.

The first of these two areas is typically acquired from Flood Insurance Rate Maps (FIRMs), which are prepared by the Federal Emergency Management Agency (FEMA) through the National Flood Insurance Program (NFIP).[1] FIRMs show flood hazard areas along bodies of water where there is a risk of flooding by a base flood, i.e., a flood having a 1-percent chance of being equaled or exceeded in any given year.[2]

---

[1] The NFIP is a voluntary program whose goal is to reduce the loss of life and the damage caused by floods, to help victims recover from floods and to promote an equitable distribution of costs among those who are protected by flood insurance and the general public. Conducting flood hazard studies and providing FIRMs and Flood Insurance Studies (FISs) for participating communities are major activities undertaken by the NFIP.

[2] The term "100-year flood" is a misnomer. Contrary to popular belief, it is not the flood that will occur once every 100 years, but the flood elevation that has a 1-percent chance of being equaled or exceeded in any given year. The "100-year flood" could occur more than once in a relatively short period of time. The flood elevation that has a 1-percent chance of being equaled or exceeded in any given year is the standard used by most government agencies and the NFIP for floodplain management and to determine the need for flood insurance.

In addition to showing the extent of flood hazards, the maps also show base flood elevations (the heights to which flood waters will rise during passage or occurrence of the base flood) and floodways.

Floodways (which are channels of a river, creek or other watercourse) and adjacent land areas must be kept clear of encroachments so that the base flood can discharge without increasing the water surface elevations by more than a designated height.[3]

Some local jurisdictions develop and subsequently adopt flood hazard maps that are more extensive than FEMA maps. In such cases, flood design and construction requirements must be satisfied in the areas delineated by the more extensive maps. Thus, a design flood is a flood associated with the greater of the area of a base flood or the area legally designated as a flood hazard area by a community.

The NFIP divides flood hazard areas into flood hazard zones beginning with the letter "A" or "V." "A" zones are those areas within inland or coastal floodplains where high-velocity wave action is not expected during the base flood. In contrast, "V" zones are those areas within a coastal floodplain where high-velocity wave action can occur during the base flood event. Table 7.1 contains general descriptions of these zones. Such zone designations are contained in FIRMs and essentially indicate the magnitude and severity of flood hazards.

The concept of a Coastal A Zone is introduced in Chapter 5 of ASCE/SEI 7-05 and in ASCE/SEI 24-05, *Flood Resistant Design and Construction,* to facilitate application of load combinations in Chapter 2 of ASCE/SEI 7.[4] In general, a Coastal A Zone is an area located within a flood hazard area that is landward of a V Zone or landward of an open coast without mapped V Zones (such as the shorelines of the Great Lakes). Wave forces and erosion potential should be taken into consideration when designing a structure for the effects of flood loads in such zones.

To be classified as a Coastal A Zone, the principal source of flooding must be from astronomical tides, storm surges, seiches or tsunamis and not from riverine flooding. Additionally, stillwater flood depths must be greater than or equal to 1.9 ft and breaking wave heights must be greater than or equal to 1.5 ft during the base flood conditions.[5]

---

[3] Designated heights are found in floodway data tables in FISs. Also, included in FISs are the FIRM, the Flood Boundary and Floodway Map (FBFM), the base flood elevation (BFE) and supporting technical data.

[4] IBC 1612.4 references Chapter 5 of ASCE/SEI 7-05 and ASCE/SEI 24-05 for the design and construction of buildings and structures located in flood hazard areas. The requirements in these documents are covered in the next section of this publication. The NFIP regulations do not differentiate between Coastal and Non-Coastal A Zones.

[5] See Section 4.1.1 of ASCE/SEI 24. Stillwater depth is the vertical distance between the ground and the stillwater elevation, which is the elevation that the surface of water would assume in the absence of waves. The stillwater elevation is referenced to the North American Vertical Datum (NAVD), the National Geodetic Vertical Datum (NGVD) or other datum and it is documented in FIRMs.

Table 7.1 FEMA Flood Hazard Zones (Flood Insurance Zones)[6]

Zone	Description
Moderate to Low Risk Areas	
X*	These zones identify areas outside of the flood hazard area. • Shaded Zone X identify areas that have a 0.2-percent probability of being equaled or exceeded during any given year. • Unshaded Zone X identify areas where the annual exceedance probability of flooding is less than 0.2 percent.
High Risk Areas	
A, AE, A1-30, A99, AR, AO and AH	These zones identify areas of flood hazard that are not within the Coastal High Hazard Area.
High Risk – Coastal Areas	
V, VE and V1-V30	These zones identify the Coastal High Hazard Area, which extends from offshore to the inland limit of a primary frontal dune along an open coast and any other portion of the flood hazard zone that is subject to high-velocity wave action from storms or seismic sources and to the effects of severe erosion and scour. Such zones are generally based on wave heights (3 ft or greater) or wave runup depths (3 ft or greater).

* Zone B on older FIRMs corresponds to shaded Zone X on more recent FIRMs. Zone C on older FIRMs corresponds to unshaded Zone X on more recent FIRMs.

The principal sources of flooding in Non-Coastal A Zones are runoff from rainfall, snowmelt or a combination of both.

It is recommended to check with the local building official for the most current information on flood hazard areas prior to designing a structure in a flood-prone area.

7.3 DESIGN AND CONSTRUCTION

7.3.1 General

According to IBC 1612.4, the design and construction of buildings and structures located in flood hazard areas shall be in accordance with Chapter 5 of ASCE/SEI 7-05 and ASCE/SEI 24-05, *Flood Resistant Design and Construction*. Section 1.6 of ASCE/SEI 24 requires that design flood loads and their combination with other loads be determined by ASCE/SEI 7.

The provisions of ASCE/SEI 24 are intended to meet or exceed the requirements of the NFIP. Figures 1-1 and 1-2 in ASCE/SEI 24 illustrate the application of the standard and the application of the sections in the standard, respectively.

---

[6] Comprehensive definitions for each zone can be found on the FEMA Map Service Center website (http://msc.fema.gov/).

The provisions contained in IBC Appendix G, Flood-Resistant Construction, are intended to fulfill the floodplain management and administrative requirements of NFIP that are not included in the IBC. IBC appendix chapters are not mandatory unless they are specifically referenced in the adopting ordinance of the jurisdiction.

Other provisions related to construction in flood hazard areas worth noting are found in IBC Chapter 18. Section 1804.4 prohibits grading in flood hazard areas unless the specific requirements given in the section are met. Section 1805.1.2.1 requires raised floor buildings in flood hazard areas to have the finished grade elevation under the floor such as at a crawl space to be equal to or higher than outside finished grade on at least one side. The exception permits under floor spaces in Group R-3 residential buildings to comply with the FEMA technical bulletin FEMA/FIA-TB-11, *Crawlspace Construction for Buildings Located in Special Flood Hazard Areas*.[7] This bulletin provides guidance on crawlspace construction and gives the minimum NFIP requirements for crawlspaces constructed in Special Flood Hazard Areas.

7.3.2 Flood Loads

Flood waters can create the following loads, which are referenced in ASCE/SEI 5.4:

- Hydrostatic loads (ASCE/SEI 5.4.2)
- Hydrodynamic loads (ASCE/SEI 5.4.3)
- Wave loads (ASCE/SEI 5.4.4)
- Impact loads (ASCE/SEI 5.4.5)

Determination of these loads is based on the design flood, which is defined in Section 7.2 of this publication. The design flood elevation (DFE) is the elevation of the design flood including wave height. For communities that have adopted minimum NFIP requirements, the DFE is identical to the base flood elevation (BFE). The DFE exceeds the BFE in communities that have adopted requirements that exceed minimum NFIP requirements.[8]

The equations in Table 7.2 can be used to determine flood loads in accordance with Chapter 5 of ASCE/SEI 7. Figure 7.1 illustrates the relationships among the various flood parameters that are used in the equations. Additional information on each type of load follows.

Loads on walls that are required by ASCE/SEI 24 to break away (i.e., breakaway walls) are given in ASCE/SEI 5.3.3. The minimum design load must be the largest of the following loads: (1) wind load in accordance with ASCE/SEI Chapter 6, (2) seismic load in accordance with ASCE/SEI Chapter 11 through 23, and (3) 10 psf. The maximum permitted collapse load is 20 psf, unless the design meets the conditions of ASCE 5.3.3.

---

[7] FEMA/FIA-TB-11 is available from FEMA at http://www.fema.gov/library/viewRecord.do?id=1724.
[8] Communities that have chosen to exceed minimum NFIP requirements typically require a specified freeboard above the BFE.

Table 7.2 Flood Loads

Load Type			Equation No(s).	Equation*
Hydrostatic	Lateral**		---	$F_{sta} = \gamma_w d_s^2 w / 2$
	Vertical upward		---	$F_{bouy} = \gamma_w \times Volume$
Hydrodynamic	$V \leq 10$ ft/sec		5-1	$F_{dyn} = \gamma_w d_s d_h w$ where $d_h = (aV^2/2g)$
	$V > 10$ ft/sec		---	$F_{dyn} = a\rho V^2 d_s w / 2$
Wave	Breaking wave loads on vertical pilings and columns		5-4, 5-2, 5-3	$F_D = \gamma_w C_D D H_b^2 / 2$ where $H_b = 0.78 d_s$ $d_s = 0.65(BFE - G)$
	Breaking wave loads on vertical walls	Space behind the vertical wall is dry	5-5, 5-6	$P_{max} = C_p \gamma_w d_s + 1.2 \gamma_w d_s$ $F_t = 1.1 C_p \gamma_w d_s^2 + 2.4 \gamma_w d_s^2$
		Free water exists behind the vertical wall	5-5, 5-7	$P_{max} = C_p \gamma_w d_s + 1.2 \gamma_w d_s$ $F_t = 1.1 C_p \gamma_w d_s^2 + 1.9 \gamma_w d_s^2$
	Breaking wave loads on nonvertical walls		5-8	$F_{nv} = F_t \sin^2 \alpha$
	Breaking wave loads from obliquely incident waves		5-9	$F_{oi} = F_t \sin^2 \alpha$
Impact	Normal		C5-3	$F = \dfrac{\pi W V_b C_I C_O C_D C_B R_{max}}{2g(\Delta t)}$
	Special		C5-4	$F = C_D \rho A V^2 / 2$

* γ_w = unit weight of water, which is equal to 62.4 pcf for fresh water and 64.0 pcf for salt water.
V = water velocity in ft/sec; see ASCE/SEI C5.4.3 for equations on how to determine V.
Additional information on these equations is given in ASCE/SEI Chapter 5 and in this section.

** In communities that participate in the NFIP, it is required that buildings in V Zones be elevated above the BFE on an open foundation; thus, hydrostatic loads are not applicable. It is also required that the foundation walls of buildings in A Zones be equipped with openings that allow flood water to enter so that internal and external hydrostatic pressure will equalize.

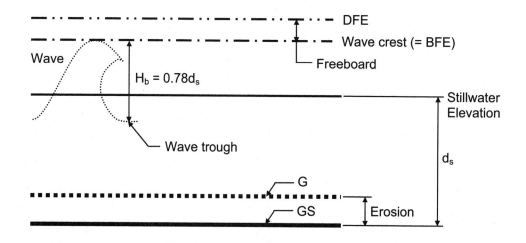

BFE = Base Flood Elevation

DFE = Design Flood Elevation

d_s = Design stillwater flood depth

G = Ground elevation

GS = Lowest eroded ground elevation adjacent to structure

H_b = Breaking wave height

Figure 7.1 Flood Parameters

Hydrostatic Loads

Hydrostatic loads occur when stagnant or slowly moving (velocity less than 5 ft/sec; see ASCE/SEI C5.4.2) water comes into contact with a building or building component. The water can be above or below the ground surface.

Hydrostatic loads are commonly subdivided into lateral loads, vertical downward loads and vertical upward loads (uplift or buoyancy). The hydrostatic pressure at any point on the surface of a structure or component is equal in all directions and acts perpendicular to the surface.

Lateral hydrostatic pressure is equal to zero at the surface of the water and increases linearly to $\gamma_s d_s$ at the stillwater depth d_s where γ_s is the unit weight of water. The total force F_{sta} on the width w of a vertical element acts at the point that is two-thirds below the stillwater surface of the water. See the second footnote in Table 7.2 for more information on lateral hydrostatic pressure in V Zones and A Zones.

Buoyant forces on a building can be of concern where the actual stillwater flood depth exceeds the design stillwater flood depth. Such forces are also of concern for tanks and swimming pools. The buoyant force F_{bouy} is calculated based on the volume of flood water displaced by a submerged object.

Hydrodynamic Loads

Hydrodynamic loads are caused by water moving at a moderate to high velocity above the ground level. Similar to wind loads, the loads produced by moving water include an impact load on the upstream face of a building, drag forces along the sides and suction on the downstream face.

For a water velocity less than or equal to 10 ft/sec, ASCE 5.4.3 permits the dynamic effects of moving water to be converted into an equivalent hydrostatic load. This is accomplished by increasing the DFE by an equivalent surcharge depth d_h calculated by ASCE/SEI Eq. (5-1) (see Table 7.2). In Eq. (5-1), V is the average velocity of water, which can be estimated by ASCE/SEI Eqs. (C5-1) and (C5-2), g is the acceleration due to gravity (32.2 ft/sec^2), and a is the drag coefficient (or shape factor) that must be greater than or equal to 1.25.[9] It is assumed in this case that the total force F_{dyn} on the width w of a vertical element acts at the point that is two-thirds below the stillwater surface of the water.

For a water velocity greater than 10 ft/sec, basic concepts of fluid mechanics must be utilized to determine loads imposed by moving water. The equation in Table 7.2 can be used to determine the total load F_{dyn} in such cases. In this equation, ρ = the mass density of water = γ_w / g and A = surface area normal to the water flow = wd_s. The recommended value of the drag coefficient a is 2.0 for square or rectangular piles and is 1.2 for round piles.[10] In this case, F_{dyn} is assumed to act at the stillwater mid-depth (halfway between the ground surface and the stillwater elevation).

Wave Loads

Wave loads result from water waves propagating over the surface of the water and striking a building or other object. The following loads must be accounted for when designing buildings and other structures for wave loads:

- Waves breaking on any portion of a building or structure
- Uplift forces caused by shoaling waves beneath a building or structure (or any portion thereof)
- Wave runup striking any portion of a building or structure

---

[9] Guidelines on how to determine a are given in ASCE/SEI C5.4.3.
[10] See Table 11.2 in *Coastal Construction Manual*, Third Edition, FEMA 55, 2000, for values of a that can be used for larger obstructions.

- Wave-induced drag and inertia forces
- Wave-induced scour at the base of a building or structure, or at its foundation

The effects of nonbreaking waves and broken waves can be determined using the procedures in ASCE 5.4.2 and 5.4.3 for hydrostatic and hydrodynamic loads, respectively.

Of the wave loads noted above, the loads from breaking waves are the highest. Thus, this load is used as the design wave load where applicable.

Wave loads must be included in the design of buildings and other structures located in both V Zones (wave heights equal to or greater than 3 ft) and A Zones (wave heights less than 3 ft). Since present NFIP mapping procedures do not designate V Zones in all areas where wave heights greater than 3 ft can occur during base flood conditions, it is recommended that historical flood damages be investigated near a site to determine whether or not wave forces can be significant.

ASCE 5.4.4 permits three methods to determine wave loads: (1) analytical procedures (ASCE 5.4.4), (2) advanced numerical modeling procedures and (3) laboratory test procedures. The analytical procedures of ASCE 5.4.4 for breaking wave loads are discussed next.

Breaking wave loads on vertical pilings and columns. The net force F_D resulting from a breaking wave acting on a rigid vertical pile or column is determined by ASCE/SEI Eq. (5-4) (see ASCE/SEI 5.4.4.1 and Table 7.2). In this equation,

C_D = drag coefficient for breaking waves
 = 1.75 for round piles or columns
 = 2.25 for square or rectangular piles or columns

D = pile or column diameter for circular sections
 = 1.4 times the width of the pile or column for rectangular or square sections

H_b = breaking wave height (see Figure 7.1)
 = $0.78 d_s = 0.51(BFE - G)$

This load is assumed to act at the stillwater elevation.

Breaking wave loads on vertical walls. Two cases are considered in ASCE/SEI 5.4.4.2. In the first case, a wave breaks against a vertical wall of an enclosed dry space (i.e., the space behind the vertical wall is dry). Equations (5-5) and (5-6) give the maximum pressure and net force, respectively, resulting from waves that are normally incident to the wall (the direction of the wave approach is perpendicular to the face of the wall). The hydrostatic and dynamic pressure distributions are illustrated in ASCE/SEI Figure 5-1.

In the second case, a wave breaks against a vertical wall where the stillwater level on both sides of the wall are equal.[11] The maximum combined wave pressure is still computed by Eq. (5-5), and the net breaking wave force F_t is computed by Eq. (5-7). ASCE/SEI Figure 5-2 illustrates the pressure distributions in this case.

Values of the dynamic pressure coefficient C_p are given in Table 5-1 based on the building category. ASCE/SEI C5.4.4.2 contains information on the probabilities of exceedance that correspond to the different building categories listed in the table.

Breaking wave loads on nonvertical walls. ASCE/SEI Eq. (5-8) can be used to determine the horizontal component of a breaking wave load F_{nv} on a wall that is not vertical. The angle α is the vertical angle between the nonvertical surface of a wall and the horizontal.

Breaking wave loads from obliquely incident waves. Maximum breaking wave loads occur where a wave strikes perpendicular to a surface. Equation (5-9) can be used to determine the horizontal component of an obliquely incident breaking wave force F_{oi} where the angle α is the horizontal angle between the direction of the wave approach and a vertical surface.[12]

Impact Loads

Impact loads occur where objects carried by moving water strike a building or structure. Normal impact loads result from isolated impacts of normally encountered objects while special impact loads result from large objects such as broken up ice floats and accumulated debris. These two types of impact loads are commonly considered in the design of buildings and similar structures.

The recommended method for calculating normal impact loads on buildings is given in ASCE/SEI C5.4.5. Equation (C5-3) can be used to determine the impact force F. The parameters and coefficients in this equation are discussed in that section.

ASCE/SEI C5.4.5 also contains Eq. (C5-4), which can be used to determine the drag force due to debris accumulation (i.e., special impact load). Additional methods to predict such loads can be found in the references provided at the end of that section.

It is assumed that objects are at or near the water surface level when they strike a building. Thus, the loads determined by Eqs. (C5-3) and (C5-4) are usually assumed to act at the stillwater flood level; in general, these loads should be applied horizontally at the most critical location at or below the DFE.

---

[11] This can occur, for example, where a wave breaks against a wall equipped with openings (such as flood vents) that allow flood waters to equalize on both sides.

[12] In coastal areas, it is usually assumed that the direction of wave approach is perpendicular to the shoreline. Therefore, Eq. (5-9) provides a method for reducing breaking wave loads on vertical surfaces that are not parallel to the shoreline.

7.4 EXAMPLES

The following sections contain examples that illustrate the flood load provisions of IBC 1612, Chapter 5 in ASCE/SEI 7-05 and ASCE/SEI 24-05.

7.4.1 Example 7.1 – Residential Building Located in a Non-Coastal A Zone

The plan dimensions of a residential building are 40 ft by 50 ft. Determine the design flood loads on the perimeter reinforced concrete foundation wall depicted in Figure 7.2 given the design data below.

DESIGN DATA	
Location:	Non-Coastal A Zone
Design stillwater elevation, d_s:	1 ft-6 in.
Base flood elevation (BFE):	3 ft-0 in.

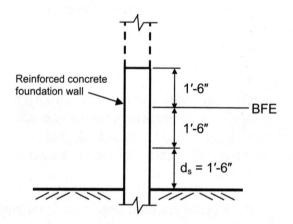

Figure 7.2 Reinforced Concrete Foundation Wall

SOLUTION

In this example, it is assumed that the BFE and the DFE are equal.

This residential building is classified as an Occupancy Category II building in accordance with IBC Table 1604.5.[13] The elevation of the top of the lowest floor relative to the BFE must be greater than or equal to 3 + 1 = 4 ft to satisfy the requirements of Section 2.3 of ASCE/SEI 24 for Category II buildings located in Non-Coastal A Zones (see Table 2-1 of ASCE/SEI 24).

---

[13] Table 1-1 of ASCE/SEI 24 is the same as Table 1-1 of ASCE/SEI 7. Where the IBC is the legally adopted building code, Table 1604.5 should be used to determine the occupancy category of the structure.

EXAMPLES 7-11

The applicable flood loads are hydrostatic, hydrodynamic, breaking wave and impact.

- Step 1: Determine water velocity V

    Since the building is located in a Non-Coastal A Zone, it is appropriate to use the lower bound average water velocity, which is given by ASCE/SEI Eq. (C5-1):

    $V = d_s /1 \text{ sec} = 1.5 \text{ ft/sec}$

- Step 2: Determine lateral hydrostatic load F_{sta}

    The equation for the lateral hydrostatic load is given in Table 7.2. Assuming fresh water,

    $$F_{sta} = \frac{\gamma_w d_s^2}{2} = \frac{62.4 \times 1.5^2}{2} = 70 \text{ lb/linear ft of foundation wall}$$

This load acts at 1 ft-0 in. below the stillwater surface of the water (or, equivalently, 6 in. above the ground surface).

- Step 3: Determine hydrodynamic load F_{dyn}

    According to ASCE/SEI 5.4.3, the dynamic effects of water can be converted into an equivalent hydrostatic load where the velocity of water is less than 10 ft/sec.

    The equivalent surcharge depth d_h is determined by ASCE/SEI Eq. (5-1):

    $$d_h = \frac{aV^2}{2g}$$

    In lieu of a more detailed analysis, the drag coefficient a is determined from Table 11.2 of the *Coastal Construction Manual*, FEMA 55. Since the building will not be completely immersed in water, a is determined by the ratio of longest plan dimension of the building to d_s : $50/1.5 = 33.3$. For a ratio of 33.3, $a = 1.5$.

    Thus, $d_h = \dfrac{1.5 \times 1.5^2}{2 \times 32.2} = 0.05 \text{ ft}$

    Using the equation in Table 7.2, the hydrodynamic load is:

    $F_{dyn} = \gamma_w d_s d_h = 62.4 \times 1.5 \times 0.05 = 5 \text{ lb/linear ft of foundation wall}$

This load acts at 1 ft-0 in. below the stillwater surface of the water (or, equivalently, 6 in. above the ground surface).

- Step 4: Determine breaking wave load F_t

Assuming that the dry-floodproofing requirements of Section 6.2 of ASCE/SEI 24 are not met, flood vents must be installed in the foundation wall (see Section 2.6.1 of ASCE/SEI 24).[14] Thus, the breaking wave load is determined by ASCE/SEI Eq. (5-7), which is applicable where free water exists behind the wall:

$$F_t = 1.1 C_p \gamma_w d_s^2 + 1.9 \gamma_w d_s^2$$

Category II buildings are assigned a value of C_p corresponding to a 1-percent probability of exceedance, which is consistent with wave analysis procedures used by FEMA (see ASCE/SEI C5.4.4.2). For a Category II building, $C_p = 2.8$ from ASCE/SEI Table 5-1. Thus,

$$F_t = (1.1 \times 2.8 \times 62.4 \times 1.5^2) + (1.9 \times 62.4 \times 1.5^2)$$

$$= 432 + 267 = 699 \text{ lb/linear ft of foundation wall}$$

This load acts at the stillwater elevation, which is 1 ft-6 in. above the ground surface.

- Step 5: Determine impact load F

Both normal and special impact loads are determined.

1. Normal impact loads

    ASCE/SEI Eq. (C5-3) is used to determine normal impact loads:

    $$F = \frac{\pi W V_b C_I C_O C_D C_B R_{max}}{2g(\Delta t)}$$

    Guidance on establishing the debris weight W is given in ASCE/SEI C5.4.5. It is assumed in this example that $W = 1000$ lb.

    It is also assumed that the velocity of the object V_b is equal to the velocity of the water V. Thus, from Step 1 of this example, $V_b = 1.5$ ft/sec.

---

[14] The design of the flood vents must satisfy the requirements of Section 2.6.2 of ASCE/SEI 24.

The importance coefficient C_I is obtained from Table C5-1. For an Occupancy Category II building, $C_I = 1.0$.

The orientation coefficient $C_O = 0.8$. This coefficient accounts for impacts that are oblique to the structure.

The depth coefficient C_D is obtained from Table C5-2 or, equivalently, from Figure C5-1. For a stillwater depth of 1.5 ft in an A Zone, $C_D = 0.125$.

The blockage coefficient C_B is obtained from Table C5-3 or, equivalently, from Figure C5-2. Assuming that there is no upstream screening and that the flow path is wider than 30 ft, $C_B = 1.0$.

The maximum response ratio for impulsive load R_{max} is determined from Table C5-4. Assume that the duration of the debris impact load is 0.03 sec (see ASCE/SEI C5.4.5) and that the natural period of the building is 0.2 sec. The ratio of the impact duration to the natural period of the building is 0.03/0.2 = 0.15. From Table C5-4, $R_{max} = (0.4 + 0.8)/2 = 0.6$.

Therefore,

$$F = \frac{\pi \times 1000 \times 1.5 \times 1.0 \times 0.8 \times 0.125 \times 1.0 \times 0.6}{2 \times 32.2 \times 0.03} = 146 \text{ lb}$$

This load acts at the stillwater flood elevation and can be distributed over an appropriate width of the foundation wall.

2. Special impact loads

ASCE/SEI Eq. (C5-4) is used to determine special impact loads:

$$F = \frac{C_D \rho A V^2}{2}$$

Using a drag coefficient $C_D = 1$ and assuming a projected area of debris accumulation $A = 1.5 \times 50 = 75$ sq ft, the impact force F is

$$F = \frac{1 \times \left(\frac{62.4}{32.2}\right) \times 75 \times 1.5^2}{2} = 164 \text{ lb}$$

This load acts at the stillwater flood elevation and is uniformly distributed over the width of the foundation wall.

The flood load effects determined above must be combined with the other applicable load effects in accordance with IBC Sections 1605.2.2 and 1605.3.1.2, which reference ASCE/SEI 7 for load combinations involving flood loads. The above flood load effects are combined with other loads in accordance with the modified strength design load combinations including flood loads in ASCE/SEI 2.3.3(2) or the modified allowable stress design load combinations including flood loads in ASCE/SEI 2.4.2(2).

The design and construction of the foundation, including the foundation walls, must satisfy the requirements of Section 1.5.3 of ASCE/SEI 24.

7.4.2 Example 7.2 – Residential Building Located in a Coastal A Zone

For the residential building described in Example 7.1, determine the design flood loads on the reinforced concrete columns depicted in Figure 7.3 given the design data below.

DESIGN DATA

Location:	Coastal A Zone[15]
Design stillwater elevation, d_s:	3 ft-6 in.
Base flood elevation (BFE):	6 ft-0 in.

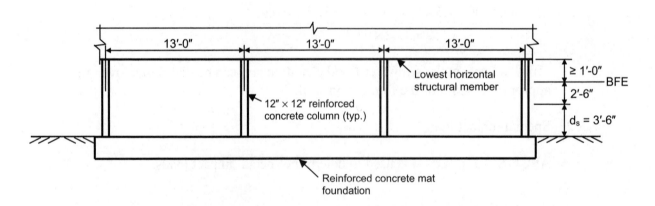

Figure 7.3 Partial Elevation of Residential Building

SOLUTION

In this example, it is assumed that the BFE and the DFE are equal.

---

[15] The provided data satisfies the criteria of Coastal A Zones given in Section 4.1.1 of ASCE/SEI 24: stillwater depth = 3.5 ft > 1.9 ft and wave height = $0.78\,d_s$ = 2.7 ft > 1.5 ft.

This residential building is classified as an Occupancy Category II building in accordance with IBC Table 1604.5. The elevation of the bottom of the lowest supporting horizontal structural member of the lowest floor relative to the BFE must be greater than or equal to 6 + 1 = 7 ft to satisfy the requirements of Section 4.4 of ASCE/SEI 24 for Occupancy Category II buildings located in Coastal A Zones (see Table 4-1 of ASCE/SEI 24).

The applicable flood loads are hydrodynamic, breaking wave and impact.

- Step 1: Determine water velocity V

    Since the building is located in a Coastal A Zone, it is appropriate to use the upper bound average water velocity, which is given by ASCE/SEI Eq. (C5-2):

    $$V = (gd_s)^{0.5} = (32.2 \times 3.5)^{0.5} = 10.6 \text{ ft/sec}$$

- Step 2: Determine hydrodynamic load F_{dyn}

    Since the water velocity exceeds 10 ft/sec, it is not permitted to use an equivalent hydrostatic load to determine the hydrodynamic load (ASCE/SEI 5.4.3).

    Use the equation in Table 7.2 to determine the hydrodynamic load F_{dyn} on one reinforced concrete column:

    $$F_{dyn} = \frac{a\rho V^2 d_s w}{2}$$

    Based on the recommendations in ASCE/SEI C5.4.3 and FEMA 55, the drag coefficient a is taken as 2.0.

    Assuming salt water, the hydrodynamic load is

    $$F_{dyn} = \frac{2.0 \times \left(\frac{64.0}{32.2}\right) \times 10.6^2 \times 3.5 \times \left(\frac{12}{12}\right)}{2} = 782 \text{ lb}$$

    This load acts at 1 ft-9 in. below the stillwater surface of the water.

- Step 3: Determine breaking wave load F_D

    The breaking wave load is determined by ASCE/SEI Eq. (5-4), which is applicable for vertical pilings and columns:

$$F_D = \frac{\gamma_w C_D D H_b^2}{2}$$

According to ASCE/SEI 5.4.4.1, the drag coefficient C_D is equal to 2.25 for square columns.

The breaking wave height H_b is determined by Eq. (5-2):

$$H_b = 0.78 d_s = 0.78 \times 3.5 = 2.7 \text{ ft}$$

Therefore, the breaking wave load on one of the columns is

$$F_D = \frac{64.0 \times 2.25 \times \left(1.4 \times \frac{12}{12}\right) \times 2.7^2}{2} = 735 \text{ lb}$$

This load acts at the stillwater elevation, which is 3 ft-6 in. above the ground surface.

- Step 4: Determine impact load F

Both normal and special impact loads are determined.

1. Normal impact loads

    ASCE/SEI Eq. (C5-3) is used to determine normal impact loads:

    $$F = \frac{\pi W V_b C_I C_O C_D C_B R_{max}}{2g(\Delta t)}$$

    Guidance on establishing the debris weight W is given in ASCE/SEI C5.4.5. It is assumed in this example that $W = 1000$ lb.

    It is also assumed that the velocity of the object V_b is equal to the velocity of the water V. Thus, from Step 1 of this example, $V_b = 10.6$ ft/sec.

    The importance coefficient C_I is obtained from Table C5-1. For an Occupancy Category II building, $C_I = 1.0$.

    The orientation coefficient $C_O = 0.8$. This coefficient accounts for impacts that are oblique to the structure.

The depth coefficient C_D is obtained from Table C5-2 or, equivalently, from Figure C5-1. For a stillwater depth of 3.5 ft in an A Zone, $C_D = (0.75 + 0.5)/2 = 0.63$.

The blockage coefficient C_B is obtained from Table C5-3 or, equivalently, from Figure C5-2. Assuming that there is no upstream screening and that the flow path is wider than 30 ft, $C_B = 1.0$.

The maximum response ratio for impulsive load R_{max} is determined from Table C5-4. Assume that the duration of the debris impact load is 0.03 sec (see ASCE/SEI C5.4.5) and that the natural period of the building is 0.2 sec. The ratio of the impact duration to the natural period of the building is 0.03/0.2 = 0.15. From Table C5-4, $R_{max} = (0.4 + 0.8)/2 = 0.6$.

Therefore,

$$F = \frac{\pi \times 1000 \times 10.6 \times 1.0 \times 0.8 \times 0.63 \times 1.0 \times 0.6}{2 \times 32.2 \times 0.03} = 5212 \text{ lb}$$

This load acts at the stillwater flood elevation.

2. Special impact loads

   ASCE/SEI Eq. (C5-4) is used to determine special impact loads:

   $$F = \frac{C_D \rho A V^2}{2}$$

   Using a drag coefficient $C_D = 1$ and assuming a projected area of debris accumulation $A = 1 \times 3.5 = 3.5$ sq ft, the impact force F on one column is

   $$F = \frac{1 \times \left(\frac{64.0}{32.2}\right) \times 3.5 \times 10.6^2}{2} = 391 \text{ lb}$$

   This load acts at the stillwater flood elevation.

The flood load effects determined above must be combined with the other applicable load effects in accordance with IBC Sections 1605.2.2 and 1605.3.1.2, which reference ASCE/SEI 7 for load combinations involving flood loads. The above flood load effects are combined with other loads in accordance with the modified strength design load combinations including flood loads in ASCE/SEI 2.3.3(1) or the modified allowable stress design load combinations including flood loads in ASCE/SEI 2.4.2(1).[16]

The design and construction of the mat foundation must satisfy the requirements of Section 4.5 of ASCE/SEI 24. The top of the mat foundation must be located below the eroded ground elevation and must extend to a depth sufficient to provide the support to prevent flotation, collapse, or permanent lateral movement under the design load combinations (Section 1.5.3).

The design and construction of the reinforced concrete columns must satisfy the requirements of ACI 318-08 (Section 4.5.7.3 of ASCE/SEI 24).

7.4.3 Example 7.3 – Residential Building Located in a V Zone

For the residential building described in Examples 7.1 and 7.2, determine the design flood loads on 8-in.-diameter reinforced concrete piles given the design data below. The partial elevation of the building is similar to that shown in Figure 7.3.

DESIGN DATA	
Location:	V Zone[17]
Design stillwater elevation, d_s:	4 ft-6 in.
Base flood elevation (BFE):	7 ft-8 in.

SOLUTION

In this example, it is assumed that the BFE and the DFE are equal.

This residential building is classified as an Occupancy Category II building in accordance with IBC Table 1604.5. The elevation of the bottom of the lowest supporting horizontal structural member of the lowest floor relative to the BFE must be greater than or equal to 7.67 + 1 = 8.67 ft to satisfy the requirements of Section 4.4 of ASCE/SEI 24 for Category II buildings located in V Zones (see Table 4-1 of ASCE/SEI 24).

The applicable flood loads are hydrodynamic, breaking wave, and impact.

- Step 1: Determine water velocity V

    Since the building is located in a V Zone, it is appropriate to use the upper bound average water velocity, which is given by ASCE/SEI Eq. (C5-2):

---

[16] All of the flood loads calculated in this example will not occur on all of the columns at the same time. See Section 11.6.12 of FEMA 55 for guidance on how to apply the flood loads.

[17] The provided data satisfies the criteria of V Zones given in Section 4.1.1 of ASCE/SEI 24: stillwater depth = 4.5 ft > 3.8 ft and wave height = $0.78 d_s$ = 3.5 ft > 3.0 ft.

$$V = (gd_s)^{0.5} = (32.2 \times 4.5)^{0.5} = 12.0 \text{ ft/sec}$$

- Step 2: Determine hydrodynamic load F_{dyn}

    Since the water velocity exceeds 10 ft/sec, it is not permitted to use an equivalent hydrostatic load to determine the hydrodynamic load (ASCE/SEI 5.4.3).

    Use the equation in Table 7.2 to determine the hydrodynamic load F_{dyn} on one reinforced concrete pile:

    $$F_{dyn} = \frac{a\rho V^2 d_s w}{2}$$

    Based on the recommendations in ASCE/SEI C5.4.3 and FEMA 55, the drag coefficient a is taken as 1.2.

    Assuming salt water, the hydrodynamic load is

    $$F_{dyn} = \frac{1.2 \times \left(\frac{64.0}{32.2}\right) \times 12.0^2 \times 4.5 \times \left(\frac{8}{12}\right)}{2} = 515 \text{ lb}$$

    This load acts at 2 ft-3 in. below the stillwater surface of the water.

- Step 3: Determine breaking wave load F_D

    The breaking wave load is determined by ASCE/SEI Eq. (5-4), which is applicable for vertical pilings and columns:

    $$F_D = \frac{\gamma_w C_D D H_b^2}{2}$$

    According to ASCE/SEI 5.4.4.1, the drag coefficient C_D is equal to 1.75 for round piles.

    The breaking wave height H_b is determined by Eq. (5-2):

    $$H_b = 0.78 d_s = 0.78 \times 4.5 = 3.5 \text{ ft}$$

    Therefore, the breaking wave load on one of the piles is

$$F_D = \frac{64.0 \times 1.75 \times \left(\frac{8}{12}\right) \times 3.5^2}{2} = 457 \text{ lb}$$

This load acts at the stillwater elevation, which is 4 ft-6 in. above the ground surface.

- Step 4: Determine impact load F

Both normal and special impact loads are determined.

1. Normal impact loads

    ASCE/SEI Eq. (C5-3) is used to determine normal impact loads:

    $$F = \frac{\pi W V_b C_I C_O C_D C_B R_{max}}{2g(\Delta t)}$$

    Guidance on establishing the debris weight W is given in ASCE/SEI C5.4.5. It is assumed in this example that $W = 1000$ lb.

    It is also assumed that the velocity of the object V_b is equal to the velocity of the water V. Thus, from Step 1 of this example, $V_b = 12.0$ ft/sec.

    The importance coefficient C_I is obtained from Table C5-1. For an Occupancy Category II building, $C_I = 1.0$.

    The orientation coefficient $C_O = 0.8$. This coefficient accounts for impacts that are oblique to the structure.

    The depth coefficient C_D is obtained from Table C5-2. For a V Zone, $C_D = 1.0$.

    The blockage coefficient C_B is obtained from Table C5-3 or, equivalently, from Figure C5-2. Assuming that there is no upstream screening and that the flow path is wider than 30 ft, $C_B = 1.0$.

    The maximum response ratio for impulsive load R_{max} is determined from Table C5-4. Assume that the duration of the debris impact load is 0.03 sec (see ASCE/SEI C5.4.5) and that the natural period of the building is 0.2 sec. The ratio of the impact duration to the natural period of the building is 0.03/0.2 = 0.15. From Table C5-4, $R_{max} = (0.4 + 0.8)/2 = 0.6$.

    Therefore,

$$F = \frac{\pi \times 1000 \times 12.0 \times 1.0 \times 0.8 \times 1.0 \times 1.0 \times 0.6}{2 \times 32.2 \times 0.03} = 9366 \text{ lb}$$

This load acts at the stillwater flood elevation.

2. Special impact loads

   ASCE/SEI Eq. (C5-4) is used to determine special impact loads:

   $$F = \frac{C_D \rho A V^2}{2}$$

   Using a drag coefficient $C_D = 1$ and assuming a projected area of debris accumulation $A = 0.67 \times 4.5 = 3.0$ sq ft, the impact force F on one pile is

   $$F = \frac{1 \times \left(\frac{64.0}{32.2}\right) \times 3.0 \times 12.0^2}{2} = 429 \text{ lb}$$

   This load acts at the stillwater flood level.

The flood load effects determined above must be combined with the other applicable load effects in accordance with IBC Sections 1605.2.2 and 1605.3.1.2, which reference ASCE/SEI 7 for load combinations involving flood loads. The above flood load effects are combined with other loads in accordance with the modified strength design load combinations including flood loads in ASCE/SEI 2.3.3(1) or the modified allowable stress design load combinations including flood loads in ASCE/SEI 2.4.2(1).[18]

The design and construction of the foundation must satisfy the requirements of Section 4.5 of ASCE/SEI 24. Requirements for pile foundations are given in Section 4.5.5.

The design and construction of the reinforced concrete piles must satisfy the requirements of ACI 318-09 (Section 4.5.5.8 of ASCE/SEI 24). Additional design provisions are given in Section 4.5.6.

---

[18] All of the flood loads calculated in this example will not occur on all of the piles at the same time. See Section 11.6.12 of FEMA 55 for guidance on how to apply the flood loads.

2009 INTERNATIONAL CODES ON I-QUEST

ICC offers an electronic option just right for you

This powerful tool contains the complete searchable text of the 2009 International Codes®.

I-Quest's Microsoft Fast Folio® search engine allows you to:
- Enjoy multiple search capabilities
- Utilize hyperlinks to sections and tables within each code
- Copy and paste text, figures, and tables
- Highlight and bookmark frequently used sections
- Add your own searchable notes to text or hide them
- Use the "Update" feature to keep your code current
- Available versions include:
  - The Complete Collection containing all 13 I-Codes plus ICC/ANSI A117.1-2003
  - IBC, IRC, or IFC individually

Please note, I-Quest is not MAC-compatible.

I-QUEST, SINGLE USER
Enables one code user to load CD content on one computer.

2009 COMPLETE COLLECTION
SINGLE | #8005Q09

2009 IBC®
SINGLE | #8000Q09

2009 IRC®
SINGLE | #8100Q09

2009 IFC®
SINGLE | #8400Q09

All versions include new Update feature!

I-QUEST, NETWORK
An ideal selection for code users working on a shared network. I-Quest Network allows a set number of concurrent users to access content.

2009 COMPLETE COLLECTION
FIVE USER | #8005Q509

2009 IBC®
FIVE USER | #8000Q509

2009 IRC®
FIVE USER | #8100Q509

2009 IFC®
FIVE USER | #8400Q509

10, 25, and 50 user versions are also available.
Call 1-888-422-7233, x33822 for details.

NEW!
I-QUEST, VOLUME LICENSING
For organizations with staff on the go, ICC offers the ideal solution. Purchasing a volume license enables your firm, office or jurisdiction to download CD content to a set number of computers. Whether the seats are fixed desktops in the office, laptops travelling the globe, or tablet PCs in the field, ICC's new Volume Licensing has you covered.

2009 COMPLETE COLLECTION
FIVE SEAT | #8005QL509

2009 IBC®
FIVE SEAT | #8000QL509

2009 IRC®
FIVE SEAT | #8100QL509

2009 IFC®
FIVE SEAT | #8400QL509

10, 25, and 50 seat versions are also available.
Call 1-888-422-7233, x33822 for details.

ORDER YOURS TODAY! 1-800-786-4452 | www.iccsafe.org/store
FOR MORE ELECTRONIC OPTIONS: www.eCodes.biz

Don't Miss Out On Valuable ICC Membership Benefits. Join ICC Today!

Join the largest and most respected building code and safety organization. As an official member of the International Code Council®, these great ICC® benefits are at your fingertips.

EXCLUSIVE MEMBER DISCOUNTS
ICC members enjoy exclusive discounts on codes, technical publications, seminars, plan reviews, educational materials, videos, and other products and services.

TECHNICAL SUPPORT
ICC members get expert code support services, opinions, and technical assistance from experienced engineers and architects, backed by the world's leading repository of code publications.

FREE CODE—LATEST EDITION
Most new individual members receive a free code from the latest edition of the International Codes®. New corporate and governmental members receive one set of major International Codes (Building, Residential, Fire, Fuel Gas, Mechanical, Plumbing, Private Sewage Disposal).

FREE CODE MONOGRAPHS
Code monographs and other materials on proposed International Code revisions are provided free to ICC members upon request.

PROFESSIONAL DEVELOPMENT
Receive Member Discounts for on-site training, institutes, symposiums, audio virtual seminars, and on-line training! ICC delivers educational programs that enable members to transition to the I-Codes®, interpret and enforce codes, perform plan reviews, design and build safe structures, and perform administrative functions more effectively and with greater efficiency. Members also enjoy special educational offerings that provide a forum to learn about and discuss current and emerging issues that affect the building industry.

ENHANCE YOUR CAREER
ICC keeps you current on the latest building codes, methods, and materials. Our conferences, job postings, and educational programs can also help you advance your career.

CODE NEWS
ICC members have the inside track for code news and industry updates via e-mails, newsletters, conferences, chapter meetings, networking, and the ICC website (www.iccsafe.org). Obtain code opinions, reports, adoption updates, and more. Without exception, ICC is your number one source for the very latest code and safety standards information.

MEMBER RECOGNITION
Improve your standing and prestige among your peers. ICC member cards, wall certificates, and logo decals identify your commitment to the community and to the safety of people worldwide.

ICC NETWORKING
Take advantage of exciting new opportunities to network with colleagues, future employers, potential business partners, industry experts, and more than 50,000 ICC members. ICC also has over 300 chapters across North America and around the globe to help you stay informed on local events, to consult with other professionals, and to enhance your reputation in the local community.

JOIN NOW! 1-888-422-7233, x33804 | www.iccsafe.org/membership

People Helping People Build a Safer World™

Innovative Building Products

Make sure they are up to code with ICC-ES Evaluation Reports

The ICC-ES Solution

ICC Evaluation Service® (ICC-ES®), a subsidiary of ICC®, was created to assist code officials and industry professionals in verifying that new and innovative building products meet code requirements. This is done through a comprehensive evaluation process that results in the publication of ICC-ES Evaluation Reports for those products that comply with requirements in the code or acceptance critera. Today, more code officials prefer using ICC-ES Evaluation Reports over any other resource to verify products comply with codes.

FREE Access to ICC-ES Evaluation Reports!

ICC-ES Evaluation Report

ESR-4802
Issued March 1, 2008
This report is subject to re-examination in one year.

www.icc-es.org | 1-800-423-6587 | (562) 699-0543   A Subsidiary of the International Code Council®

DIVISION: 07—THERMAL AND MOISTURE PROTECTION
Section: 07410—Metal Roof and Wall Panels

REPORT HOLDER:

ACME CUSTOM-BILT PANELS
52380 FLOWER STREET
CHICO, MONTANA 43820
(808) 664-1512
www.custombiltpanels.com

EVALUATION SUBJECT:

CUSTOM-BILT STANDING SEAM METAL ROOF PANELS: CB-150

1.0 EVALUATION SCOPE

Compliance with the following codes:
- 2006 *International Building Code®* (IBC)
- 2006 *International Residential Code®* (IRC)

Properties evaluated:
- Weather resistance
- Fire classification
- Wind uplift resistance

2.0 USES

Custom-Bilt Standing Seam Metal Roof Panels are steel panels complying with IBC Section 1507.4 and IRC Section R905.10. The panels are recognized for use as Class A roof coverings when installed in accordance with this report.

3.0 DESCRIPTION

3.1 Roofing Panels:

Custom-Bilt standing seam roof panels are fabricated in steel and are available in the CB-150 and SL-1750 profiles. The panels are roll-formed at the jobsite to provide the standing seams between panels. See Figures 1 and 3 for panel profiles. The standing seam roof panels are roll-formed from minimum No. 24 gage [0.024 inch thick (0.61 mm)] cold-formed sheet steel. The steel conforms to ASTM A 792, with an aluminum-zinc alloy coating designation of AZ50.

3.2 Decking:

Solid or closely fitted decking must be minimum $^{15}/_{32}$-inch-thick (11.9 mm) wood structural panel or lumber sheathing, complying with IBC Section 2304.7.2 or IRC Section R803, as applicable.

4.0 INSTALLATION

4.1 General:

Installation of the Custom-Bilt Standing Seam Roof Panels must be in accordance with this report, Section 1507.4 of the IBC or Section R905.10 of the IRC, and the manufacturer's published installation instructions. The manufacturer's installation instructions must be available at the jobsite at all times during installation. The roof panels must be installed on solid or closely fitted decking, as specified in Section 3.2. Accessories such as gutters, drip angles, fascias, ridge caps, window or gable trim, valley and hip flashings, etc., are fabricated to suit each job condition. Details must be submitted to the code official for each installation.

4.2 Roof Panel Installation:

4.2.1 CB-150: The CB-150 roof panels are installed on roof shaving a minimum slope of 2:12 (17 percent). The roof panels are installed over the optional underlayment and secured to the sheathing with the panel clip. The clips are located at each panel rib side lap spaced 6 inches (152 mm) from all ends and at a maximum of 4 feet (1.22 m) on center along the length of the rib, and fastened with a minimum of two No. 10 by 1-inch pan head corrosion-resistant screws. The panel ribs are mechanically seamed twice, each pass at 90 degrees, resulting in a double-locking fold.

4.3 Fire Classification:

The steel panels are considered Class A roof coverings in accordance with the exception to IBC Section 1505.2 and IRC Section R902.1.

4.4 Wind Uplift Resistance:

The systems described in Section 3.0 and installed in accordance with Sections 4.1 and 4.2 have an allowable wind uplift resistance of 45 pounds per square foot (2.15 kPa).

5.0 CONDITIONS OF USE

The standing seam metal roof panels described in this report comply with, or are suitable alternatives to what is specified in, those codes listed in Section 1.0 of this report, subject to the following conditions:

5.1 Installation must comply with this report, the applicable code, and the manufacturer's published installation instructions. If there is a conflict between this report and the manufacturer's published installation instructions, this report governs.

5.2 The required design wind loads must be determined for each project. Wind uplift pressure on any roof area must not exceed 45 pounds per square foot (2.15 kPa).

6.0 EVIDENCE SUBMITTED

Data in accordance with the ICC-ES Acceptance Criteria for Metal Roof Coverings (AC166), dated October 2007.

7.0 IDENTIFICTION

Each standing seam metal roof panel is identified with a label bearing the product name, the material type and gage, the Acme Custom-Bilt Panels name and address, and the evaluation report number (ESR-4802).

ICC-ES Evaluation Reports are not to be construed as representing aesthetics or any other attributes not specifically addressed, nor are they to be construed as an endorsement of the subject of the report or a recommendation for its use. There is no warranty by ICC Evaluation Service, Inc., express or implied, as to any finding or other matter in this report, or as to any product covered by the report.

© 2008 Copyright

Page 1 of 1

William Gregory
Building and Plumbing Inspector
Town of Yorktown, New York

"We've been using ICC-ES Evaluation Reports as a basis of product approval since 2002. I would recommend them to any jurisdiction building department, particularly in light of the many new products that regularly move into the market. It's good to have a group like ICC-ES evaluating these products with a consistent and reliable methodology that we can trust."

Becky Baker, CBO
Director/Building Official
Jefferson County, Colorado

"The ICC-ES Evaluation Reports are designed with the end user in mind to help determine if building products comply with code. The reports are easily accessible, and the information is in a format that is useable by plans examiners and inspectors as well as design professionals and contractors."

VIEW ICC-ES EVALUATION REPORTS ONLINE!
www.icc-es.org